PASSIVE AND LOW ENERGY COOLING OF BUILDINGS

PASSIVE AND LOW ENERGY COOLING OF BUILDINGS

***

BARUCH GIVONI

JOHN WILEY & SONS, INC.

New York • Chichester • Weinheim • Brisbane • Singapore • Toronto

This book is printed on acid-free paper. ♾

Published simultaneously in Canada

Library of Congress Cataloging-in-Publication Data:

Givoni, Baruch.
Passive and low energy cooling of buildings / Baruch Givoni.
p. cm.
Includes bibliographical references and index.
ISBN 0-471-28473-4
1. Air conditioning. 2. Ventilation. 3. Buildings—Energy conservation. I. Title.
TH7687.5.G54 1994
697.9' 3—dc20 94-11136

Printed in the United States of America

10 9 8 7 6 5 4 3 2

CONTENTS

PREFACE

My objectives in writing this book were twofold. On one hand it is a summary of my personal research during the last 30 years in the area of passive and low energy cooling of buildings in regions with hot summers, including minimizing the needs for cooling by architectural means. At the same time it is an attempt to present the state of the art in this area by summarizing relevent research by the other researchers working in this area.

The book discusses a number of cooling techniques utilizing natural, renewable energies as well as low energy systems consuming far less electricity than conventional air conditioning: daytime and nocturnal ventilative cooling, radiant cooling utilizing longwave radiation to the sky as the main heat loss avenue, direct and indirect evaporative cooling, and the utilization of cooled soil as a heat sink for buildings. A special chapter deals with cooling options for outdoor spaces.

The book was written with several audiences in mind: prospective home owners and architects designing buildings in regions with hot summers interested in minimizing the consumption of conventional energies for cooling; scientists doing research in any one of the cooling techniques covered in the book; utilities interested in reducing the peak demand for electricity (for air conditioning); teachers and students in architectural schools in which this subject is taught; and any other interested individuals.

To be of value to the different audiences, the book deals both with the scientific fundamentals relevant to the various cooling techniques as well as with the practical and architectural design issues involved in their application.

Acknowledgments

The Academic Senate of UCLA has supported the preparation of the figures for the book. Mr. Tarek Labib, a Ph.D. student at UCLA, has prepared many of the figures.

Thanks are due to the Graduate School of Architecture and Urban Planning of UCLA for enabling me to use the computer and printer facility of the School in writing this book. Special thanks to Mrs. Patricia Victorin who helped in printing the manuscript.

1

OVERVIEW:
THE VARIOUS PASSIVE COOLING SYSTEMS AND THEIR APPLICABILITY TO DIFFERENT CLIMATES AND BUILDING TYPES

INTRODUCTION

This book discusses properties and application problems of various passive cooling systems that utilize natural energies. It is intended as a guide for architects and other building-design practitioners who are engaged in designing buildings in hot regions and are interested in providing comfortable indoor conditions while minimizing the consumption of conventional exhaustible energy sources.

The term "passive cooling systems" needs some clarification. In the context of this book it applies to various simple cooling techniques that enable the indoor temperatures of buildings to be lowered through the use of natural energy sources. The term "passive" does not exclude the use of a fan or a pump when their application might enhance performance. In most cases the cold collecting and storage elements of such a system are an integral part of the building itself—the roof, structural materials, or soil under the building, for example.

Even when mechanical devices are used, their coefficient of performance (COP)—the energy input per unit of cooling that is obtained within the building—is much higher than the value common in air conditioning. Usually the COP is above five, and often it is above ten.

While solar energy can be utilized for cooling, as it is in solar-absorption cooling systems, such systems are not discussed in this book. The reason is that with present-day technology these systems are quite complex and their application consumes a high proportion of parasitic energy.

The material presented in the book is based to a large extent on personal research done by the author, but it also includes results of research on various passive cooling systems that was conducted by other scientists.

Passive Cooling and Bioclimatic Architecture

The design and application of passive cooling systems is often confused with appropriate architectural design for hot regions, currently termed "tropical bioclimatic architecture." The distinction between these two concepts is not always clear, but it is important. The following definitions present the author's personal interpretation of the two concepts.

Any building is heated up during the day, and it cools down during the night as a result of heat loss to the outdoor air by convection and of radiant loss to the sky. These natural cooling processes occur even when the building is not provided with specialized passive cooling systems. The daytime heating and night cooling result in an indoor *average* temperature that is somewhat above the outdoor air temperature average, owing to solar radiation that is absorbed in the building's envelope and penetrates through the windows. All the heat gained during the day is lost during the night. Specifically, this pattern occurs when the climate is stable, with about the same diurnal average over a period of several days (a steady periodic pattern).

Thus, in reality, even if the windows are effectively shaded and the envelope has a reflective (white or light) color, the indoor average would be above the outdoor average. This is true even in an unoccupied building, and it is a result of some unavoidable solar energy absorption in the envelope. Higher solar gain, when the envelope has darker colors and there is internal heat generated by occupants (through cooking, lighting, and so on), increases this indoor temperature elevation. The overall elevation of the indoor average temperature caused by the direct and indirect solar energy gain is referred to as the "sol-air temperature elevation."

From the thermal performance point of view, in the context of cooling in summer, bioclimatic architectural design therefore can only minimize the sol-air elevation—that is, the elevation of the diurnal indoor average temperature above the outdoor average. Any lowering of the indoor average below the outdoor level requires the input of cooling energy which, in passive systems, is obtained from natural renewable sources. This is the operational definition of passive cooling systems applied in this book.

Bioclimatic architecture in hot regions, by this definition, involves architectural design and choice of materials aiming at providing comfort while minimizing the demand for energy used to cool a building. It involves minimizing heat gain by the building, minimizing solar heating of the envelope and solar penetration through windows, providing comfort by natural ventilation, and so on.

Architectural means for achieving these aims include such conventional design elements as the layout of the building; its orientation; the number, size, location, and details of its windows; the shading devices that surround it, and the thermal resistance and heat capacity of its envelope. Appropriate application of these design elements can bring the average indoor temperature close to the level of the outdoor average but usually not below this level.

On the other hand, the various passive cooling systems discussed in this book are capable of transferring heat from a building to various natural heat sinks. To achieve this end special design details must be applied to the building.

In a way, passive cooling systems provide "active" cooling through the use of passive processes, which often use heat-flow paths that do not exist in buildings that lack these systems. The application of a passive cooling system usually requires specialized details in the design and construction of various building components, including the roof and other internal structural elements.

Appropriate architectural bioclimatic design in a region with hot summers can be considered as a precondition for the application of passive cooling systems, and the two approaches supplement and reinforce one another. Therefore, although the subject of bioclimatic architecture is dealt with in other books by this author (Givoni 1976, 1994), the second chapter of this book presents a summary of the architectural means available to minimize the cooling needs of a building.

Scope of this Chapter

This first chapter is an overview of passive cooling systems. It is intended mainly for architects who are considering their application to a specific building in a given climate. It summarizes the detailed information that is presented in the other chapters of the book, each of which deals with a particular type of passive cooling. This chapter emphasizes issues of applicability of the various systems in different climates and for various building types. The figures mentioned in the text of this chapter are presented in the respective chapters dealing with each system.

A recent book edited by Jeff Cook (1989) also provides a comprehensive review of passive cooling techniques. However, the approach and emphasis in this book is different than that in Cook's book. Cook's book deals mainly with the physics of the heat loss to the various natural heat sinks. Its conclusions are based mainly on mathematical modelling of the various cooling systems. In the present book I have tried to put more emphasis on the architectural and climatic application issues involved in the utilization of the same natural heat sinks. Different techniques or systems by which the various natural heat cooling sources can be utilized are described. Emphasis is given to available experimental data on the performance of the different passive cooling systems under various climatic and testing conditions.

Classification of Passive Cooling Systems

Buildings can be cooled by passive systems through the utilization of several natural heat sinks such as the ambient air, the upper atmosphere, water, and the undersurface soil. Each of these cooling sources can be utilized in various ways, resulting in different systems. The various passive cooling systems are thus classified according to the major, or obvious, natural source from which the cooling energy is derived:

- **Comfort Ventilation:** providing direct human comfort, mainly during the daytime (Chapter 3).
- **Nocturnal Ventilative Cooling:** cooling the structural mass of the building interior by ventilation during the night and closing the building during the daytime, thus lowering the indoor daytime temperature (Chapter 3).
- **Radiant Cooling:** transferring into the building cold energy generated during the night hours by radiant heat loss from the roof, or using a special radiator on the roof, with or without cold storage for the daytime (Chapter 4).

- **Direct Evaporative Cooling:** mechanical or nonmechanical evaporative cooling of air. The humidified and cooled air is then introduced into the building (Chapter 5).
- **Indirect Evaporative Cooling:** evaporative cooling of the roof, for example by roof ponds. The interior space is cooled without elevation of the humidity (Chapter 5).
- **Soil Cooling:** cooling soil below its natural temperature in a given region and utilizing it as a cooling source for a building (Chapter 6).
- **Cooling of Outdoor Spaces:** Cooling techniques that are applicable to outdoor spaces, such as patios, that are adjacent to a building (Chapter 7).

The applicability of a given cooling system depends on the specified limits (boundaries) of the indoor climate. It varies with the type of the building and the local climate. Some cooling systems may be applied only to specific types of buildings or only under specific climatic conditions.

The boundaries of the climatic conditions under which a given cooling system can be applied are not the same in different countries. People living in hot regions, especially in developing countries, usually accept higher temperatures and/or humidity levels because of lower expectations and natural acclimatization. Therefore the upper limits suggested below for the applicability of various cooling systems are given as a range rather than as a fixed value. The subject of comfort is discussed in Chapter 3.

History of Passive Cooling

Systematic research on systems of passive cooling is a relatively recent phenomenon. Individual studies on various elements of passive cooling have been carried out in various research institutions during the past 25 years. But these were scattered efforts and did not produce a comprehensive and synoptic body of knowledge.

When the energy crises of the 1970s aroused interest in the use of natural renewable energies as substitutes for conventional fuels in buildings, this interest was directed to the use of solar energy for heating. This led to the development of "active" solar heating systems. In time the focus in solar space heating shifted to passive systems, but still the emphasis was on heating. Only in about 1978 did more worldwide interest arise and systematic research start in passive cooling systems.

There seem to be several reasons for the recent emerging interest in passive cooling techniques. The first is obviously the rising cost of electricity, especially during peak demand times. Peak demand is often caused by the increased use of air conditioning in hot summer days. In addition to consumer concern with the rising cost, there is an institutional interest in flattening the demand for energy by cutting down peak demand, which determines the need for generation capacity.

In the United States there is another factor that has focused interest on passive cooling. This is the internal migration of population to the southern and southwestern states. These states have hotter climates than the northeastern states, and their expanded population increases the demand for cooling. More recently there has been a global interest in reducing the emission of "greenhouse gases" suspected of causing global warming. Reducing power generation by lowering the demand for air conditioning can be helpful also in this respect.

COMFORT VENTILATION

The simplest strategy for improving comfort when the indoor temperature, under still air conditions, seems to be too warm is by daytime ventilation—providing comfort through higher indoor air speeds. The flow of outdoor air at a given speed through a building extends the upper limit of the comfort zone beyond the limit for still air conditions, and it may provide a direct physiological cooling effect even when the air is rather warm, up to about 30°C. Hence the term "comfort ventilation." This is particularly the case when the humidity is high and so the higher air speed increases the rate of sweat evaporation from the skin, thus minimizing the discomfort from the sensation of wet skin.

However, when a building is cross-ventilated during the daytime the temperature of the indoor air and surfaces closely follow the ambient temperature. Therefore there is a point in applying daytime ventilation only when **indoor comfort** can be experienced at the **outdoor air temperature** (with acceptable indoor air speed).

The distinction between comfort ventilation and nocturnal ventilative cooling as two distinct cooling systems is suggested because some building elements, such as the structural materials, and especially the amount and thermal conductivity of the building interior mass, require different designs for the optimization of each one of these ventilation strategies. However, other building design factors, such as the location and details of the openings, may be the same in these two cases.

The effect of the outdoor air flow on the indoor daytime temperature depends on the temperature that the interior would have without the ventilation, which, in turn, depends on the design details of the building, its internal heat generation, and the amount of penetrating solar energy (e.g. for daylighting).

In buildings that are well protected from solar radiation and that have high insulation of the envelope and high thermal mass, the indoor daytime temperature, in the absence of ventilation, could be well below the outdoor level. In this case daytime ventilation would raise the indoor air and the radiant temperature of surfaces. This occurs most often in residential buildings with low internal heat generation. The specific climatic conditions under which daytime ventilation is advisable, and the building design details that can maximize the beneficial effect derived from it, are discussed in detail in Chapter 3 and are summarized below.

When a building is naturally ventilated through open windows and the wind reaches even moderate speeds, the indoor air temperature tends to approach the outdoor temperature level. A high indoor air speed also increases the rate of heat exchange between the indoor air and the interior mass of the building. In most regions the wind speed during the daytime is much higher than at night, when it is often almost imperceptible. In the absence of wind a much greater difference may exist between indoor and outdoor temperatures. Consequently the indoor air speed, even when the windows are open, is significantly lower at night, and the indoor–outdoor temperature difference is greater than during the daytime.

Building materials absorb heat from the warm ventilation air during the day. The rate of heat absorption is enhanced by the higher convective heat transfer coefficient, resulting from the high daytime indoor air speed in a cross-ventilated building. Because of the thermal time lag of the building materials the indoor surfaces are cooler than the indoor air in the morning hours and warmer in the afternoon and evening, when heat flows from the interior mass to the indoor air. Thus the indoor air temperature in a high mass building reaches its maximum in the evening. Therefore, even if a building is ventilated continuously, day and night, for comfort ventilation, the indoor temperature is close to the outdoor level during the daytime but is higher than the outdoor during the night hours. As a result of the combination of low air speed and warm temperature the indoor environment is often most uncomfortable during the evening hours.

Climatic Applicability of Comfort Ventilation

Figure 3.1 shows the suggested boundaries of the outdoor temperature and humidity within which indoor comfort can be provided by natural ventilation during the day with indoor air speed of about 2 m/s (a very light breeze).

Comfort ventilation would be the simplest strategy in warm, humid regions. As mentioned above, during the day with effective cross ventilation, accompanied by a relatively high indoor air speed the indoor air closely follows the outdoor level. Therefore the temperature limit of applicability of comfort ventilation is the comfort limit at the enhanced air speed. Assuming an indoor air speed of 1.5–2.0 m/s, comfort ventilation is applicable mainly in regions and seasons when the outdoor maximum temperature does not exceed about 28 to 32°C, depending on the acclimatization of the population, and where the diurnal temperature range is less than about 10°C. These conditions are typical of warm humid climates. In these regions the relatively small diurnal temperature range does not produce a significant reduction of the indoor daytime temperature below the outdoor level, even in a closed, high-mass, well insulated building. (See the section on nocturnal convective cooling.) Daytime ventilation is therefore needed to minimize the physiological effect of the high humidity and to enhance the convective heat loss from the body.

Building Design for Comfort Ventilation

When comfort ventilation is chosen as the main strategy, the building design should aim at achieving high air speed and fast cooling of the interior during the evening hours, when the wind usually subsides. This calls for relatively large but well shaded windows. Materials should not absorb and store too much heat during the daytime hours. Therefore the preferable structural materials for buildings relying on comfort ventilation would be lightweight—for example, wood, lightweight concrete, or perforated bricks.

Comfort ventilation can be applied to all types of buildings. In regions with moderate to high daytime wind speeds the ventilation can be natural. In places without sufficient wind speeds and/or in buildings where effective cross ventilation is not possible due to the design of the interior, the indoor air can be exhausted by a fan, with outdoor air entering the building through all the open windows.

As the main function of the air flow in buildings designed for comfort ventilation is to provide direct physiological cooling, the design of the windows should direct the flow toward the areas and the vertical zone where the inhabitants are staying. The effect of window design on the internal air flow pattern in discussed in Chapter 3. More detailed discussion of this subject is presented in Givoni (1976, 1993) and in Chandra et al. (1983).

Nocturnal Ventilative Cooling

When a building is ventilated at night its structural mass is cooled by convection from the inside, bypassing the thermal resistance of the envelope. During the daytime the cooled mass, when it has sufficient amount and surface area and if it is adequately insulated from the outdoors, can serve as a heat sink. By radiation and natural convection it can absorb the heat penetrating into it, mainly through the windows, along with the heat generated inside. To this effect, the building should be closed (unventilated) during the daytime to prevent the interior being heated by the hotter outdoor air. In this respect daytime comfort ventilation and nocturnal ventilative cooling are mutually exclusive.

There is a point in designing a building for ventilative cooling only if comfort conditions can be maintained during the daytime hours in an unventilated building. Attainment of such performance depends both on the climatic conditions and on the design details of the building. From the climatic aspect the main relevant factors affecting the performance of convective cooling are the diurnal temperature range and the typical maximum temperature during the hottest months. A large diurnal range is needed because the achievable drop of the indoor maximum below the outdoor maximum is, for a building designed for this purpose, roughly proportional to the outdoor temperature range. For a high mass, well insulated, and well shaded building a drop of the indoor maximum below the outdoor maximum of about 35 to 45% of the outdoor range (depending on the actual thermal resistance, heat capacity, and solar protection levels) is possible when the building is *unventilated* day and night. The indoor minimum temperature will be higher than the outdoor minimum also by about 35 to 45% of the outdoor range. A typical indoor temperature swing for such a high mass building is about 10 to 20% of the outdoor range.

Ventilating the indoor space during the night lowers the indoor temperatures. It is possible to lower the indoor minimum more than the maximum. With nocturnal ventilation rates practical at night in residential buildings the indoor minimum in a building ventilated at night could be lowered below the level of an unventilated building by about half of the difference between the minimum of a closed building and the outdoor minimum. The drop of the indoor maximum by the nocturnal ventilation is smaller, about one half of the minimum drop (see the experimental studies below).

Ventilative cooling is applicable mainly in arid and desert regions, which have a large diurnal temperature range (about 12 to 15°K or more) and where the night minimum temperature in summer is below about 20°C. In such regions it is possible to store the coolness of the night air in the structural mass of the building. The flow of outdoor air at night through the building can be induced naturally by the wind (where wind speed at night at the building site is sufficient—above 2–3 m/s) or mechanically by an exhaust fan. During the following day the cooled mass serves as a heat sink, maintaining the indoor temperatures well below the outdoor level, if possible within the comfort range. This temperature reduction can be achieved only when the building is well insulated, with the insulation external to the structural mass, and if the building is not ventilated by the hot outdoor air during the daytime hours.

Experimental Studies of Convective Cooling

The quantitative effect of convective cooling was studied extensively by Givoni at the Institute for Desert Research of Ben Gurion University in Israel and at the University of California, Los Angeles. Figure 3.8 in Chapter 3 shows temperatures that were measured in Israel in two identical high mass rooms. One room was closed day and night while the other room was closed during the day and ventilated naturally during the night by opening a door.

The temperature pattern of the closed room illustrates how it is possible to achieve daytime indoor temperatures well below the outdoor level in regions with a large temperature range. It also demonstrates the need to prevent ventilation during the daytime. Even without night ventilation the indoor temperature was kept below the outdoor maximum by 5 to 8°K—that is, by more than ⅓ of the outdoor range. This was achieved by the thermal inertia of the room and by minimizing the solar gain. Ventilation during the daytime would, of course, bring the indoor temperature closer to the outdoor level and therefore is not advisable.

The ventilation of the building during the nighttime cools the interior mass. The cooled mass absorbs heat during the following daytime hours and lowers the indoor temperatures. The drop of the indoor daytime temperature of a high-mass building below the outdoor maximum by night ventilation is roughly proportional to the outdoor temperature range.

As a rule of thumb it can be estimated that in arid and desert regions, with a diurnal temperature range of 15–20°K, the expected reduction of the average daytime indoor temperature is 2–3°K below the level of similar buildings that do not have night ventilation. Indoor maximum temperature can be lowered by about 7–8°K below the outdoor maximum. On very hot days, which usually have a larger diurnal range, the drop of the indoor temperature during the time of the outdoor maximum may be up to about 10°K.

Applicability of Nocturnal Ventilative Cooling

Nocturnal ventilative cooling would be the preferable strategy in regions where the diurnal temperature range in summer is large enough to enable a significant reduction of the indoor air temperature below the outdoor maximum. Nocturnal ventilative cooling as a building design strategy is preferable to comfort ventilation in regions where the daytime temperatures in summer are above the upper limit of the comfort zone (with air speed of about 1.5 m/s) and it is applicable mainly in arid regions where the daytime temperature is between 30 and 36°C and the night temperatures below about 20°C. In this situation daytime ventilation is not desirable because it would raise the indoor temperature. The large diurnal temperature range, typical of arid regions, allows the indoor daytime temperature to be reduced significantly below the outdoor level by night ventilation. However, in arid regions with daytime temperatures above 36°C, night ventilation would not maintain the indoor temperature at an acceptable level. Other passive cooling systems should be considered during the too hot hours as supplements to convective cooling, systems such as evaporative cooling, earth cooling, or compression air conditioning. But even when an additional cooling system is provided the use of convective nocturnal cooling can significantly reduce the time when the additional cooling systems will be needed.

During very hot days, when the indoor temperature exceeds the upper level of comfort in still air, comfort can be improved by increasing the indoor air speed with a fan while preventing ventilation with outdoor air. Only when the outdoor temperature drops below the indoor level is it advisable to open the windows and ventilate the building.

Some practical problems may limit the applicability of nocturnal convective cooling. These problems include the need to open and close windows at prescribed times and issues of security, especially in public buildings.

A graphic illustration of the climatic boundaries (in terms of the ambient maximum temperatures) for the application of nocturnal ventilative cooling is presented in Figure 3.3.

Nocturnal ventilative cooling can be applied to all types of buildings, from single family homes to multistory apartment buildings to nonresidential buildings, provided that they are well insulated and their interior mass permits effective thermal storage of the night's coolness. Low-mass buildings, even if ventilated at night, can not retain enough "coolth reserve" to reduce significantly the rate of temperature rise during the daytime.

Radiant Cooling

Any ordinary surface that "sees" the sky looses heat by the emission of longwave radiation toward the sky and can be regarded as a heat radiator. Although the radiant heat loss takes place day and night, only during the night is the radiant balance negative. During the daytime the absorbed solar radiation counteracts the cooling effect of the longwave emission and produces a net radiant heat gain.

Roofs are usually insulated to minimize heat loss in winter and heat gain in summer. As the radiant loss takes place at the external surface of the roof the insulation minimizes the actual cooling that a building can utilize from the nocturnal radiation, unless specialized designs (radiant cooling systems) are applied. Several such systems are discussed below.

Massive Roofs with Movable Insulation as Radiators

The simplest concept of radiant cooling is that of a heavy and highly conductive roof (for example, one made of dense concrete) exposed to the sky during the night but highly insulated externally by means of operable insulation during the daytime. Such roofs, a conventional construction common in many countries, can be very efficient in losing heat at night, both by longwave radiation to the sky and by convection to the outdoor air, which cools down faster than the massive roof. During the daytime the (installed) external insulation minimizes the heat gain from solar radiation and from the hotter ambient air. The cooled mass of the roof can then serve as a heat sink and absorb, through the ceiling, the heat penetrating into and generated inside the building interior during the daytime hours.

The performance of this concept was tested by Givoni (1981) at the Institute for Desert Research of Ben Gurion University in Sede Boqer, Israel, with small lightweight test cells (55×55×50 cm). The average results are summarized in the following table:

	Minimum	*Maximum*	*Average*
Outdoor temperature	14.5	27.0	20.7
Indoor—experimental	14.0	21.3	17.2
External roof surface	12.0	19.3	15.2
Indoor—control	19.4	25.6	22.5

Thus, the roof radiator has lowered the indoor average temperature by 3.5°C and the maximum by 5.7°C, below the corresponding outdoor air temperatures. Compared with the control test cell the reductions in the average and maximum temperatures were 5.3 and 4.3°C, respectively.

Etzion and Dover, at the same institute, later tested this concept with larger test cells. Figure 4.11 shows diurnal patterns of the outdoor and the interior temperatures during a sequence of four days in July, 1987. It is interesting to note that in these independent studies the indoor maximum temperatures were lowered below the outdoor maxima by a similar fraction of the outdoor's range (45–50%).

Although it has been demonstrated that concrete roofs, which are very common in many countries, can provide effective radiant cooling, this can be achieved only with the application of movable insulation. At present, in the opinion of the author, practical and inexpensive movable insulation is not available. To the extent that simple movable insulation will be developed, concrete roofs could provide simple, inexpensive, and effective radiant cooling in many hot regions of the world.

The "Skytherm" System

Harold Hay (1978) has developed a passive solar heating and radiant cooling system, the "Skytherm." In this system the roof is made of structural steel deck plates. Above the metal deck plastic bags filled with water are placed, and above them insulation panels can be moved by a motor to cover or to expose the bags. In winter the water bags are exposed to the sun during the day and covered by the insulation panels during the nights. In summer, when cooling is needed, the water bags are exposed and cooled during the night and insulated during the daytime. The cooled water bags are in direct thermal contact with the metal deck and thus the ceiling serves as a cooling element over the whole space below.

Several buildings utilizing this system have been built in the United States. Some of them were experimental buildings that were subjected to detailed studies, and their physical performance and the operational problems encountered with the movable insulation were evaluated. See more details in Chapter 4.

The main problem encountered with the Skytherm system concerns the movable insulation. On the basis of research reports it is the opinion of the author that the availability of a simple, inexpensive, convenient, and trouble-free system of movable insulation is still in question. For example, Clark (1989), in summarizing the experience gained with the different buildings that have applied the "Skytherm" system, notes, "As a one-of-a-time kind installation, the conventional horizontally rolling panels have also been expensive and mechanically unreliable."

Specialized Long-Wave Radiators

In order to utilize the cooling effect of nocturnal radiation by buildings with ordinary insulated roofs, the cold produced at the external surface, above the insulation, should be transferred into the building's interior. Usually this "cold transfer" is provided by air flow under the radiating element, cooling it to a temperature below the ambient. The cooled air is blown through the interior space to cool the mass of the building, in a similar way to nocturnal ventilative cooling discussed above, but with a temperature below the level that can be achieved by direct ventilation with the outdoor air. The cooled mass then serves during the following day as a sink for heat penetrating into and generated inside the building.

Any painted metallic layer placed over the roof, with an air space of about five to ten cm beneath it, could serve as a radiator. The radiator can serve also as a rainproofing element. When air flows under the radiator at night it is cooled, while the radiator's temperature rises. Metals have low emissivity and thus are poor radiators. Therefore the external surface of the metal should be painted so that the paint would provide an emissive layer in thermal contact with the conductive metal underneath.

The temperature drop attained by a radiator without air flow *underneath* is the "stagnation temperature drop." To be of any value as a cooling system the radiator's stagnation temperature should be lower than the ambient air by some minimum temperature drop—at least 5°C. If it is not, the simpler and less expensive nocturnal ventilative cooling could be applied. Ambient air drawn under the radiator can be cooled by about one third to two thirds of the stagnation depression achieved by the radiator, depending on the flow rate.

Two approaches were considered and tested in order to augment the temperature drop and the effective cooling that can be obtained by metallic radiators. One approach minimized the convective heat gain through the use of a polyethylene film serving as a wind screen, while the other applied a selective surface to the radiator. These methods are discussed in Chapter 4.

Applicability of Radiant Cooling

The roof is the only element of a building that has much exposure to the sky and therefore it is the natural location for a nocturnal radiator. High-mass roofs with operable insulation, either of concrete or with roof ponds, provide the functions of cold collection and of storage in one element. Therefore, these types of radiant cooling are effective in providing *daytime cooling* in almost any region with low cloudiness at night, regardless of the air humidity.

A radiant cooling system that uses a metallic radiator with a fan-driven air flow underneath can cool *the night air* below the ambient level. The main climatic requirement is again low cloudiness during the nights. Humidity is less important as long as the nights are clear. In arid regions a temperature drop of about 3 to 5°K, depending on the flow rate of the air exiting from the radiator, can be expected. In humid regions with clear skies the expected temperature drop would be about 2 to 3°K. In humid regions, moisture would be condensed out of the air while flowing under the radiator, thus lowering both the indoor humidity and the temperature. This cooled and partly dehumidified air can be used directly to cool a building *during the evening and night hours* in regions where the outdoor air is too warm to provide comfort by direct ventilation.

EVAPORATIVE COOLING

Outdoor air can be cooled by evaporating water before it is introduced into the building. The air flow can be induced mechanically or passively—for example, evaporative cooling towers that humidify the ambient air can be used. This is direct evaporative cooling. However, passive evaporative cooling can also be indirect—the roof can be cooled with a pond and the ceiling transformed into a cooling element, which cools the space below by convection and long-wave radiation without raising the indoor humidity.

In desert regions with too low humidity (from the comfort viewpoint) direct evaporative cooling can be an inexpensive system that is desirable physiologically as well. In more humid regions the efficiency of the system is reduced and the higher humidity may be undesirable from the comfort aspect. Indirect evaporative cooling would then be an appropriate solution.

Direct Evaporative Cooling by Cooling Towers

Two types of passive cooling towers will be described below: one developed by Cunningham and Thompson and the other, an inertial "shower," developed by Givoni.

The Cunningham and Thompson System

A passive evaporative air cooling tower attached to a building has been developed and tested by Cunningham and Thompson in Tucson, Arizona. The system consists of a down-draft tower that has at its top vertical wetted cellulose pads. Water is distributed at the top of the pads and collected at the bottom by a sump and recirculated by a pump. The tower was attached to a building of about 100 m^2, which was cooled by the air exiting from the tower.

The measured performance of the system was very impressive. When the outdoor temperature maximum was 40.6°C and the Wet Bulb Temperature (WBT) 21.6°C the tower exit air temperature was 23.9°C. The corresponding speed of the exit air at that time was 0.75 m/sec.

Givoni has developed a simple mathematical model of this system, which is described in Chapter 5, predicting the performance of such a tower and also the indoor temperatures of a building cooled by it. The model was validated by comparing measurements of the cooled building performance with predictions by the model. Figures 5.8, 5.26, and 5.27 show diurnal patterns of the measured indoor temperatures in the Tucson building during sequences of five days in June, July, and August, 1985, and the indoor temperatures calculated by the model. It can be seen that, overall, the predicted curves are very close to the measurements.

Inertial Convective/Evaporative "Shower" Cooling Tower

This cooling system has been developed by Givoni while serving as a consultant on cooling outdoor rest areas for the '92 EXPO in Seville, Spain. However, this system can also be applied to cool buildings.

When fine drops of water, which have a very large surface area, are sprayed vertically downward, like a shower, from the top of an open shaft, the falling water contains a large volume of air. The water is collected in a small pond at the bottom of the shaft and is pumped back to the shower head. The momentum of the falling water is thus transmitted to an air stream, creating an inertial air flow down the shaft. The evaporation from the fine drops cools the water, as well as the air in the shaft, to a level close to the ambient Wet Bulb Temperature (WBT). A wind catcher can be placed above the shower head to supplement the inertial air flow by wind effect. Because any evaporation from the droplets takes place in the free air stream any type of water, even brackish water or sea water, if available, could be used for this system.

The performance of this system was tested at UCLA with a tower that was two meters high. Figure 5.11 shows the temperature of the air exiting from the tower, together with the ambient DBT and the WBT, measured on October 8, 1992.

The air temperature drop, as a function of the WBT depression with both fresh water and sea water, is shown in Figure 5.10. The data with the fresh water and the sea water are marked differently. With both types of water the relative temperature drop was about the same, as is shown by the almost identical regression lines. The air temperature drop is about 72% of the WBT depression (72% saturation efficiency).

The fact that the shower system provides the same cooling performance with sea water as with fresh water has profound implications for applicability in arid and desert coastal regions, where fresh water is in short supply while brackish water and sea water often are plentiful. In addition to being cooled the air emerging from the tower is also cleaned of dust and the drops can absorb, and remove from the air stream, some other air pollutants.

Indirect Evaporative Cooling by Roof Ponds

Instead of using evaporation to humidify and cool the ambient air that is introduced into a building, it is possible to use evaporation to cool the roof of the building by placing a cooled pond over its roof. The building then is cooled by conduction across the roof. This is an indirect evaporate cooling, which lowers indoor air and radiant temperatures without elevating the indoor vapor content.

The performance of two test rooms, cooled by two different types of roof ponds, was investigated by the author at the Institute for Desert Research of Ben Gurion University in Israel. The ponds had two different means for preventing solar heating and for providing winter insulation:

- **A** An insulated fixed shade over the pond, with wind flow between the insulation and the water.
- **B** A pond with floating insulation and with water circulation at night over the insulation (invented by Dick Bourne of Davis, California).

The measured water, ceiling, and indoor air temperatures, and the indoor air of a "control" test room with highly insulated roof, are shown in Figure 5.14. The thermal performance of the two types of ponds was very similar, in spite of the basic differences in their design details and cooling processes. In the pond with fixed insulation and continuous evaporation the cooling is produced only by evaporation, which takes place day and night and consumes a large quantity of water. In the pond with floating insulation the cooling effect is produced only at night, mainly by outgoing longwave radiation, and water consumption by evaporation is very small.

Furthermore, the pond with fixed insulation requires the construction of two roofs—the structural roof supporting the pond and a second, insulated roof to shade the pond and provide insulation during the winter. On the other hand the pond with floating insulation requires only the structural roof to support it. When repair is needed the insulating panels can be removed. It seems that this type is the preferable one, considering the simplicity of construction, the lower cost, and the very significant conservation of water.

The water temperature in any one of these types of roof ponds would be about 1 to 2°K above the average ambient Wet Bulb. When the roof is made of materials with high thermal conductivity, such as concrete, the combined water and roof structure serves as an integrated "cold" storage. The ceiling temperature, in the case of a concrete roof over a well insulated building, would be about 2°K above the water temperature. In this way the soffit of the system, the ceiling, serves as a passive cooling panel for the space below, cooling it both by radiation and by natural convection.

Thus the heat exchange between the ceiling and the indoor space is maximized, enabling effective cooling even with a small temperature difference of about 2 to 3°K between the indoor air and the ceiling. This cooling is achieved without elevating the indoor humidity and is accompanied by some reduction of the indoor radiant temperature. These factors enable application of indirect evaporative cooling with higher WBT and maximum air temperatures than direct evaporative cooling.

Applicability of Evaporative Cooling

As the performance of both direct and indirect evaporative cooling depends on the wet bulb temperature this is the main climatic criterion for their applicability. The temperature of the humidified air is above the ambient WBT by about 25% of the WBT depression.

The indoor air temperature in a residential building cooled by a direct evaporative cooling system is about 2 to 4°K above the temperature of the humidified air, depending on the cooled air flow rate, the thermal quality of the building, and its solar load. These factors serve as the basis for suggesting the applicability limits for direct evaporative cooling.

Indoor air speed can be enhanced in a closed building by the use of fans. Consequently, indirect evaporative cooling can be applied in places where the WBT is 24°C and the maximum DBT is 44°C.

THE EARTH AS A COOLING SOURCE

The earth mass under, around, and sometimes also above a building can serve in most climatic regions as a natural cooling source for the building, either in a passive or an active way. In regions with a temperate climate the natural temperature of the soil in summer, at a depth of two to three meters, may be low enough to serve as a cooling source. In hot regions, where cooling is more of interest, the "natural" temperature of the soil in summer is usually too high for it to serve as a cooling source.

However, it is possible by very simple means to lower the earth temperature well below the "natural" temperature characteristics of a given location. Experiments in Israel and North Florida have demonstrated that it is possible to lower the earth surface temperature by about 8 to 10°K below the summer temperature of the exposed soil.

Two methods have been used successfully to lower the earth temperature by shading it while permitting water to evaporate from the surface:

- **A** Covering the soil with a layer of mulch, for example gravel or wood chips, at least 10 cm thick and, in regions with dry summers, irrigating it.
- **B** Raising the building off the ground and allowing water that was provided by either summer rains or irrigation to evaporate from the shaded soil surface.

Once the soil surface temperature in summer is lowered the temperatures of the layers below the surface are also lowered. Figure 6.2 shows diurnal temperature patterns of "natural" and cooled soil in Israel. Figures 6.6 and 6.7 show seasonal and diurnal temperature patterns of cooled soil in Tallahassee, Florida. These figures demonstrate that the difference between the outdoor maximum air temperature and the cooled earth temperature in midsummer can be up to about 12 to 14°K in arid regions and up to 10 to 12°K in some hot humid regions. With such a temperature difference the soil can provide a heat sink for a building, especially in cooling the ventilation air.

Figure 6.6 demonstrates that even in a hot humid region such as Florida it is possible to cool the soil in summer to a level below the outdoor minimum temperature. The difference between the outdoor maximum and the cooled soil is greatest during periods of ambient heat waves.

Options for Utilizing the Cooling Potential of the Earth

When the soil temperature is cool enough, it is possible to utilize it for cooling buildings by several methods. In the case of earth-integrated buildings, in which the walls are bermed and the roof covered by earth, cooling of the earth mass adjacent to the building provides direct passive conductive cooling for the building. This approach would be most suitable in hot, dry regions with mild winters. In such places this direct conductive cooling of the building will be very effective. Concerning the water availability issue, it should be noted that by effectively shading the soil surface with a layer that is thermally insulative but permeable to vapor (for example, with wood bark mulch), effective soil cooling can be achieved with water consumption much lower than what would be needed to water a lawn, for instance.

In hot regions with cold winters the direct-conductive coupling of the indoor space with the surrounding soil, through highly conductive walls and roof, may be effective during the summer. However, this method may be undesirable since it will cause a high rate of heat loss in winter.

Another approach, an active one, is to insulate the building and to install air pipes in the cooled soil, circulating through them the air from the building or the ventilation air. The cooler earth mass then serves as a heat sink to cool the air that is introduced into the building. Heat transfer to the cool soil can be made through an array of pipes, usually plastic, often PVC. The air circulation may be through a closed circuit, or it may be outdoor air that is used for ventilation. Circulating the indoor air through air tubes embedded in the cool soil can keep the indoor temperature about 10°K below the outdoor average maximum air temperature. During a "heat wave" with abnormally high outdoor temperatures the cooling effect of the tubes, embedded in the more stable earth, can be even greater.

In hot, wet regions with summer rains it is possible to raise the building off the ground, thus both minimizing floor wetness problems and providing permanent shading of the soil under the building. The rains would keep the shaded soil moist and the water evaporation would lower its temperature well below the natural soil, as was demonstrated in the study in Florida.

Summary: Applicability of the Various Passive Cooling Systems

This section summarizes the applicability of the various passive cooling systems to different climatic regions.

Comfort Ventilation

Comfort ventilation is the simplest strategy in warm, humid regions. With daytime-effective cross ventilation the indoor air temperature closely follows the outdoor level if it is accompanied by a relatively high indoor air speed. Therefore, the temperature limit of applicability of comfort ventilation is the comfort limit at the enhanced air speed, at any region or season. Consequently, assuming an indoor air speed of 1.5 to 2.0 m/s, comfort ventilation is applicable in regions and seasons when the outdoor *maximum* air temperature does not exceed about 28 to 32°C, depending on the acclimatization and comfort expectations of the population. These conditions are typical of summer in many temperate and warm humid regions.

Nocturnal Ventilative Cooling

From the climatic aspect nocturnal ventilative cooling as a cooling strategy, and the appropriate building design details needed for its satisfactory performance, would be preferable to comfort ventilation in regions where the daytime temperatures in summer are above the upper limit of the comfort zone (with air speed of about 1.5 m/s).This strategy is applicable mainly in arid regions where the daytime temperatures are between 30 and 36°C and the night temperatures are at or below 20°C (to enable sufficient nocturnal cold storage). In this situation daytime ventilation is not desirable, even if it could provide conditions acceptable in the daytime, because additional heat will be stored in the structural mass and raise the indoor temperatures at night, when the wind usually subsides. With outdoor temperatures above about 32°C the indoor temperature in a ventilated building would rise above the comfort limit, even with the higher air speed.

In arid regions with daytime temperatures above 36°C, night ventilation would not maintain the indoor daytime conditions at acceptable levels even in a high-mass building. Other cooling systems should be considered in these regions, at least for operation during the too-hot hours. These systems include evaporative cooling, air flow through earth cooling pipes, and compression air conditioning. But even if an additional cooling system has to be provided at some times, the use of ventilative cooling can significantly reduce the length of the periods when these systems must be operated.

Radiant Cooling

High-mass roofs with operable insulation, either of concrete or metallic with water bags, are effective in providing daytime cooling for single story buildings or the upper floors of multistory buildings in almost any region. The problem with such systems is the cost and the technical problems inherent in the movable insulation. Radiant cooling utilizing metallic specialized radiators with fan-driven air flow can cool *the night air* below the ambient level. The main climatic requirement is low cloudiness during the nights. Humidity is less important as long as the nights are clear. In arid regions a temperature drop of about 3 to 5°K of the air exiting from the radiator can be expected. In humid regions with clear skies the expected temperature drop would be about 2 to 3°K and moisture might be condensed out of the air as it flows under the radiator, thus also lowering the indoor humidity. The cooled air can be used directly to cool a building *during the evening and night hours* in regions where the outdoor air is too warm to provide comfort by direct ventilation. If the building contains enough thermal storage the daytime temperatures also will be lowered.

Comparing Nocturnal Ventilative and Radiant Cooling with Specialized Radiators

These two cooling systems have many similar features. Both collect the cold energy by nocturnal air flow and both systems need interior thermal storage to store the nocturnal cold energy for daytime use. The differences between them are in the complexity and cost on one hand and in the "operating" temperatures of the air on the other hand. The air in a radiant cooling system is cooler by a few degrees than the ambient air used in ventilative cooling.

However, ventilative cooling may be accomplished without any specialized equipment, except for an exhaust fan in regions lacking wind during the night, and it can be applied to any type of building. Radiant cooling is applicable only to a single story and a radiator must be constructed. The provision of a fan is indispensable.

Evaporative Cooling

Direct Evaporative Cooling

Direct evaporative cooling is applicable only in arid regions, and where high quality water is available in abundance. The maximum WBT is the main climatic criterion for the applicability of this system. The indoor maximum air temperature in reasonably insulated residential buildings, cooled by a direct evaporative cooling system, would be about 3 to 4°K above the maximum ambient WBT.

Taking into account the effects on comfort of the higher indoor air speed and the higher humidity associated with direct evaporative cooling, it can be suggested that direct evaporative cooling is advisable only where and when the WBT maximum in summer is not higher than about 22 or 24°C and the DBT is not higher than 42 or 44°C, in developed and in developing countries, respectively.

Indirect Evaporative Cooling by Roof Ponds

The water temperature of a shaded roof pond, and that of the ceiling, follow approximately the *average* ambient WBT. The lowered ceiling temperature lowers the average indoor *radiant* temperature. For these reasons it is possible to apply roof pond cooling in places where the *maximum* WBT and DBT are higher than the applicability limits of direct evaporative cooling. The suggested maximum WBT for application of indirect evaporative cooling is 25°C and the maximum DBT is 46°C.

In buildings cooled by a roof pond the indoor air speed is very low. If the indoor temperature is too high for comfort with still air, a higher indoor air speed can be provided by interior fans, thus improving the thermal comfort of the occupants.

Roof pond cooling can be applied only to a single story building or to the top floor of a multistory building.

The Soil as a Cooling Source

In regions with temperate climate the natural temperature of the soil in summer, at a depth of two to three meters, may be low enough to serve as a cooling source. In hot regions, on the other hand, the "natural" temperature of the soil in summer is usually too high to serve as a cooling source. However, it is possible by very simple means to lower the earth temperature well below the "natural" temperature characteristics of a given location.

When the soil temperature is cool enough, it is possible to use it to cool buildings by several methods. In the case of earth-covered buildings the cool earth mass provides direct conductive passive cooling. This approach would be most suitable in hot, dry regions with mild winters. In regions with hot summers but cold winters the direct-conductive coupling of the indoor space with the surrounding soil may cause a high rate of heat loss in winter. In these regions, indirect "active" coupling of the building to the cool soil, by circulating air through pipes imbedded in the soil, can provide the required cooling.

References

Chandra, S., P. Fairey, and M. Houston. 1983. *A Handbook for Designing Ventilated Buildings.* Cape Canaveral, Florida: Florida Solar Energy Center.

Clark, G. 1989. "Passive Cooling Systems." In J. Cook, Ed., *Passive Cooling,* pp. 347–538.

Cook, J., Ed. 1989. *Passive Cooling.* Cambridge, MA: MIT Press.

Givoni, B. 1976. *Man, Climate and Architecture,* 2nd ed. London: Applied Science Pub.

——— 1994. *Urban and Building Design Guidelines for Different Climates.* In preparation for Publication by Van Nostrand Reinhold.

2

MINIMIZING COOLING NEEDS BY BUILDING DESIGN

Architectural means for minimizing the heat gain of buildings, and consequently their cooling needs, generally would be less expensive than the application of cooling systems, even passive ones. Therefore, there is no point in applying passive cooling systems in a hot climate to a building that does not have an appropriate design for that climate. The subject of building design from the climatic aspect in different climate types is covered in great detail in other books (Givoni 1976, 1994). Here the subject of minimizing cooling needs in hot climates is only briefly discussed. The rest of the book deals with different passive cooling systems.

Minimizing the cooling needs of a building by appropriate architectural design mainly means minimizing the solar load and the conductive daytime heat gain through the envelope. In hot regions the summer is the most stressful season and dealing with summer performance issues may govern the climatic design of the building. The design should therefore aim at lowering the indoor temperatures and enabling effective natural ventilation when ventilation is advisable. However, some regions with hot summers also experience cold winters. In such regions winter performance also should be considered carefully.

The following architectural design features that affect the solar load on a building in summer are discussed below:

— Building layout
— Orientation of main rooms and windows
— Window size, location, and details
— Shading devices for windows
— Color of the building's envelope
— Vegetation near the building

The optimal design of these features may be different in different types of hot climates—for example, in dry and in humid regions.

The thermal properties of building materials and their effect on the conductive heat gain and the cooling needs of buildings are discussed at length in other books (Givoni 1976, 1994). They are not covered in the present book.

BUILDING LAYOUT (SHAPE)

The appropriate layout of buildings in hot regions depends on whether the building is to be air-conditioned most of the time or if it is to rely on natural ventilation whenever ventilation can provide indoor comfort. In the first case, in hot, dry as well as in hot humid regions, the building should be compact, with a minimum of envelope surface area relative to occupied space, in order to reduce the heat gain and the resulting load on the cooling equipment.

On the other hand, when a building is designed to enjoy as much natural ventilation as possible in order to minimize the need for air conditioning, a very compact share is not the best design. In this case the details of a specific optimal shape are somewhat different in hot dry and in hot humid climates.

Building Layout in a Hot Dry Climate

During the summer in a hot dry climate it is both desirable and possible to lower the indoor temperature significantly below the outdoor level during the daytime hours by minimizing the heat gain from the outdoor air. To this end the building should be compact—the surface area of its external envelope should be as small as possible to minimize the heat flow into the building. The ventilation rate should be kept to the minimum required for health (about 0.5 air change per hour in residential buildings) in order to minimize the heating of the interior by the hotter outdoor air. When the construction is not completely airtight, as is the case with the stud wall construction common in the United States, keeping the surface area of the walls and roof small also helps in reducing daytime infiltration.

However, during summer evenings the outdoor temperature in many hot dry regions (and in almost all hot regions during some months) drops down rapidly and reaches a level below the indoors, within or even below the comfort zone. This situation changes the desired climatic performance of the building. The objective in the summer evenings then would be to speed up the cooling rate of the interior as much as possible. From the layout aspect this calls for a spread-out building with greater exposure to the outdoor air. The need to enhance the rate of cooling in the evenings is of special importance in the case of high mass buildings, which naturally have a very slow response to changes in the outdoor temperature.

Using the appropriate architectural design details described below it is possible to change the effective surface area of the building's envelope, on a seasonal, daily, or even an hourly basis, and thus to increase the exposure of the building mass to the cooler evening air while maintaining a compact configuration during the hot hours.

When the building facades are "indented" by deep and narrow porches the surface area of the envelope is greatly increased. However, these porches can be equipped with closeable insulated shutters along the lines of the adjoining walls. When these shutters are closed they become an integral part of the building's envelope. These shutters could be in the form of insulated doors or they could take other forms. An illustration of this configuration is shown in Figure 3.2.

When the insulated shutters are closed the porches, and the openings of the adjoining rooms leading to them, are within an interior or semi-interior space. They are thus protected from hot air, solar radiation, and dust. During the evening and night hours, on the other hand, the insulated shutters should be opened. The porches are then exposed directly to the outdoor air, increasing the surface area of the envelope and the openings through which the building can be cooled by ventilation.

With such geometrical configuration and operating procedures the building is compact during the summer daytime hours (as well as during the winter) and widespread during the nights. The changeable surface area of the effective envelope minimizes the rate of daytime heating and increases the rate of cooling in the evenings.

The area of the porches can be usable year round and thus forms an integral part of the effective floor area of the dwelling.

Building Layout in Hot Humid Climate

In hot humid climate ventilation is the most effective way to minimize the physiological effect of the high humidity. The small diurnal temperature range characteristic of such regions does not enable, in any way, a significant reduction of the indoor daytime temperature below the outdoor level without some cooling system. A spread-out building allows better natural cross-ventilation than a compact one by providing more wall areas, and in more directions, for catching the winds.

Once the building is cross-ventilated during the daytime hours its indoor temperature tends to follow the outdoor pattern. In this case, the heat flow through the envelope is small and the larger surface area does not significantly affect the daytime indoor temperature. On the other hand, during the evening and night hours, when winds usually subside, the envelope's larger area permits faster cooling.

Air conditioning, of course, is too expensive for the vast majority of people in the developing countries, which cover most of the hot, humid regions. In these countries, a building layout that provides good potential for cross-ventilation would be the optimal one.

Building's Orientation (Solar and Wind Considerations)

In considering the orientation of a building the main issue is the orientation of the windows. Solar energy penetration through large windows in summer can elevate a building's indoor temperature high above the outdoor daytime level and thus cause significant thermal stress, as well as increasing the building's cooling load. The problems of indirect solar gain through walls, on the other hand, can be minimized effectively through the use of a reflective (white) color or through shading by plants, as well as by adequate thermal resistance (insulation) of the walls and the roof.

The potential of solar penetration through windows in summer, and its effect on the elevation of the indoor temperature, depends greatly on the orientation of the windows. This is illustrated in Figure 2.1, which shows the indoor temperatures of unventilated thermal models with unshaded windows facing in the four cardinal directions. It can be seen that with an outdoor maximum temperature of about 26°C the indoor maximum in the model with the east window reached about 33°C (elevation of about 9°C above the outdoor temperature at that time). The model with the western window has reached a maximum of about 38°C (12°C above the corresponding outdoor temperature). The maximum temperatures of the models with the south and the north windows were about 30°C, only 4 degrees above the outdoor maximum temperature.

Indoor temperature elevation caused by solar energy penetrating through southern

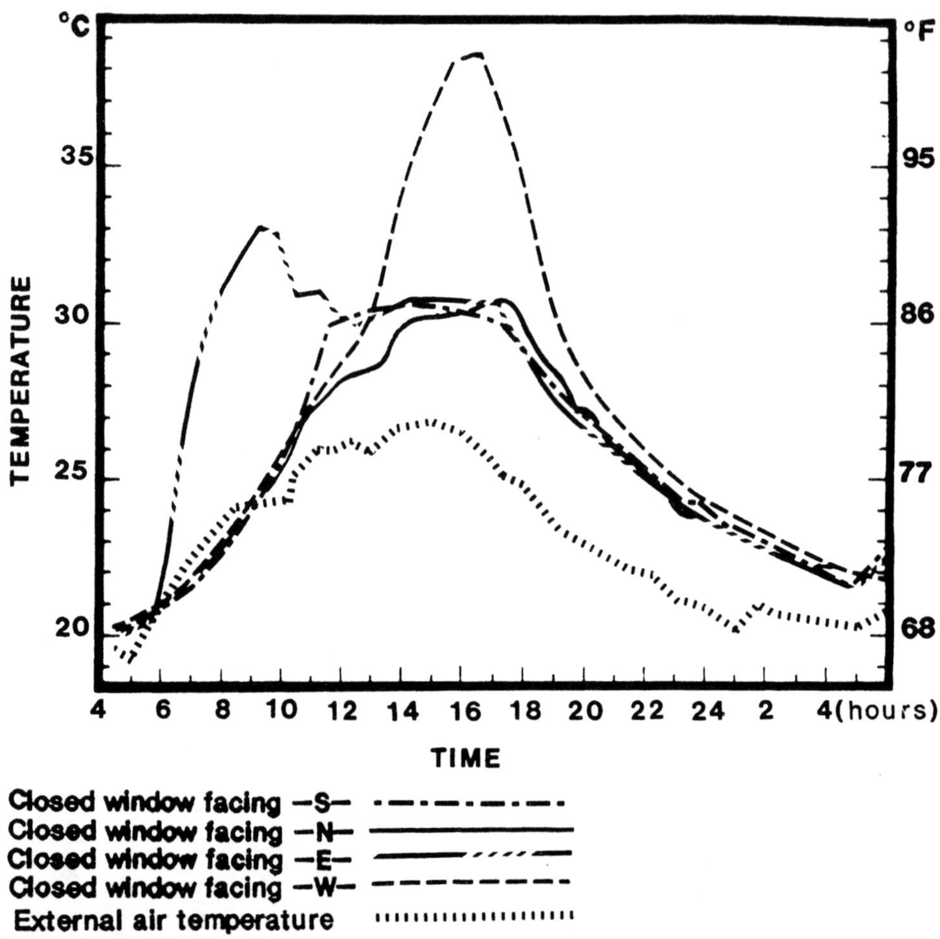

FIGURE 2.1 Indoor temperatures of unventilated thermal models with unshaded windows facing the four cardinal directions.

windows in winter is, of course, desirable. In fact, it is the basis for the "Direct Gain" passive solar heating system.

The main considerations affecting the orientation of the main facades and windows of a building are to some extent different in arid and in humid regions because of the different relative importance of solar and ventilation issues.

Orientation Considerations in Hot Dry Regions

Most of the world's hot dry areas are located in subtropical latitudes where the highest intensities of the impinging solar radiation in summer fall on the eastern and western walls in the summertime and on the southern wall during the winter.

This pattern of solar irradiation on the different walls results in a clear preference in subtropical hot dry regions for north–south orientations for the main facades, and especially for the windows. Such orientation enables easy and inexpensive shading of the southern windows in summer, and of the wall in general, by horizontal overhangs. In the northern hemisphere southern overhangs can effectively block the rays of the summer sun high in the sky (the solar noon altitude of 70 to 80°) while irradiating the southern wall in winter.

The main objective in hot dry regions is to minimize the impact of the sun on the building in summer and this is the main consideration affecting orientation. Although ventilation in the evening hours is also very important in hot dry regions this factor is secondary to the solar aspect in the choice of orientation.

Orientation Considerations in Hot Humid Regions

In hot humid regions the provision of effective cross ventilation under the local wind direction is the major factor that may affect the building's orientation. In view of the primary importance of natural ventilation in hot humid climates a building's relationship to the wind direction should be a major consideration in determining the location of the main rooms—the living and sleeping rooms—during the design stage.

Of course, minimizing solar penetration through the windows and solar absorption at the surfaces of the walls and the roof also is very important in humid regions, but when wind and solar considerations conflict while decisions about building orientations are made, ventilation should be the primary factor. As will be discussed below, the issues of orientation with respect to the sun can be taken care through such measures as shading devices and the color of the envelope. Plants around the building, in particular, can be very effective in humid regions in providing shading both for windows and walls (Givoni 1991-b, Akbari et al. 1992).

Orientation for ventilation does not imply that the building should be perpendicular to the wind direction. Oblique winds at angles between 30 and 120 degrees to the wall can provide effective cross ventilation if openings are provided in the windward and leeward walls. However, the wind must be able to flow inside the building from the inlet to the outlet opening—there should be as few obstacles as possible in its path.

Many hot humid regions at low latitudes have winds mainly from the east—the Trade Wind Belt. In these areas there is a conflict between the best orientation for the solar aspect (south–north) and the optimal orientation for ventilation. This apparent conflict can be resolved by suitable design details.

Solar radiation on a building can be controlled by effective shading of the openings and by the color of the opaque walls (see below). A white wall, or a wall shaded by vegetation, is effectively exposed to a low level of radiation, even when it faces east or west. Similarly, eastern and western windows, if equipped with appropriate *operable* shutters, can be protected from the sun while the building takes advantage of an easterly or westerly wind for ventilation.

Size of Windows

The functions of windows are numerous in any climate—among other things, they provide daylight, visual contact with the outdoors, sun penetration in winter, and ventilation in summer. Here only the impact of window's design on the ventilation and cooling needs of a building will be discussed. A comprehensive review of the various roles of windows and the effects of their design details is presented in a chapter written by Givoni, "Windows," in the book edited by H. J. Cowan (1991).

The demands placed on daytime ventilation are very different in hot dry and hot humid climates and the thermal effect of windows is also very different in these two types of hot climates. Consequently, the issues of a window's size and its effect on the cooling load of buildings are treated separately.

Hot Dry Regions

In hot dry regions, the daytime temperature in properly designed buildings is much higher outdoors than indoors. Heat gain through windows, per unit area, is much higher than through walls or the roof. Conventional windows tend therefore to raise the indoor temperature, and the larger the window area the greater its heating effect, especially when sun penetration is not effectively prevented by shading or orientation. Even when sun penetration is effectively eliminated and the windows are closed, their small thermal resistance and the air infiltration through the surrounding cracks makes them the weakest point from the aspect of heat gain. Sunlight in hot dry regions is very intense, and large windows may cause glare discomfort, reinforcing the notion that small windows are more suitable for deserts than larger ones. This view was supported by the observation that vernacular buildings in desert regions, built mainly from compacted mud, adobe blocks, or stone, usually have very small windows.

However, it should be pointed out that the traditional way of life of the inhabitants of vernacular houses in hot dry regions was to sleep on the roof or in a courtyard, testifying to the uncomfortable indoor conditions that prevailed indoors at night. In fact, good ventilation is of primary importance in hot dry climates during summer evenings and nights. Ventilation can enhance physiological comfort as well as increase the rate of cooling of the building's structural mass. Without effective nighttime cross ventilation a building's interior may be unbearably hot at night, just when comfort is essential for restful sleep.

With special design details, in these regions large windows can provide thermal advantages. When highly insulated shutters are added to large openable windows, their thermal effect can be adjusted to varying needs, both diurnally and annually. In summer the shutters can be closed during the hot hours. Then, light will filter into the house only through the small areas provided by the shutters. In the evening, the shutters and the windows can be opened, increasing the rate of cooling of the interior.

In winter, large southern windows can provide significant direct solar heating of the interior. Closing the insulated shutters during the night traps the heat indoors and reduces the rate of cooling. This helps to maintain comfortable indoor night temperatures.

Hot Humid Regions

When a building is cross ventilated the indoor temperature follows the outdoor pattern closely. If penetration of solar radiation is prevented no significant heat flow through the envelope in the conventional sense takes place and thus well shaded large windows do not affect the cooling needs of the building.

Openings in a hot humid climate play a major role in determining the thermal comfort of the occupants as their location and size determine the ventilation conditions of the building. In this respect, large openings in all the walls can provide the design solution for effective cross ventilation. However, solar radiation can penetrate directly through unshaded openings into the interior of the building and elevate the indoor temperature above the outdoor level. Therefore, utmost care should be taken in ensuring that all openings in the envelope of the building are effectively shaded.

It is desirable to ensure independent cross ventilation to every individual room in the building. This means that each room will have at least two openings, in different walls, preferably one of them in a wall facing the direction of the wind. In practice it is difficult in many cases to have independent cross ventilation of every individual room in the building, especially in large apartment buildings or even in townhouse rows. In such cases, it is important to make sure that air can flow in and out of every room, passing through a series of rooms or a hall in the building on its way to the outlet openings.

When the wind direction is at a very small angle to the wall—that is, when it is nearly parallel to it—it is still possible to create effective cross ventilation. To do so, there must be at least two windows in the windward wall, and each one must have a single "wing wall"—a single vertical projection on one side of the window (Givoni 1969, Chandra 1989). In each one of these windows the projection should be installed on alternate left and right sides. The windows should preferably be vertical, that is, narrow and high. With the wind coming from any direction across an angle of down to about 15° from the wall, one window will be in a wind pressure zone, acting as inlet, and the other window will be in the suction zone, acting as outlet. (See more details on wing walls in Chapter 3).

SHADING OF WINDOWS

Solar gain through windows is a major component of the total heat gain of a building. Minimizing this heating source through the use of shading devices is therefore of primary importance in all types of hot climates. The effectiveness of shading in minimizing the solar effect, as well as minimizing the effect of window's orientation, is illustrated in Figure 2.2 (Givoni 1976). The figure shows the same thermal models shown in Figure 2.1, but here all the windows are shaded by venetian blinds that have been placed outside the glazing. It can be seen that the differences in the indoor temperatures of the models with the different orientations are very small. The temperatures in all models closely follow the outdoor pattern.

Shading devices can be of two major types: fixed or adjustable. Adjustable shading devices can be applied either outside the glazing (external shading) or inside (internal shading). Each one of these types has different performance characteristics and different implications from the architectural viewpoint. In hot dry regions the differences in performance between the different types of shading devices is even more significant than in other climates.

Fixed Shading Devices

Fixed shading devices are major architectural features and they provide one of the main ways to articulate the facades of a building. Many architects have used fixed shading devices as the main determinant of architectural expression. However, the choice of a given shading type is often based not on its thermal effect but rather on stylistic considerations.

There is a consensus with respect to the appropriate fixed shading for southern and northern windows. In the northern hemisphere a horizontal overhang above a southern window, extending sufficiently on both sides, can be designed to provide complete

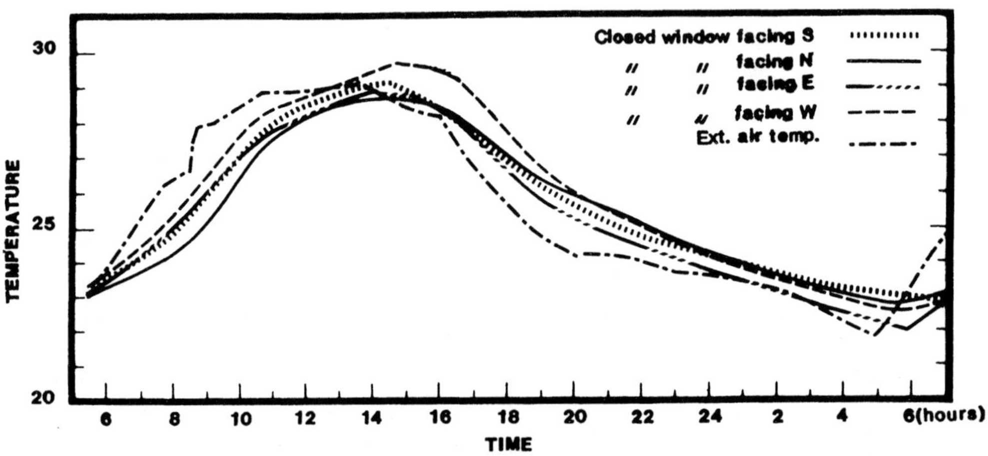

FIGURE 2.2 The same thermal models shown in Figure 2.1, but with all the windows shaded by venetian blinds placed outside the glazing.

shading during midsummer (April through August) and to permit solar penetration in winter (October through February). Vertical fins protect northern windows in summer from the low sun in the morning and late afternoon hours.

On the other hand, there is a common misconception concerning the appropriate shading for eastern and western windows. In many books and articles it is stated that for these orientations vertical shading devices (fins) are more effective than horizontal overhangs of the same depth. In fact, even for eastern and western windows fixed horizontal overhangs are more effective for summer shading, although neither can provide complete shading in these orientations.

The only way an eastern or western window can be shaded effectively by fixed devices is by a "frame" consisting of a horizontal overhang and a vertical fin oblique at 45° towards the south (Givoni 1976).

External Operable Shading Devices

Operable external shading devices can be of many types, including shutters (hinged, sliding, etc.) rotatable fins, horizontal plates, retractable venetian blinds, or canvas awnings. They can be made of many materials, such as wood, metals, asbestos-cement, fabrics, and so on. The common feature to all operable devices is that they can be adjusted at will to either exclude or admit solar radiation. As with fixed devices, they intercept the sun's rays before the rays hit the glazing. Many of the operable types can also intercept solar radiation reflected from the ground, in addition to intercepting the direct and most of the diffused radiation from the sky. Operable shading devices can admit all of the solar radiation when this is desirable, as it is in winter. Therefore, they are inherently more effective than the fixed shading. Operable external shading devices can reduce solar heat gain through windows and other glazed areas down to about 10 to 15% of the radiation impinging on the wall.

The color of external operable shading devices may have little effect on their thermal performance unless they are completely closed. With dark shading most of the solar radiation is absorbed at the outer surface of the shades and then mostly dissipated to the outdoors by convection. With white shading most of the radiation is reflected away but part is reflected inward. The net gain is almost the same in these two cases.

However, the solar radiation reflected indoors by white shutters provides much more daylighting than is allowed indoors through dark shutters. Therefore, from the daylighting point of view, white shutters are preferable.

Internal Operable Shading

Internal shading devices can be in many forms, including venetian blinds, roller blinds, and curtains. All of them intercept the sun's rays after the rays are transmitted indoors through the glazing. In consequence, all the different types of internal shading have low efficiency from the thermal point of view, although they can effectively control indoor lighting conditions.

Although white internal shades reflect more radiation outward than dark shades do, the overall heat gain of internal shading devices, even white ones, is much more than that of external shading. Therefore, internal shading generally is not suitable in hot regions, especially when applied to large windows and other (fixed) glazed areas.

COLOR OF THE ENVELOPE

The color of the external surfaces of the walls and the roof has a tremendous effect on the impact of the sun on the building and on the indoor temperature of un-conditioned buildings, particularly in desert regions where solar intensity is higher than in other regions. It may affect the practicality of applying passive cooling to a given building even in climates where the potential for passive cooling does exist.

The effect of the external color of walls and roofs was studied extensively by Givoni and Hoffman (1968) at the Technion in Haifa, Israel (a coastal city with a humid Mediterranean climate), with different types of thermal models. The research covered the effect of colors of roofs and walls. The effect of roof color was tested with lightweight roofs made of insulated panels (plywood/Polystyrene sandwiches) and of Ytong (lightweight concrete, 600 Kg/m^3). The effect of wall color was studied with test cells built of different materials (1.35×1.47×1.35 meters). Each test cell had two small windows (25×40 cm) that could be closed by insulated panels. The roofs were white in all the experiments that tested the effect of wall color.

Here the effect of roof color will be summarized only with respect to Ytong roofs. The effect of wall color will be shown with unventilated test cells built of 22 cm thick walls of various materials. A more detailed discussion of the results of this research, including the interaction between the effects of color and those of ventilation, materials, and wall thickness, is presented in another book (Givoni 1994).

Effect of Roof Color

Figure 2.3 shows the ceiling temperatures of Ytong roofs, 7, 12, and 20 cm thick, painted gray, together with the outdoor temperature and the average external surface temperature of the roofs. With outdoor air maximum temperature of about 31°C the average maximum external surface temperature was about 69°C. The maximum ceiling temperatures were affected, of course, by the roof thickness: they were about 45, 39 and 33°C for the 7, 12, and 20 cm thick roofs, respectively.

Figure 2.4 shows the respective temperatures when the roofs were painted white. With maximum air temperature of about 27°C the average external maximum surface temperature was about 27.5°C. The ceiling maximum temperatures were about the same for all roofs, around 25.5°C, with a time lag of the maxima increasing with roof thickness. In fact, the diurnal average of the external surface temperature was *lower* than the air average, and the ceiling's minima were *lower* as the roof was *thinner*, indicating that the 24-hour longwave

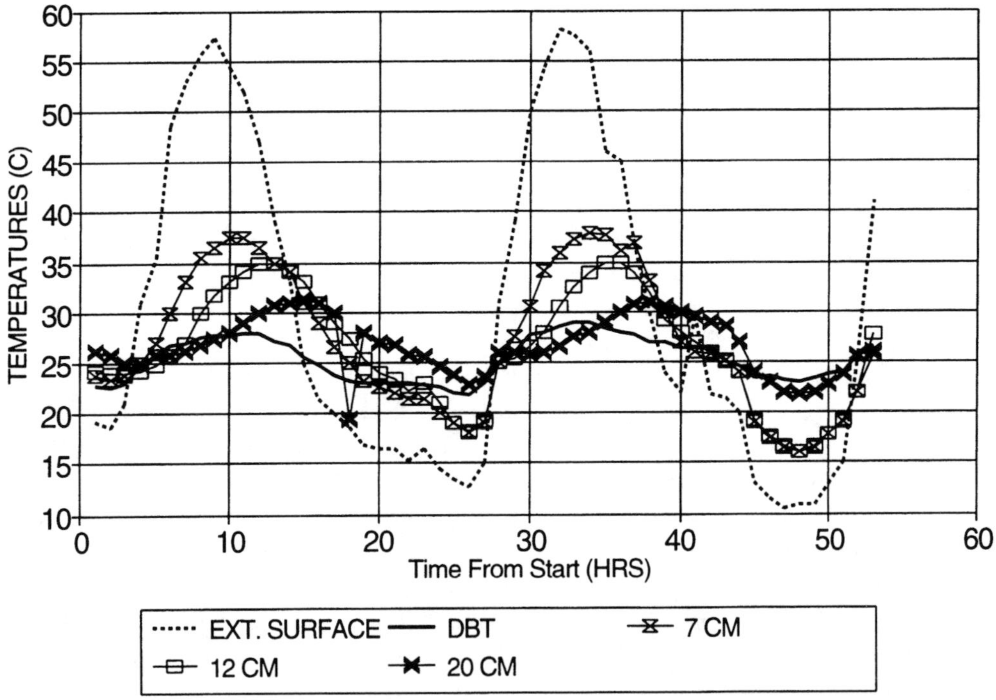

FIGURE 2.3 Ceiling temperatures of Ytong roofs, 7, 12, and 20 cm thick, painted gray, together with the outdoor temperature and the average external surface temperature of the roofs.

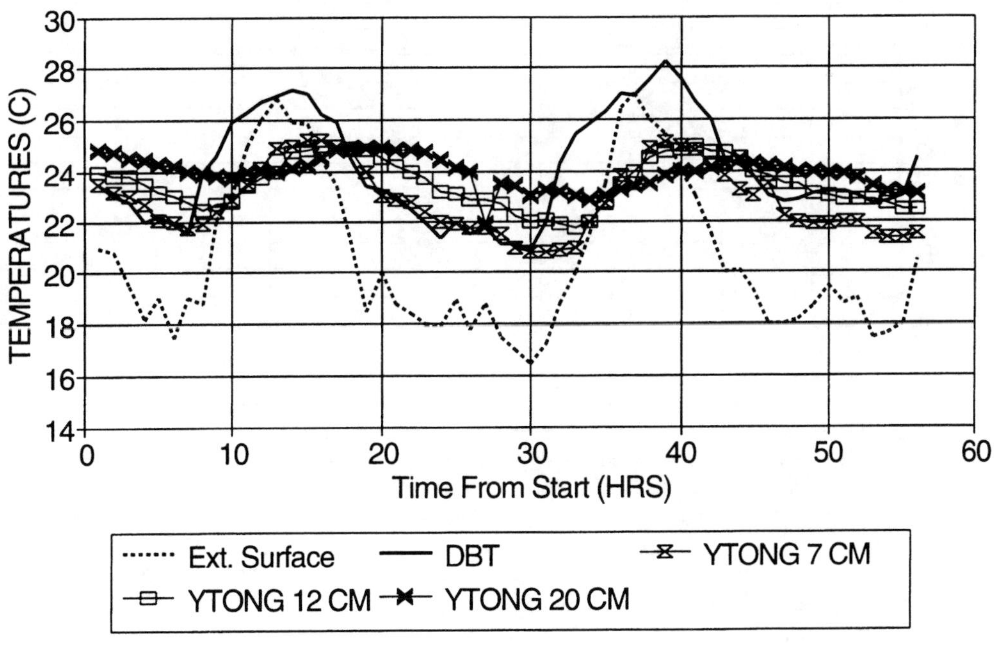

FIGURE 2.4 Ceiling temperatures of Ytong roofs, 7, 12, and 20 cm thick, painted white, together with the outdoor temperature and the average external surface temperature of the roofs.

radiant loss was greater than the solar energy absorbed in the white roofs. This phenomenon, of white roofs being at lower diurnal temperatures than the ambient air, was observed repeatedly, with different roof types, even in midsummer on very clear days.

Effect of Wall Color

Figure 2.5 shows the temperatures of the indoor air of four test cells in Haifa (solid concrete and Ytong 12 and 22 cm thick), when the walls were painted gray and the windows were closed. The indoor temperatures were constantly above the outdoor. The average indoor air temperatures of the cells with the thick walls were about 4.5°K above the outdoor average and those of the cells with the thin walls about 4.4°K. It should be noted that this elevation was caused only by the solar energy absorbed in the walls—the roofs were kept white. The indoor maxima were above the outdoor maximum. In the case of the cell with the 12 cm concrete walls the temperature difference was about 8°K.

Figure 2.6 shows similar temperature patterns when the walls were painted white. During most of the daytime hours the indoor temperatures were below the outdoor. The daily average elevations above the outdoor average air temperature of the indoor air was about 1.1°K. The indoor maxima of the cells with the thick walls were about 2°K below the outdoor maximum and the maxima of the cells with the thin walls were at about the same level.

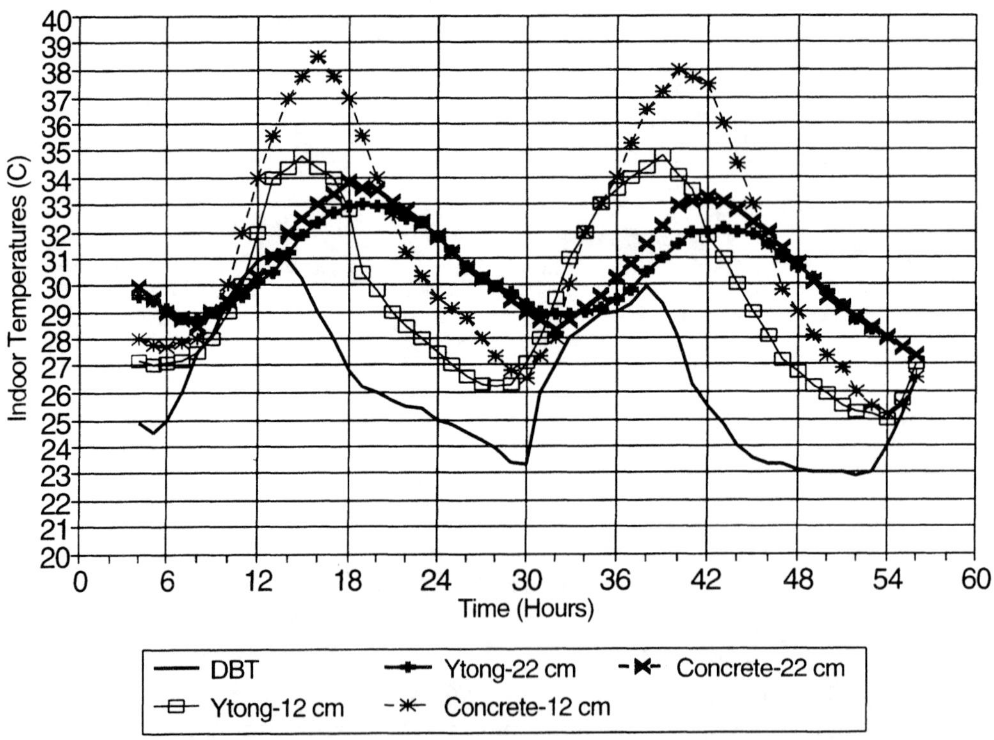

FIGURE 2.5 Temperatures of indoor air and of interior surfaces of the western walls, of three test cells (solid concrete, hollow concrete block and Ytong). Walls gray and windows closed.

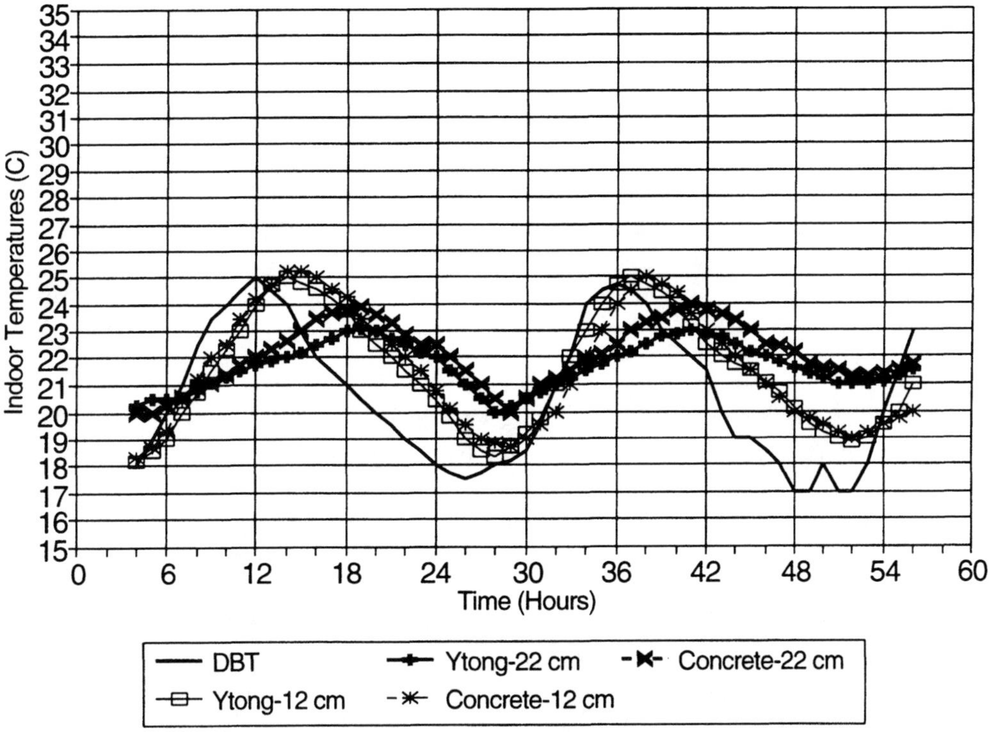

FIGURE 2.6 Temperatures of indoor air and of interior surfaces of the western walls, of three test cells (solid concrete, hollow concrete block and Ytong). Walls white and windows closed.

It should be noted that these measurements were taken in a coastal city with a humid climate characterized by small outdoor diurnal temperature swings. The depression of the daytime indoor temperature of white painted buildings below the outdoor level is much greater in arid regions, with a larger temperature swing. More data of the indoor temperatures of white painted buildings in an arid climate is presented in Chapter 3.

As the choice of color does not involve extra cost, a reflective color of the building envelope—white—is a very powerful architectural climatic control feature, and it is the most cost-effective way to minimize the building's heat load in summer, as was demonstrated in this research.

Because of the different intensities of solar radiation incidence on the roof and on walls with different orientations, the importance of color as a controller of the indoor temperature is variable. For the roof, the influence of color is at a maximum. The difference in the maximum external surface temperature between a white roof and a black one in a desert in the summer can be 30 to 40°C. The resulting heat gain to the building interior depends on the thermophysical properties of the roof, but in general it is quite significant. A western wall is almost as sensitive to the effect of color as the roof.

Eastern and southern walls are also very sensitive to their external color, while the northern wall is the least sensitive. The southern wall presents a special case because it receives most radiation in winter, when heating may be desirable.

Climatic Impact of Plants Around Buildings

Plants, in the form of trees, vines, high shrubs, and pergolas, can provide effective shading for walls and windows. They can even shade the roofs of small low-rise buildings. A detailed discussion of the effects of plants is provided in Givoni 1991-b and Givoni 1994. Here only the issue of shading by plants and its effect on the cooling load will be summarized briefly.

Plants can affect the indoor temperature and the cooling load of buildings in several ways:

- **A** Trees with high canopies and pergolas near walls and windows provide shade and reduce the solar heat gain with relatively small blockage of the wind (shading effect).
- **B** Vines climbing over walls and high shrubs next to the walls, while providing shade, also reduce the wind speed next to the walls appreciably (shading and insulation effects).
- **C** Plants near a building can lower the air temperature next to the building's skin, as has been measured by UCLA student Al Hemiddi (1991), thus reducing the conductive and infiltration heat gains. In winter of course, they reduce the desired solar gain and may increase wall wetness after rains.
- **D** Ground cover by plants around a building reduces the reflected solar radiation and the longwave radiation emitted toward the walls from the surrounding area, thus lowering the solar and longwave heat gain in summer.
- **E** If the ambient temperature around the condenser of a building's air conditioning unit can be lowered by surrounding plants, the COP of the system can be improved, so that less electrical power is consumed for a given cooling energy delivered to the building.

Although a lot has been written on the thermal effect of plants in urban areas, not many experimental studies were done on the subject. Parker (1983, 1987, 1989) reported on research done in Miami, Florida, in which the effect of landscaping on wall temperature was measured. On hot sunny late-summer days the average temperature of walls shaded by trees or by a combination of trees and shrubs was reduced by 13.5 to 15.5°C. Climbing vines reduced the surface temperature by 10 to 12°C.

Parker (1983) has also measured the effect of landscaping on cooling energy consumption. The building was an insulated double-width mobile home serving as a day-care center for children. Energy consumption by the air conditioners was compared during days with similar weather conditions in two periods, before and after landscaping. The landscaping consisted of trees and shrubs planted around the building. Average daily rate of energy consumption for air conditioning on hot summer days in the period without the plants' shading was 5.56 Kw. After landscaping, it was reduced to 2.28 Kw. The effect of planting was even more marked during the afternoon hours (the peak load period): the average rate of energy use was reduced from 8.65 to 3.67 Kw.

McPherson et al. (1989) measured the effect of landscaping on the cooling energy consumption of three ¼ scale models of a building. The landscape treatments consisted of: (1) Bermuda grass turf around the building and no shade, (2) rock mulch around the buildings and the walls shaded by shrubs, and (3) rock mulch with neither grass nor shade.

The noon-time surface temperature of grass turf was lower by about 15°C than that of the rocks. The air temperature, about 0.5 m. above the turf, was lower by about 2°C than the rocks. The model with the rock mulch consumed between 20% to 30% more cooling energy than the models with the turf and the shrubs. The turf apparently had reduced the longwave radiant heat gain and the ambient air temperature next to the building's skin, while the shrubs reduced the solar heat gain.

Hoyano (1988) has summarized the results of several experimental studies he has performed in Japan, dealing with the effects of plants on surface and air temperatures.

A vertical vine was installed in front of the south-west veranda of a house and a comparison was made with an unscreened veranda with the same orientation. The screen was effective in providing shade—the temperature inside the screened veranda was significantly lower than in the unscreened veranda. However, the leaf temperature was higher than the ambient air. The air temperature inside the screened veranda was higher than the ambient air but lower than inside the unscreened veranda.

The vertical vine screen significantly reduced the wind speed through the window in the screened veranda and the cross ventilation of the room behind it, as compared with the unscreened veranda. Without the screen the air speed at the center of the window was, on average, about 45% of the outdoor wind, while with the vine screen in the veranda it was about 17%. Thus the overall effect of a vertical screen on comfort in a hot humid climate may be negative, due to the reduction of the indoor air speed.

The effect of Japanese ivy covering a bare west concrete wall was evaluated by comparing the temperatures across the wall before and after the ivy cover. Without the ivy the exterior surface maximum temperature was about 10°C above the ambient air while with the ivy cover it was about 1°C below it. The average surface temperature of the exposed wall was about 3°C above the ambient air while with the ivy it was only 1°C above the outdoor average temperature.

Under the direction of the author a UCLA graduate student, Nasser Al Hemiddi, has measured the effect of plants (Al Hemiddi 1991). In this study the surface and air temperatures (at about 1 meter height) above land areas with different ground treatments were measured in the UCLA campus around noon time, during different periods over a whole year, for a total of about 70 days. The ground treatments included pavements exposed or shaded by trees, an exposed lawn, and a space between a high and dense shrub fence and a building.

During clear days, especially in summer, differences in the air temperatures of up to about 3°K were often observed between the air above an exposed pavement and the space behind the shrubs. Such differences in the air temperature next to the building skin, in addition to the shading effects of the plants, can significantly reduce the heat gain through walls and the resulting energy consumption for air conditioning.

A review of measured data on the cooling savings from landscaping is also presented in Meier (1990).

References

Akbari, H., S. Davis, S. Dorsano, J. Huang, and S. Winnett. 1992. *Cooling Our Communities.* US Environmental Protection Agency—PM-221.

Al Hemiddi, N. 1991. "Measurements of Surface and Air Temperatures Over Sites with Different Land Treatments." Submitted to the PLEA'91 Conference in Seville, Spain.

Chandra, S., P. Fairy, and M. Houston. 1983. *A Handbook for Designing Ventilated Buildings.* Cape Canaveral, Florida: Florida Solar Energy Center.

Clark, G. 1989. "Passive Cooling Systems." In J. Cook, Ed., *Passive Cooling,* pp. 347–538.

Cook, J., Ed. 1989. *Passive Cooling.* Cambridge, MA: MIT Press.

Cowan, H. J. 1992. Encyclopedia of Building Technology. New York: Van Nostrand Reinhold.

Givoni, B. 1976. *Man, Climate and Architecture.* 2nd Ed., Applied Science Pub., London.

——— 1987. "Passive Cooling—State of the Art." Portland, OR: Proc, 12th Passive Solar Conf., pp. 11–19.

——— 1991a. "Performance and Applicability of Passive and Low Energy Cooling Systems." Energy and Buildings, V-17, pp. 177–199.

——— 1991b. "Impact of Planted Areas on Urban Environmental Quality." *Atmospheric Environment,* 25B, pp. 289–300.

——— 1992. *Windows in Buildings.* Chapter 18 in Cowan, 1992.

——— 1994. *Building and Urban Design Guidelines for Different Climates.* New York: Van Nostrand Reinhold.

Givoni, B., and M. E. Hoffman. 1968. *Effect of Building Materials on Internal Temperatures.* Technion: Israel Institute of Technology: Research Report, Building Research Station.

Hoyano, A. 1988. "Climatological Uses of Plants for Solar Control and the Effects on the Thermal Environment of a Building." In *Energy and Buildings,* V-11, pp. 181–199.

McPherson, E. G., J. R. Simpson, and M. Livingston. 1989. "Effect of Three Landscape Treatments on Residential Energy and Water Use in Tucson, AZ." *Energy and Buildings,* V-13 (2), pp. 127–138.

Meier, A. K. 1990. "Measured Cooling Savings From Vegetative Landscaping." Washington, D.C. Proc, *ACEEE 1990 Summer Study on Energy Efficiency in Buildings—Environment,* pp. 4.133–4.143.

Parker, J. H. 1983. "The Effectiveness of Vegetation on Residential Cooling." *Passive Solar Journal,* pp. 123–132.

——— 1987. "The Use of Shrubs in Energy Conservation Plantings." *Landscape Journal,* V-6, pp. 132–139.

——— 1989. "The Impact of Vegetation on Air Conditioning Consumption." Proceedings, *Conf. on Controlling the Summer Heat Island.* Lawrence Berkeley Laboratory, Berkeley, California. LBL-27872, pp. 46–52.

3

VENTILATIVE COOLING

There are two ways in which ventilation can improve comfort. One is a direct physiological effect. For example, opening the windows to let the wind in, and thus providing a higher indoor air speed, makes people inside a building feel cooler. This approach is termed **comfort ventilation.** The other way is an indirect one: to ventilate the building only at night and thus to cool the interior mass of the building. During the following day the cooled mass reduces the rate of indoor temperature rise and thus provides a cooling effect. This strategy is termed **nocturnal ventilative cooling.** From the building design viewpoint it calls for some different details than comfort ventilation, and it is discussed in the second part of this chapter.

When a building is cross-ventilated during the daytime the temperature of the indoor air and surfaces closely follow the ambient temperature. Therefore there is a point in applying daytime ventilation only when **indoor comfort** can be experienced at the **outdoor air temperature** (with acceptable indoor air speed). Furthermore, in unventilated buildings the indoor temperature during the daytime often can be appreciably below the outdoor temperature. In this case daytime ventilation actually raises the temperature of the indoor air. In this situation nocturnal ventilation may be more advisable.

Because buildings are usually occupied during the daytime hours, nocturnal ventilative cooling can be effective only to the extent that it can lower the indoor temperatures during the following day. It is particularly important that the indoor maximum temperature the next day be lowered. Such lower indoor temperatures can be achieved only if the building is closed and unventilated during the daytime hours. In this respect daytime comfort ventilation and convective cooling are mutually exclusive. At any given place or on any given day one or the other should be considered as the best approach to provide daytime comfort.

Daytime ventilation cannot bring the indoor average temperature below the outdoor level. In this respect it is not a passive cooling system according to the definition presented in Chapter 1, but it is part of building climatology and of a climatically appropriate design. Still, in many cases it can make mechanical cooling unnecessary and therefore it is dealt with in this book.

This chapter discusses the two modes of ventilative cooling. Short sections deal also with ventilation requirements from the thermal comfort aspect and with building design factors that affect natural ventilation. In another book (Givoni 1994) these two subjects are dealt with much more comprehensively.

SECTION 1: COMFORT VENTILATION

The simplest strategy for improving comfort when the indoor temperature, under still air conditions, is felt as too warm is to open the window and to enhance comfort by ventilation: providing comfort through higher indoor air speeds. Introducing outdoor air with a given speed into a building may provide a direct physiological cooling effect even when the indoor air temperature is actually elevated. This is particularly the case when the humidity is high, as the higher air speed increases the rate of sweat evaporation from the skin, thus minimizing the discomfort people feel when their skin is wet.

Such comfort ventilation may be desirable, from the physiological viewpoint, even when the outdoor air temperature is higher than the indoor air temperature, because the upper temperature limit of comfort is shifted upward with a higher air speed. Therefore, even if the indoor temperature is actually elevated by ventilation with the warmer outdoor air, the effect on the comfort of the occupants, up to a given temperature limit, might be beneficial. This is especially the case in hot, humid regions.

COMFORT STANDARDS AND VENTILATION REQUIREMENTS

The delineation of the boundaries of the climatic conditions under which ventilation enhances comfort and can be applied instead of cooling systems that actually reduce the indoor temperature are closely related to the issue of the upper temperature and air speed limits of the comfort zone, that is, to comfort standards. In particular, specifying the limit of acceptable indoor air speeds from the comfort aspect can determine the climatic applicabilities of comfort ventilation on the one hand and of nocturnal ventilative cooling on the other hand.

Comfort standards, such as the ASHRAE comfort zone in the United States, were developed mainly for use by the air conditioning industry. They specify the boundaries of air temperature and humidity for sedentary people, within which the **mechanical system** has to maintain the indoor climate. They were constructed mainly for use in air conditioned office buildings but they are also widely used in evaluating the indoor climate in residential buildings. The ASHRAE comfort zone is also used extensively as the basis for structuring bioclimatic charts.

The Comfort Zone in the ASHRAE Guide (1985) extends between two fixed levels of vapor content, 4 and 12 g/kg. For still air conditions (0.15 m/s in winter and 0.25 m/s in summer) the ASHRAE boundaries of acceptable temperature in summer extends from 23 to about 26.5°C. With higher air speed the upper temperature limit is elevated by 1°C for an increase of 0.275 m/s, up to a temperature of 28°C at air speed of 0.8 m/s, which is the highest indoor air speed allowed by the ASHRAE Guide. The acceptable upper humidity limit is not affected at all by the higher air speed in the ASHRAE Standard.

Some problems exist when these comfort standards, or similar standards developed by other societies of air conditioning engineers, are used in buildings that are not air conditioned, especially in developing countries with hot climates. One issue concerns the boundaries of acceptable climatic conditions in buildings that are not air conditioned. Another problem concerns the application of the ASHRAE humidity and air speed limits in hot humid locations, taking into account the actual acclimatization and comfort expectations of residents in such places and the role of higher air speeds in enhancing comfort at high humidity.

Indoor climate in un-conditioned buildings responds to the variations in outdoor climate and residents usually experience a wider diurnal (and annual) climatic range than they would in air conditioned buildings. For example, indoor temperatures ranging from 20°C in the morning to 28°C in the afternoon on summer days are common in un-conditioned, naturally ventilated, buildings. The indoor air speed in cross ventilated buildings is often around 1 to 2 m/s. People who live in naturally ventilated buildings usually accept such a wider range of temperatures and air speeds as normal.

It is also reasonable to assume that people in developing hot countries living mostly in un-conditioned buildings are acclimatized to, and would accept, higher temperature and/or higher humidity.

Research on the effect of acclimatization and standard of living on comfort sensations and expectations in residential buildings in hot developing countries is limited, although some studies can be mentioned (Webb 1959, Nicol 1974, Humphreys 1975, and Tanabe 1988). These studies suggest that there is a real rise in the temperature that people consider acceptable as the local average annual temperature of the place is higher. Humphreys (1975), in summarizing previous studies on comfort in different countries, has derived a formula correlating the "neutral" temperatures observed in the different studies (T_n) with the mean air temperature of the location during the experimental period of each study (T_m):

$$T_n = 2.6 + 0.831\,T_m;\ (^\circ C)$$

No studies are known to the author that deal with the relationship between acclimatization to hot humid climate and the subjective response to high humidity, especially the discomfort from wetness of the skin.

In addition to the humidity issue, the subject of acceptable **indoor air speed** in un-conditioned buildings deserves special attention. A higher air speed is the most common "remedy" to minimize discomfort from high temperature, and especially from high humidity. The ASHRAE limit of 0.8 m/s seems therefore to be far too restrictive when dealing with providing comfort in un-conditioned residential buildings in countries with hot summers.

Comfort ventilation can be provided at any time when it is physiologically beneficial, during day time and/or evening and night hours, in contrast with nocturnal ventilative cooling. With a high ventilation rate the indoor *air* temperature tends to approach the outdoor level, although the *radiant* temperature of indoor surfaces may be at a significantly different, and usually higher, level, especially during the evening and night hours.

The radiant temperature of an *unventilated* indoor space, in the absence of solar penetration, is close to the average temperatures of the surfaces surrounding the space, weighted according to their respective areas. During daytime hours, mainly around noon and in the afternoon, the temperatures of the internal surfaces of the external walls and of the ceiling tend to be higher than the indoor air temperature, as the direction of the heat flow across the envelope is *inward.* The difference between the indoor air temperature and the radiant temperature increases as the walls (and the roof) are of lower thermal resistance, of higher heat capacity (mass), and with darker external colors.

A strategy relying on comfort ventilation, which is common mainly in warm, humid regions, calls for design details that will maximize the comfort achievable with natural ventilation. The relevant design factors are discussed in the next section. When dealing with ventilation in the context of its effect on comfort, it should be remembered that the important factor is the air speed over the human body. This air speed can be increased by opening the windows for natural ventilation with outdoor air, and also by the use of such devices as ceiling fans in closed buildings. The later option is often preferable when nocturnal ventilative cooling is the chosen strategy.

Climatic Applicability of Comfort Ventilation

Providing a higher indoor air speed in hot regions by natural ventilation is desirable when the outdoor air is at a lower temperature than the indoor air or when it prevents elevation of the indoor vapor content of the air. This situation is common in hot, humid regions. On the other hand, appropriate building design in hot, dry regions can keep the indoor temperature during the daytime at significantly lower levels than the outdoor maximum temperature. The vapor content of the indoor air in such regions is, in most cases, low enough so that indoor vapor generation does not raise it to levels producing discomfort due to sensible perspiration. In this case, when a higher indoor air speed is desired, it may be preferable to provide it by fans rather than to ventilate the building with warmer outdoor air by opening the windows.

Assuming an indoor air speed of 1.5 to 2.0 m/s, comfort ventilation is applicable mainly in regions and seasons when the outdoor *maximum* air temperature does not exceed about 28 to 32°C, depending on the humidity level and the acclimatization of the population, and where the diurnal temperature range is less than about 10°C (Givoni 1991).

These conditions are typical in warm humid climates. In these regions the relatively

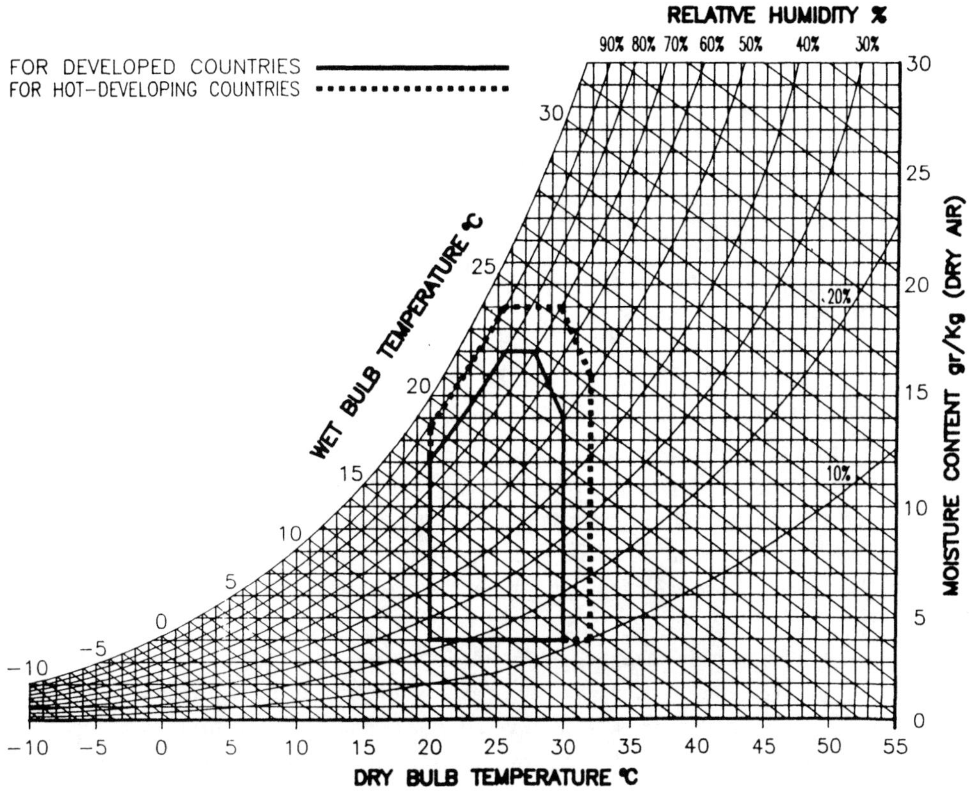

FIGURE 3.1 Suggested boundaries of outdoor temperature and humidity within which indoor comfort can be provided by natural ventilation. Assumed air speed 2 m/s.

small diurnal temperature range does not enable a significant reduction of the indoor daytime temperature below the outdoor level, even in a closed, high-mass, well insulated building.

Figure 3.1 (Givoni 1992a) shows the suggested boundaries of the outdoor temperature and humidity within which indoor comfort can be provided by natural ventilation during the day, with indoor air speed of about 2 m/s (a very light breeze). Two sets of boundaries are suggested: one for developed countries and the other for developing countries, especially those with hot climates.

Building Design for Natural Ventilation

The rate of air flow through a naturally ventilated building and the distribution of air velocities throughout the ventilated space depend both on the wind conditions around the building and on design details of the building itself. This section discusses the effects of the design of the building itself and, in particular, the effect of the design of the openings through which ventilation takes place on indoor ventilation.

The wind conditions around the building under given regional wind speed and direction depend in general on the planning of the town and, in particular, of the neighborhood where the building is situated. This will be discussed in the subsequent section. More detailed discussion of the effect of building and urban design on ventilation is presented in another book (Givoni 1994).

The main details of building design that affect indoor ventilation conditions are:

- Geometrical configuration of the building's envelope: its shape and profilation (projection, recesses, etc.).
- Location of openings with respect to wind direction.
- Total area of openings in the pressure and suction regions of the building's envelope.
- Type of windows and details of their opening.
- Vertical location of openings.
- Interior obstructions to airflow from the inlet to the outlet openings, particularly when the air is flowing through more than a single room. This includes information such as the size and location of the openings connecting the adjacent rooms—the doors.
- Presence or absence of fly screens in the openings

The major impact of these design details depends on the extent to which they enable or inhibit cross ventilation in the building as a whole and in its individual rooms. Cross ventilation is defined as the situation in which outdoor air can flow from openings on one side of the building (the inlet openings, or pressure zones around the building) through the building and out via openings on the suction sections of the building.

Criteria for Evaluating Indoor Ventilation Conditions

In evaluating the ventilation from the human comfort point of view, attention should be paid not only to the overall amount of airflow but also to the distribution of air velocities throughout the ventilated space. In any discussion concerning ventilation, quantitative criteria by which indoor ventilation conditions can be evaluated are needed. Examples of such criteria, as used by various investigators, are:

- Air speed at the inlet opening.
- Maximum speed at any point in the space.
- Average air speed in the space.
- Average speed at occupancy level (that is, at about 1 m above the floor).

It should be pointed out that there is no unique relationship between air speeds at various points in a given room. The airflow may be concentrated in a stream flowing directly from the inlet to the outlet opening, leaving the rest of the room's space at a much lower velocity. This pattern is typical when the wind is perpendicular to a small inlet opening and the outlet is larger and is located in the direct line of the flow. In this case the air speed at the inlet is maximized, while the average speed over the entire room is much lower than the maximum speed. On the other hand, when the wind is oblique or when the outlet is not in line with the initial flow, the air has to change direction within the room.

This change of direction sets off a turbulent flow in the whole room. While the overall airflow is reduced, the distribution of air velocities in the room is better and the *average* velocity may be higher.

Effect of Specific Design Factors on Natural Ventilation

Detailed discussion of the effect of building design on its ventilation is presented in another book (Givoni 1994). Here this subject is discussed only briefly.

Type and Shape of the Building

The type of a building, whether a detached house, a townhouse, a high-rise apartment house with single-loaded or double-loaded corridors, or a square building with four units, determines in practice the possibilities of providing effective cross ventilation. The main issue to consider is the possibility of providing, in the same residential unit, openings on both the windward and leeward sides. Detached houses with external walls that offer the possibility of openings all around may be thought of as the most advantageous construction. In practice, however, it is possible to design other types of buildings, such as townhouses and apartment buildings, with good cross ventilation.

In some building types it may be impossible, or quite complicated, to provide effective natural ventilation, at least for some of the dwelling units. In some cases special design details can solve the problem:

1. Long apartment buildings with central double-loaded corridors, which are common in many United States housing projects, rarely can provide cross ventilation in the dwelling units. Such apartments generally have to use mechanical air conditioning for most of the summer, even in places where natural ventilation can provide satisfactory indoor comfort.
2. Long apartment blocks with single-loaded corridors can have cross ventilation when windows are provided in the wall facing the corridor. However, this compromises residents' privacy—anyone walking in the corridor can hear and see what is going on in the apartments when the windows are open. This problem can be solved by special details that enable cross ventilation without loss of privacy. For example, if the rooms have a ceiling height of about 3 meters and the corridors can have a ceiling height of about 2.3 meters, it is possible to provide a breezeway above a false ceiling in the corridors, permitting airflow through the apartments to the other side of the building.
3. A multistory square building with a central staircase or elevator surrounded by four units cannot provide good ventilation conditions for at least one of the units in each floor, the one located on the leeward side of the building.
4. Blocks of townhouses, two or three stories high, can enable good ventilation conditions for all units, as each unit has at least two external walls, and thus it is possible to provide windows on opposite sides of the building.
5. Medium-height apartment buildings with only two or three apartments on each floor per staircase or elevator can provide good cross ventilation to all dwelling units, as each apartment has two or three external walls.

Geometrical Configuration

The term "geometrical configuration" is used here to describe such features as the compactness of the buildings, the ratio of external walls to floor area, and the existence of porches and balconies projecting outward or recessed inward from the main facade. Such features affect the options for providing openings as well as the details of the pressure distribution along the various facades of the building. These two factors, in turn, determine the potential for effective cross ventilation.

Generally speaking, the more spread out the building and the more irregular its shape, the better the potential for cross ventilation. Two different factors contribute to this effect.

As the area of the external walls for a given floor area increases, there are more opportunities to provide openings that will catch wind from different directions and provide direct and independent ventilation to the various rooms in the building. In this respect, there might be a conflict between ventilation and energy conservation when the building is either heated or air conditioned—if it is air conditioned, a compact configuration is more desirable.

It is possible, however, to design a building with special details that will enable it to be compact when it is heated and to be widespread and irregular in shape when it is naturally ventilated. For instance, it is possible to design a building with porches recessed inward and flanked by the adjoining rooms. The porches should be equipped with operable large windows and/or insulated shutters, as is illustrated in Figure 3.2. When windows and shutters are open, the porch is open, providing a ventilation option for the adjoining rooms through windows or doors. On the other hand, when the windows and/or shutters of the porch are closed, the facade becomes smooth, reducing the area of heat loss.

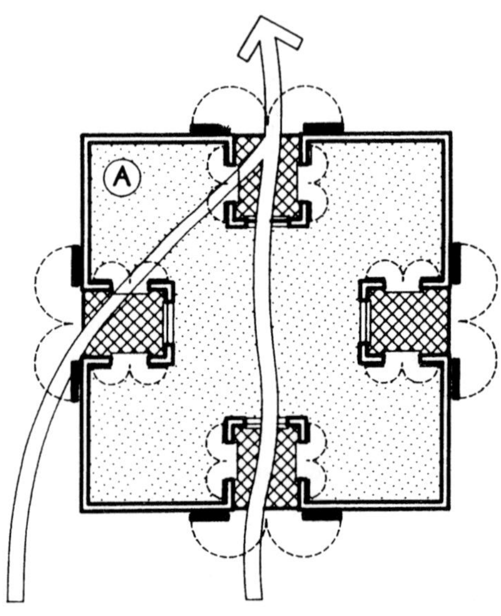

FIGURE 3.2A A building with recessed porches and insulating shutters open. The porch functions as an open space, providing ventilation for adjoining rooms.

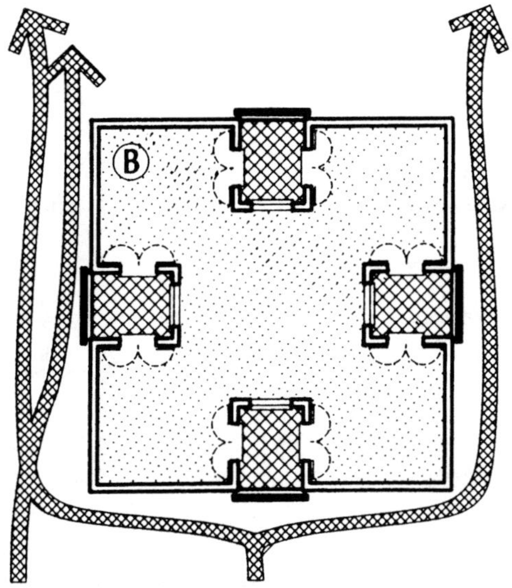

FIGURE 3.2B When windows and/or shutters of the porch are closed, the facade becomes smooth, reducing the area of heat loss.

Orientation of Openings with Respect to Winds

It is a common belief that for good ventilation in elongated buildings the main walls should be perpendicular to the prevailing wind direction. With such an orientation the largest pressure differential is created between the windward and the leeward walls. It is assumed that this orientation provides the best ventilation.

In reality, however, the situation is often different. Buildings that are exposed to oblique winds, with angles of 30 to 60° away from the normal, can provide better ventilation conditions in individual rooms and in the dwelling as a whole. When the wind is oblique to the building, a pressure gradient is created along the windward walls. If two windows are provided in a given room along the windward wall, the upwind window is at a higher pressure than the downwind one. Thus, air enters the room through the upwind window and leaves through the downwind aperture. When the wind is perpendicular to the wall, the two openings are exposed to the same pressure and this reduces the ventilation of the room.

When the wind is oblique to the wall, it is possible to greatly increase the pressure difference between the two windows by adding a single wing wall (a vertical projection on one side of the window). If such a wing wall is placed downwind of the first window, high pressure is created in front of it. A wing wall upwind of the second window creates a suction in front of it.

The idea of using such a wing wall was first developed by the author in 1962 (Givoni 1962, 1976). The quantitative effect of this design detail is shown in Figure 3.3. Later, the performance of wing walls was further studied by S. Chandra et al. (1983) at the Florida Solar Energy Center, both in models and in a full-scale experimental building (see Figure 3.4). The effectiveness of the wing walls was verified in both cases.

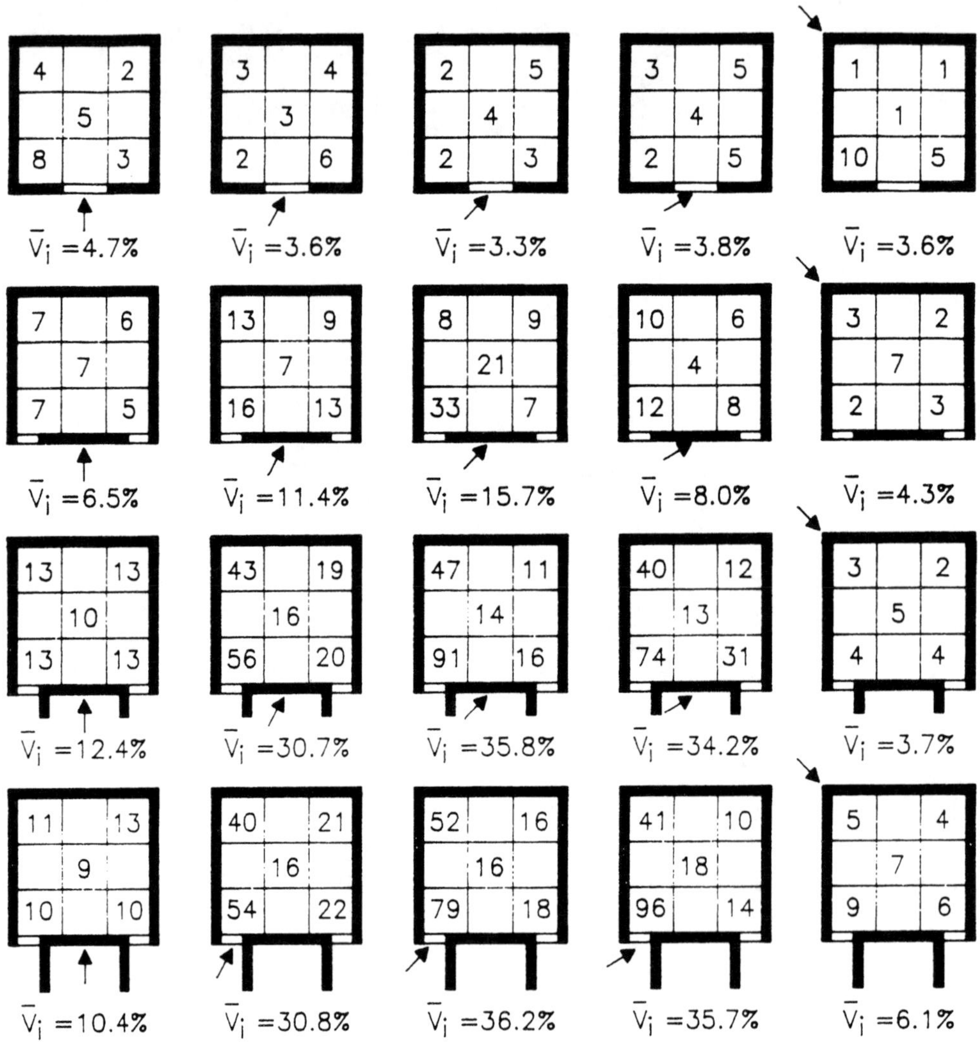

FIGURE 3.3 Effect of wing walls on indoor air speed (percent of outdoor speed at the same level). Data from Givoni (1976).

Strong pressure differences can be created not only by wing walls but also by such design elements as rooms projecting slightly from the main wall or recessed or projecting porches, as is illustrated in Figure 3.5.

Window Size

The effect of window size depends greatly on whether or not the room is cross ventilated. In rooms with a single window, the size of the window will have only a small effect on internal airflow, especially if the wind is perpendicular to the wall.

When the window faces an oblique wind, increasing its width provides a somewhat larger pressure gradient and increases indoor air speed. When a room or a complete

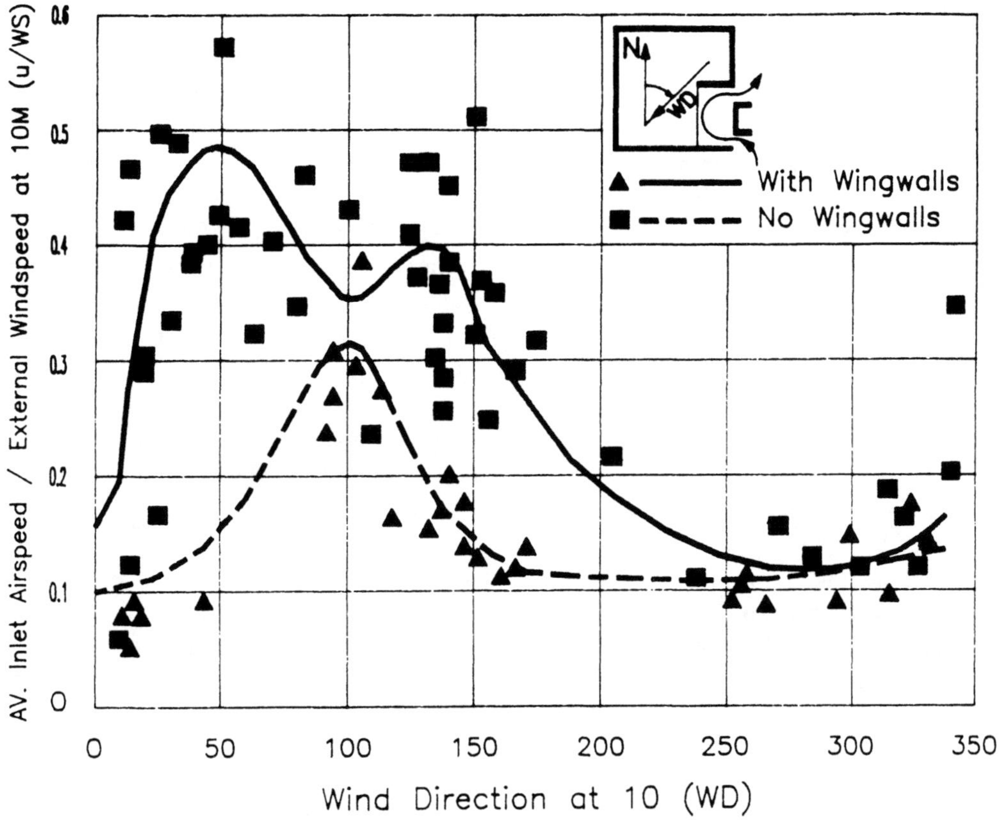

FIGURE 3.4 Effect of wing walls on indoor air speed. Data from Chandra et al (1983).

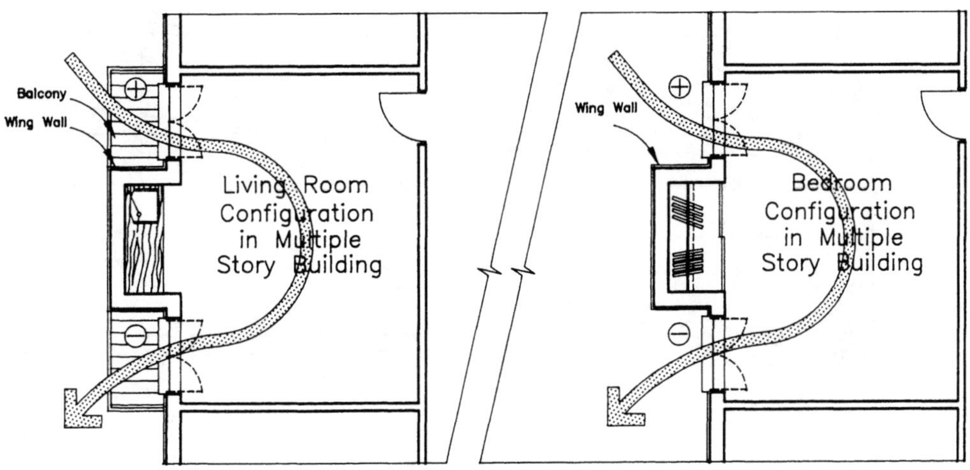

FIGURE 3.5 Architectural features acting as wing walls.

building is cross ventilated, enlarging its windows increases the airflow rate and interior air speed. See Figure 3.6 (Givoni 1962).

For a given overall area of windows, having equal areas of inlet and outlet openings produces the highest flow rate and indoor air speeds. When inlet and outlet openings are not of equal area, the flow rate is determined mainly by the smallest one, either the inlet or the outlet.

Increasing the area of inlet openings without a corresponding increase of outlets, or vice versa, produces only a small increase in interior air speed. When the inlet is small and the outlet is large, *maximum* indoor speed is much higher and average speed only slightly higher, compared with the situation when the inlet is large and the outlet small. On the other hand, the distribution of speeds in the room is much more nearly equal when the inlet is large and the outlet is smaller.

Taking into account these characteristics, it can be suggested that small inlets may be suitable in rooms where the place of occupancy is defined and is close to the inlet—for example, in a bedroom with the bed next to the window. In this case, even a small inlet, if placed over the bed, can provide a good airflow over it. It should be possible, of course, to direct the flow, by the design details of the window, away from the bed (upward) when high speed is not desirable. On the other hand, in a living room in which any spot may be occupied (at a level of 0.5 to 1.0 m above the floor), a large inlet would be most suitable even if the outlet is small.

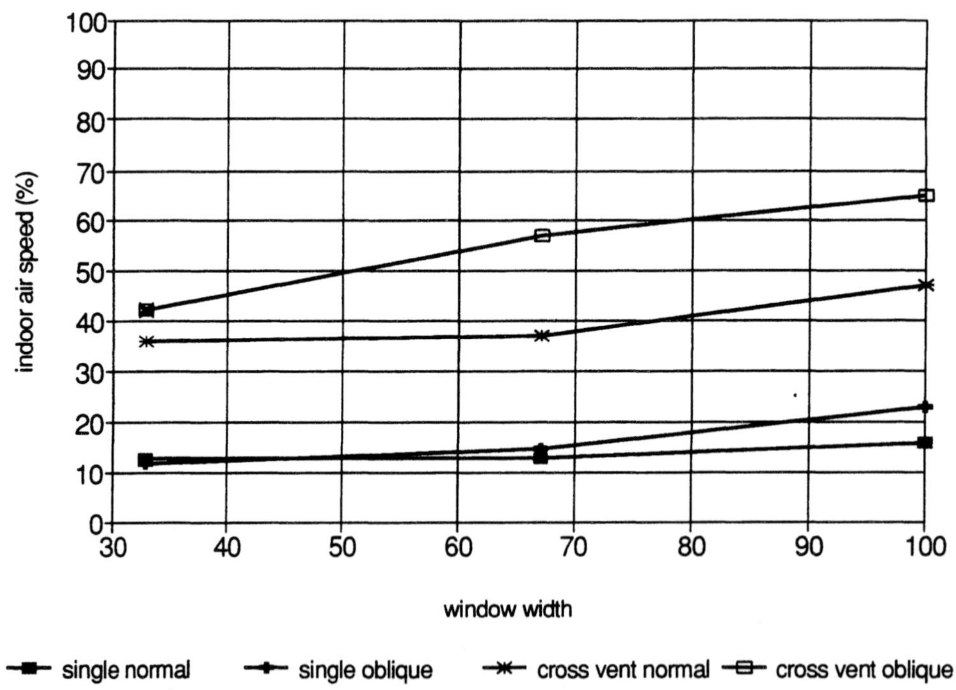

FIGURE 3.6 Effect of window size (width) and wind direction (normal or oblique to the wall) on indoor air speed.

Vertical Location of Windows

Wind direction is much more stable in the vertical plane than in the horizontal plane. The reason is that the wind flow over and within an urban built-up area is, in most cases, nearly horizontal. Therefore, the vertical components of the wind within the built-up space are more constant than the horizontal components.

The indoor flow pattern is determined at the inlet opening because of the inertia of the air; therefore, the vertical distribution of the indoor airflow, when the inlet is an ordinary opening located at a given height, is more constant than the horizontal distribution. Thus, the height of the inlet opening may determine the level of the main indoor airflow. At levels below the sill of the inlet there is a sharp drop in air speed unless this is prevented by special design features.

In some cases, architectural, functional, or privacy requirements call for placing openings near the ceiling, with their sills high above the level of occupancy. In this case, poor ventilation conditions may exist in the occupied zone of the room unless the flow is directed downward.

It is possible to direct the airflow in any desired direction by the opening details of the inlet. If the inlet window is hinged at the top and opens upward, it forces the flow downward because the open sash is in an inclined position. The actual angle of the sash then determines the angle of the flow in the vertical plane. This point is of special importance in earth-integrated buildings with walls bermed up the sill of high clerestories, which otherwise may suffer from poor ventilation conditions when comfort ventilation is desired. See Chapter 6.

Window Types and Ways of Opening

Different types of windows, when serving as inlets, produce different patterns of indoor airflow and provide different options for controlling the direction and level of the flow.

Double-hung windows, by their height, determine the vertical level of airflow but not its direction and pattern. The maximum free opening is less than one-half of the total area of the sashes, a factor that limits the effective ventilation rate.

Horizontally sliding windows also provide less than half of the free-opening area. They allow less control of the indoor flow pattern than double-hung windows because horizontal variations in the flow direction are much greater than in the vertical plane as a result of changes in wind direction.

Casement windows opened to the outside can serve as wing walls, creating an elevated pressure when one sash, the downwind one, is opened, or creating a suction zone when the upwind sash is opened. However, when both sashes are opened, they may provide a smaller airflow than when only the downwind sash is open, because when both are open there is interference in the flow.

Horizontal center pivot hung windows permit control of the vertical pattern of airflow (either upward or downward) if the sashes can be made to open downward on the room side, 10° below the horizontal. Experiments by Givoni (1976) have demonstrated that by altering the angle to which the sash is opened it is possible to modify and alter flow patterns and distribution of velocities throughout the indoor space.

In some cases—for example, at table level in offices and classrooms—it is desirable to have a sharp drop in indoor air speed at some level. This also can be accomplished by the detailing of the inlet windows.

Subdivision of Interior Space

Whenever the air has to pass through more than one room on its way from the inlet opening to the outlet, it encounters additional resistance. Actual resistance depends on the size of interior openings (doors, for example) through which the air flows. In addition, changes in the flow direction along the way, as well as the contraction and expansion of the airstream as it passes through interior passages, create turbulence, increase flow resistance, and reduce the flow rate. Therefore, if internal subdivision of the building allows independent cross ventilation of individual rooms, better overall ventilation of the building is produced.

It should be noted that if the aperture connecting two rooms, through which the air flows, is smaller than either the inlet or the outlet, the smallest aperture will determine the actual flow rate. Thus, in many apartment building designs two bedrooms are located on opposite sides of the building with a corridor separating them. The planned ventilation path is from the inlet in the room on the windward side of the building, through shutters in the doors leading to the corridor from both rooms (because the doors may be closed for privacy), through the room on the leeward side of the building, and out through the outlet opening. The bottleneck in this path are the door shutters in or above the doors. The flow resistance provided by these shutters is the limiting factor for the whole airflow.

Fly Screens and Ventilation

In many hot regions fly screens in the openings are essential. However, from the point of view of ventilation, they are a mixed blessing because they reduce the airflow rate.

It would be desirable, of course, to provide protection from insects and at the same time minimize interference with ventilation. On theoretical grounds it can be argued that if the area of a fly screen were much greater than the area of the opening itself—for example, if a screen would be placed over a porch in front of the openings—the reduction in airflow would be minimized.

To test this idea, a series of experiments was conducted with models in a wind tunnel by Givoni (1968). The models had one outlet and one inlet opening at the center of the walls. The area of the inlets and outlets were identical—one-ninth of the area of the walls. The following conditions were tested:

- No fly screens
- Screens only on inlet openings
- Screens on both inlet and outlet openings
- Screens over the front balcony
- Screens over both front and rear balconies

Figure 3.7 shows the data of the tests with one inlet and one outlet when wind direction was either perpendicular or oblique to the building: the indoor average velocity in percent of external wind speed.

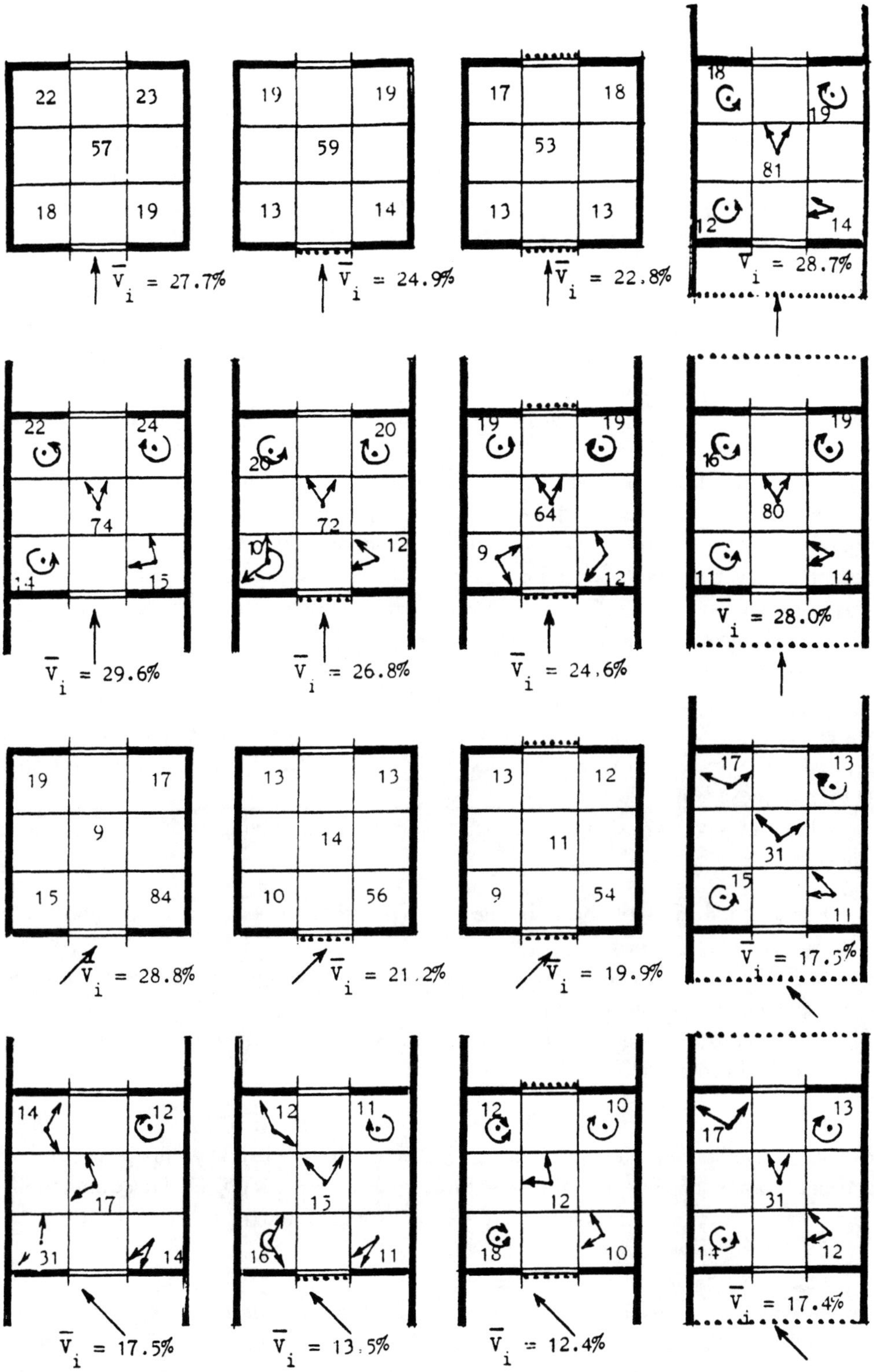

FIGURE 3.7 Effect of fly screens on indoor air speed (percent of outdoor speed at the same level).

It can be seen that putting the fly screens in front of the balconies reduced the interference of the screens with the air flow. This design solution also protects the balconies from the presence of insects and allows their use in the evening.

SECTION 2. NOCTURNAL VENTILATIVE COOLING

The term "nocturnal ventilative cooling" applies here to the strategy of keeping the building closed (unventilated) during the hot daytime hours and cooling the structural mass at night by circulating outdoor air, either through the indoor space (whole space ventilation), or through air passages within the mass elements. Another option is to cool at night by ventilation a specialized storage mass, such as a rock-bin or a water tank with embedded air tubes. The cooled mass serves during the following day as a heat sink, absorbing the heat penetrating the building as well as the heat generated inside the building. Heat flow during the daytime, from the indoor air into the cold-storage mass, can occur passively by natural convection and long-wave radiation (when the interior structural mass is the cold storage medium) or by an induced airflow, between the indoor space and a specialized cold-storing mass, in a closed loop.

From the building-design viewpoint, some building elements, such as openings, can be designed to permit either comfort ventilation or convective cooling, as is desirable at any given time. On the other hand, other building elements, such as structural details and especially the amount of the building mass, call for quite different designs for optimization of each of these ventilation strategies.

There are several design options to provide the mass serving as the nocturnal cold storage:

1. Structural mass of the building: walls, partitions, floors, etc, cooled by whole space ventilation.
2. Embedded air spaces (passages) within floors, ceilings and/or walls, through which outdoor air is circulated.
3. Specialized storage: a rock bed or a water tank with embedded air tubes, cooled at night by outdoor air.

The applicability of nocturnal ventilative cooling is limited to a certain range of conditions. Limitations on applicability are posed on the one hand by climatic conditions and on the other hand by the comfort and functional needs of occupants. These limitations particularly affect the decision to leave windows open throughout the night to obtain effective nocturnal ventilative cooling.

The climatic parameters that determine whether ventilative cooling can be applied are the ambient-air *minimum* temperature, the temperature *range*, and the water-vapor-pressure level. The ambient minimum determines the temperature to which the storage mass potentially can be brought at the end of the night. The outdoor temperature range determines the potential for lowering the indoor maximum below the outdoor maximum. The ambient vapor pressure determines the upper temperature limit of indoor comfort without daytime ventilation, either at still air conditions or at a higher indoor air speed, which can be provided in unventilated buildings by such devices as ceiling fans.

From the climatic aspects the main limiting factor is the minimum outdoor air temperature. It should be noted that the ultimate temperature of the storage mass, at the end of the ventilation period, would be higher than the outdoor minimum temperature. In view of the human comfort range the outdoor minimum should be below about 20°C (68°F) in order to cool adequate amounts of the storage mass sufficiently. Only with such night temperatures can the mass serve effectively as a heat sink during the following day and prevent elevation of indoor air and radiant temperatures above the upper limit of the comfort zone.

With a suitable airflow rate, a large storage mass, provided it has sufficient surface area exposed to the flowing air, can be cooled at night to a temperature reasonably close—within 2 to 3°C—to the outdoor minimum air temperature. This applies in particular to a specialized storage mass such as a rock bed.

In an occupied building with structural storage mass cooled at night by whole space ventilation the elevation of the mass temperature above the outdoor minimum may be larger, about 4 to 5°C, and consequently the stored cold energy is less than what is possible from climatic considerations. This may happen as a result of reluctance of the occupants to leave windows open all through the night and/or to tolerate a high indoor air speed.

Experimental Studies in Israel on Night Ventilative Cooling

At the Institute for Desert Research in Israel measurements were taken in two similar room-size test cells. Three of the walls (20 cm thick) and the roof (15 cm thick) were built of concrete and were insulated on the exterior side of the envelope. The fourth (south) wall was changeable to enable testing of various passive solar heating systems. In the cooling experimental series the wall was built of plywood sandwiches filled with 5 cm Polystyrene. One room (the control room) was closed day and night while the other was ventilated naturally at night (by opening the door facing the prevailing wind) and closed during the daytime hours. Indoor temperature patterns in the room cooled by nocturnal ventilation and in the control room during three days are shown in Figure 3.8. Figure 3.9 shows the comparison of the temperature patterns in these two rooms in another experimental series over a period of 11 days, during which the outdoor climate has changed significantly.

The temperature patterns of the closed room in Figures 3.8 and 3.9 illustrate both the possibility of achieving daytime indoor temperatures well below the outdoor level in regions with a large temperature range, and also the need to prevent ventilation by ambient air during the daytime. Even without night ventilation the indoor temperature was kept below the outdoor maximum, and the difference increased during a heat wave period up to about 5 to 7°K—that is, by more than ⅓ of the outdoor range. This was achieved by the thermal inertia of the mass in the room and by minimizing the solar gain. Ventilation during the daytime would, of course, bring the indoor temperature closer to the outdoor level and therefore is not advisable.

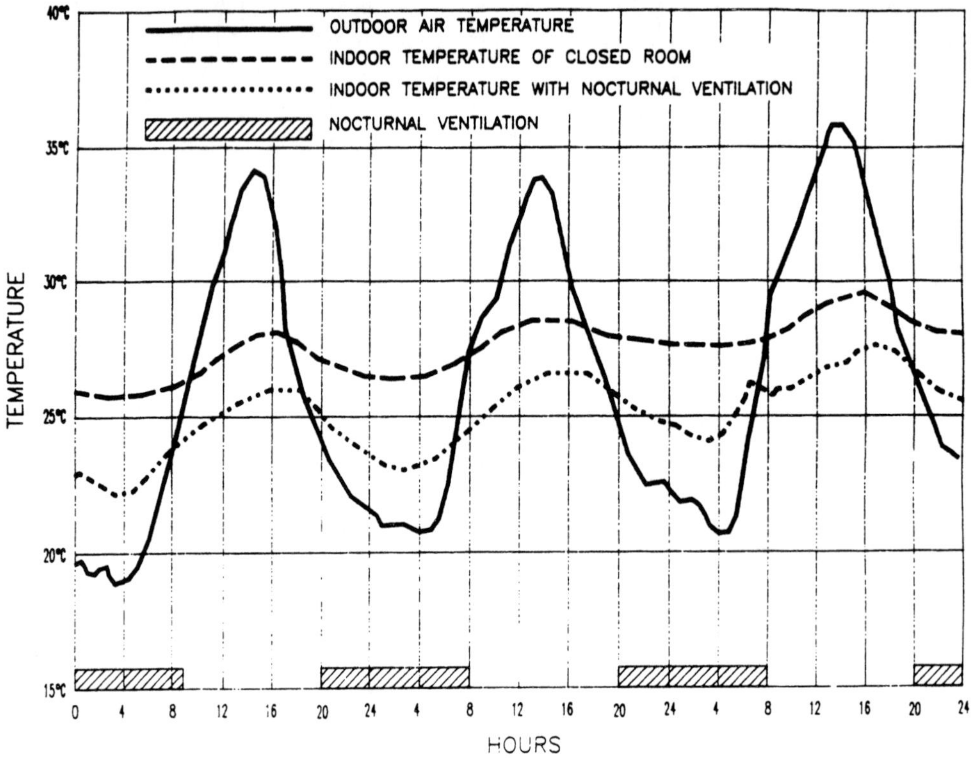

FIGURE 3.8 Study in Israel: Indoor temperature patterns in a high-mass room cooled by nocturnal ventilation and in a similar control room during a sequence of three days with similar climate.

The ventilation of the building during the nighttime lowers the indoor nighttime temperature below the level of a closed building. With a high mass building ventilated through one opening, and with the very light night winds that are common in this region, the observed temperature drop was on the average about one half of the difference between the temperatures of the closed building and the outdoor. The cooled mass also lowered the daytime temperature in the convectively cooled buildings but the reduction in the maximum temperature was about one half of the reduction in the minimum temperature.

The drop of the indoor daytime temperature of a high-mass building below the outdoor maximum by night ventilation during a period of steady outdoor temperatures is roughly proportional to the outdoor temperature range. During a period of rising outdoor temperatures the difference increases, and when the outdoor temperatures drop this difference becomes smaller. This study, lasting in several series over many weeks, has demonstrated that with structural materials of high mass such as concrete and bricks (which are the common building materials in many countries), solar control by effective shading and light external colors, nocturnal natural ventilation can lower the indoor maximum temperature and bring it close to the outdoor daily average.

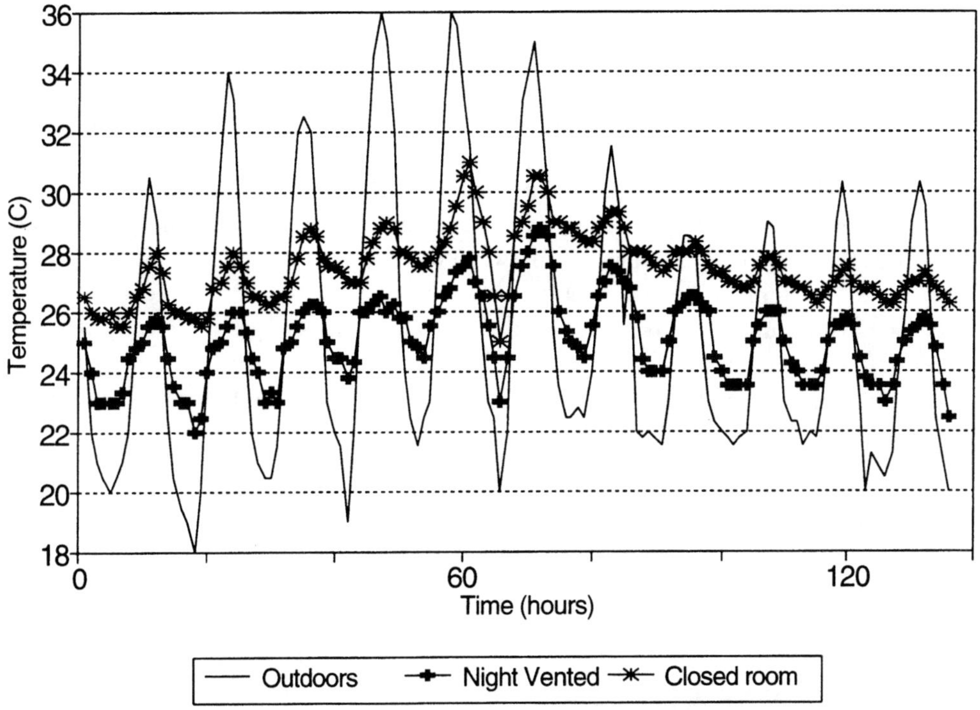

FIGURE 3.9 Study in Israel: Indoor temperature patterns in a high-mass room cooled by nocturnal ventilation and in a similar control room during a sequence of 11 days with changing outdoor climate.

EXPERIMENTAL STUDY IN CALIFORNIA ON NIGHT VENTILATIVE COOLING

An extensive study on the effectiveness of structural mass, in conjunction with night ventilation, in lowering the indoor daytime temperatures (especially the indoor maximum) was conducted in 1993 at the Pala research Facility near San Diego, California (Givoni 1994).

In this chapter the performance of two of the monitored buildings will be summarized: a "low-mass" and a "high-mass" building. Each building is about 23 m^2 (260 ft^2), built on an uninsulated concrete slab, and divided into two equally-sized (south and north) rooms with ceiling heights of 2.4 m (8 ft). The total window area of each building is about 4.6 m^2 (50 ft^2), divided about equally between the east, west, south, and north walls. At first, the windows were without any shading, but, after the first monitoring series, they were shaded by external fixed shading devices.

The color of the roofs is dark brown (absorptivity about 0.7–0.75) and that of the walls medium beige (absorptivity about 0.5–0.55). These colors determine, of course, the effective indirect solar gain of the buildings through their envelopes.

The low mass building is of conventional Californian stud-wall construction. All external walls are fiberglass insulated to a resistance of R-11 (1.94 m_2C/W). The ceiling is insulated with fiberglass in the attic to an R-19 (3.35 m_2C/W) level. The calculated diurnal building loss coefficient of the buildings, BLC is 2315 Wh/day.C or 101 Wh/m^2.day.C (4390 Btu/day.F or 16.9 Btu/ft^2.day.F).

The high mass building has solid concrete walls, 10 cm thick, and is insulated externally with rigid foam and plaster in order to have the same level of insulation as the low mass building. The partition wall between the two rooms is also of solid concrete so that the total volume of the concrete is about 5.75 m^3 or about 0.25 m^3 per each m^2 of the floor area. The interior surface area of the concrete, interacting convectively with the interior space, is about 65 m^2.

The performance of the buildings were monitored under the following conditions:

1 Windows unshaded, closed day and night
2 Windows shaded, closed day and night
3 Windows shaded, closed during the day and ventilated at night by a fan with three speeds
   A Low speed
   B Medium speed
   C High speed

Buildings Closed Day and Night

Figure 3.10 shows the daily indoor temperature patterns of the two buildings (along with the outdoor temperature pattern and daily averages) during two series: first with unshaded windows and then with windows shaded by external fixed shading devices. For a period of time, starting on July 29, 1993, the buildings were shaded but not ventilated at night.

Figure 3.11 shows the daily indoor and outdoor maximum temperatures during these two series. The daily average outdoor temperature is also shown in this figure.

It can be seen that, without shading, the indoor maximum temperatures in the two buildings were above the outdoors' maxima. With shading, the maximum of the low-mass building was still above the outdoors' maxima, but the maximum temperature of the high-mass building was below the outdoors' maxima. It can also be seen that the patterns of the indoor maxima follow closely the outdoor average but not the outdoor maximum. It means that for buildings that are closed during the daytime there is better correlation between the indoor maxima and the outdoor daily averages than with the outdoor maximum temperatures.

Buildings Ventilated at Night with Shaded Windows

When buildings are ventilated at night the indoor temperature during the night hours drops drastically, in turn lowering the indoor daily average temperature. However, the effect of the night ventilation on the indoor daytime conditions may or may not be significant.

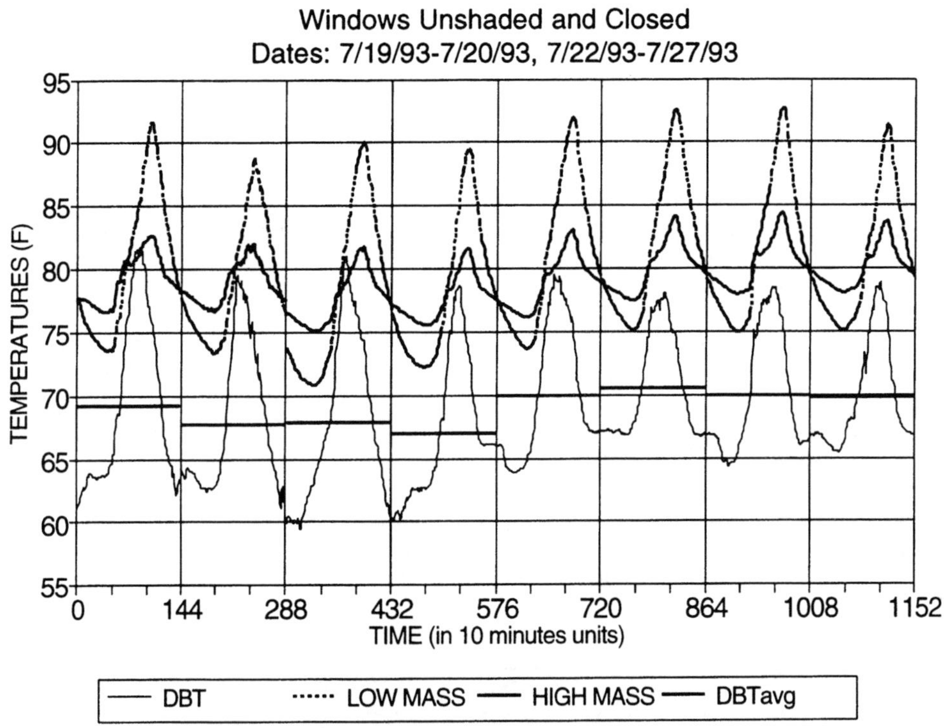

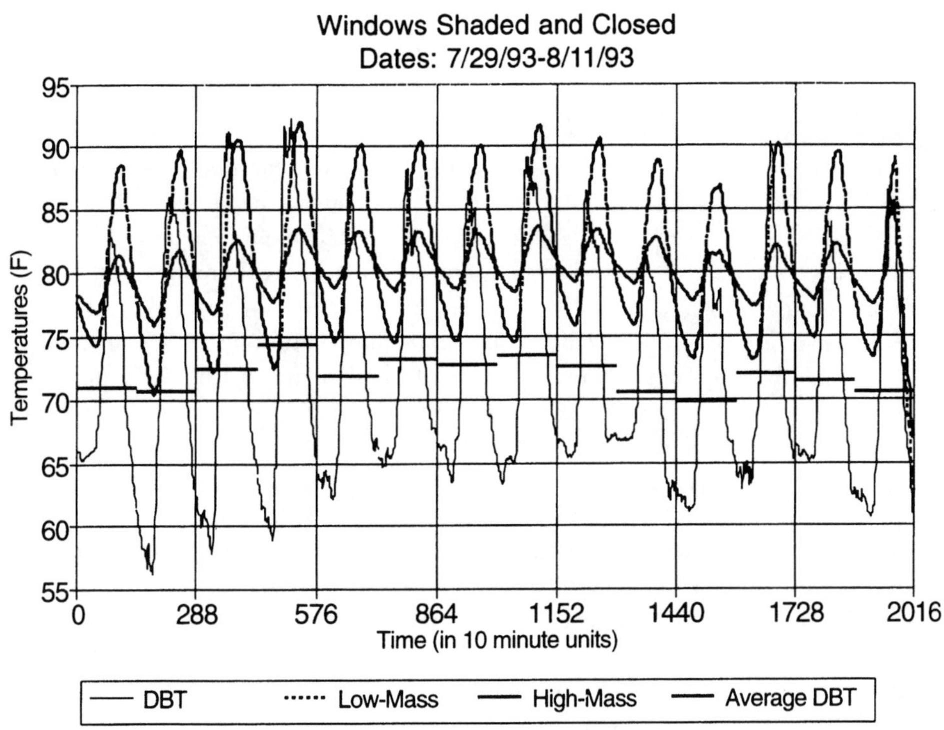

FIGURE 3.10 Study in California: Daily indoor temperatures of a low-mass and a high-mass building during two series—first with unshaded windows (top) and then with windows shaded by external fixed shading devices (bottom). Buildings were not ventilated at night.

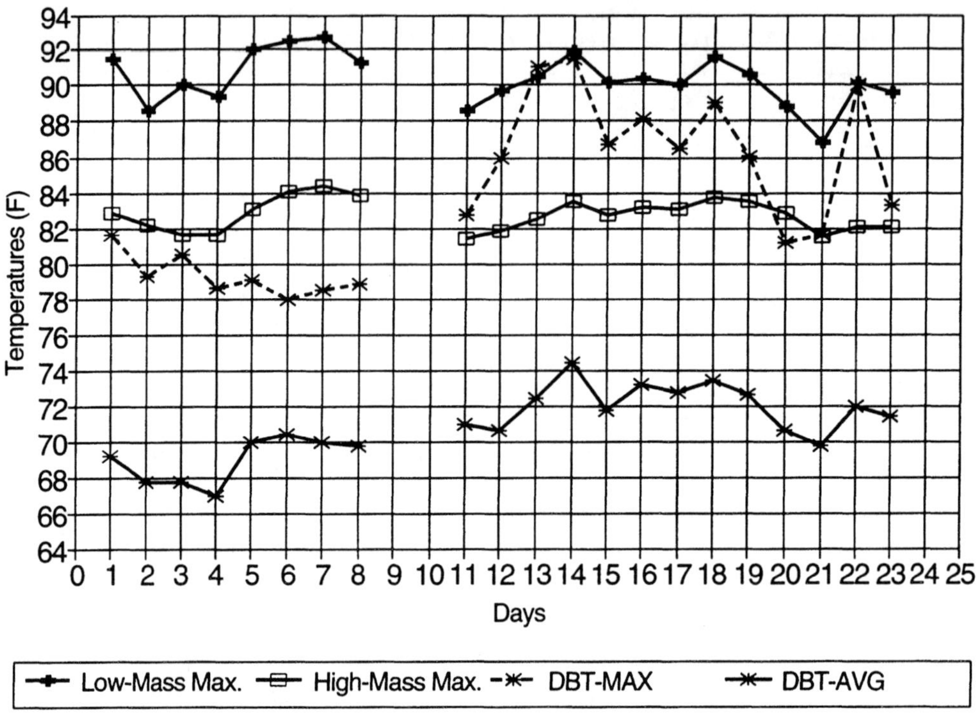

FIGURE 3.11 Study in California: Daily indoor and outdoor maximum temperatures during the two series shown in Figure 3.10.

When the chief concern revolves around daytime comfort conditions and lowering the peak demand for air conditioning, interest in nighttime indoor conditions is limited only to the extent that night ventilation significantly affects daytime temperature. Therefore, the analysis of the data obtained with the night ventilation will emphasize the indoor maximum temperatures.

Figures 3.12 and 3.13 show the indoor maximum temperatures of the low-mass and the high-mass buildings, respectively, together with the outdoor maxima, minima, and average, during the period when the buildings where ventilated at night. The outdoor climate was very unstable during this period. There were very sharp rises and falls in the maxima, with lowest maximum being about 72°F and the highest maximum, just 5 days later, being about 100°F. At the same time, the outdoor minima have changed sharply.

In Figure 3.12 it can be seen that the indoor maximum temperature in the low-mass building was rather close to the outdoor maximum, except during days with a sharp rise in the outdoor maximum. During the cooler days, the indoor maximum was even above the outdoor maximum, in spite of the night ventilation. Thus, in the case of low-mass buildings, nocturnal ventilation is quite ineffective in lowering the daytime temperatures and improving the daytime comfort conditions.

In the case of the high-mass building, as shown in Figure 3.13, the night ventilation has lowered the indoor maximum temperature consistently below the outdoor maximum.

The daily rates of change in the indoor maxima of the low-mass building were close to the changes in the outdoor average temperature. The daily rates of change in the indoor

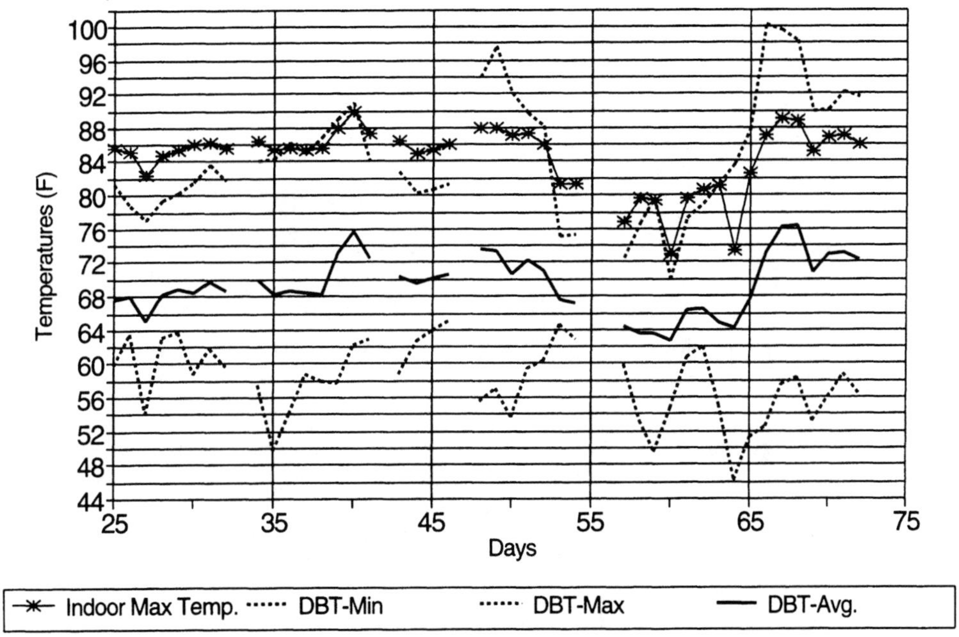

FIGURE 3.12 Study in California: Indoor maximum temperatures of the low-mass building, together with the outdoor maxima, minima, and average, during the period in which the buildings were ventilated at night.

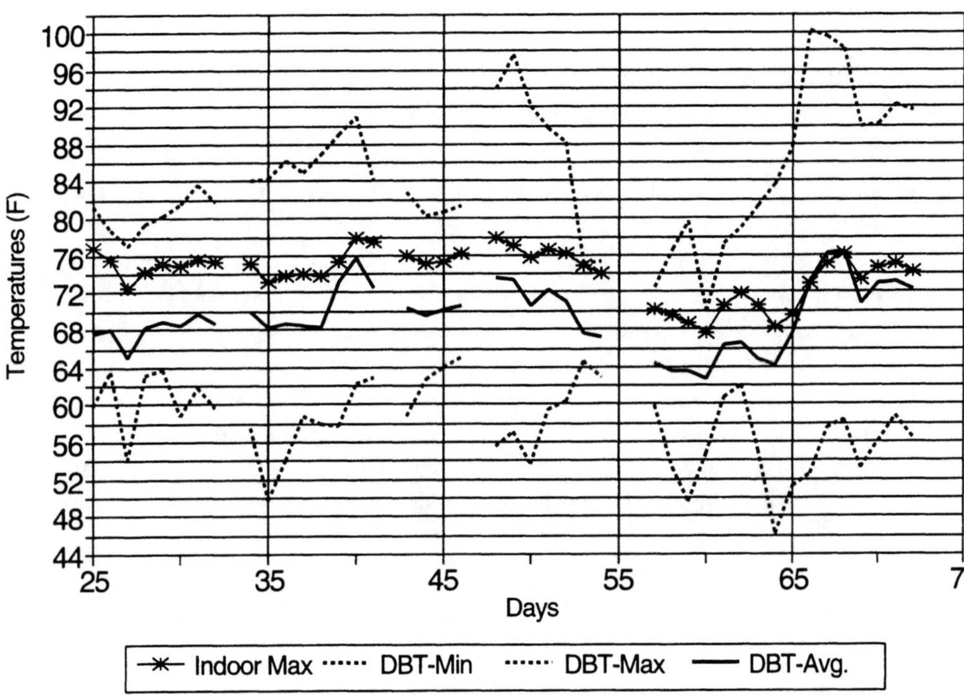

FIGURE 3.13 Study in California: Indoor maximum temperatures of the high-mass building, together with the outdoor maxima, minima, and average, during the period in which the buildings were ventilated at night.

maxima of the high-mass building were smaller (about one half) in comparison with the changes in the outdoor average temperature.

This observation provided the basis for the development of the predictive formula for the expected indoor maximum temperatures, which is described next.

Experimental Formula Predicting the Expected Indoor Maximum Temperature

Analysis of the relationship between the indoor maxima and various parameters of the outdoor climate has demonstrated that the best correlation exists between the outdoor **average** and the indoor **maximum.** The following preliminary formula has been developed to express the indoor maximum as a function of the outdoor average, the shading of the windows and the ventilation conditions—no ventilation or night ventilation.

At the present state of development, the formula is limited to buildings insulated at least as the ones in this study—either with mass at levels as low as the conventional stud-wall construction or at least as high as the high-mass building in this research (insulated 10 cm. concrete walls, light weight ceiling).

Observations on which the Formula is Based

The predictive formula of the expected indoor maximum temperature is based on the following observations in the experimental research:

- A The climatic parameter best correlated with the indoor maximum is the outdoor average.
- B Under steady climate conditions, when the outdoor average is about constant, the elevation of the indoor maximum above the outdoor average depends on the shading conditions of the windows and the color of the opaque envelope, as well as on the thermal mass of the building.
- C Under dynamic climate, when the outdoor average rises or falls, the indoor maximum of the high-mass building changes at a rate about one half of the change in the outdoor average. The rate of change of the low-mass building maximum is about 0.8 of the changes in the outdoor average.

The Experimental Predictive Formula for Indoor Maximum Temperature

The above observations are expressed mathematically in the general form of the predictive formula:

$$T_{max} = GT_{avg} + DelT + k^*(T_{avg} - GT_{avg})$$

where:

T_{max} = Indoor maximum temperature in a particular day
GT_{avg} = "Grand average" of the outdoor temperature
DelT = Average elevation of the indoor maximum above the outdoor average. Its value depends on the mass, shading and ventilation conditions, as shown in Table 3-1.
T_{avg} = Outdoor temperature average in a particular day
k = Ratio of the rates of daily changes of the indoor maximum to the rate of change of the outdoor average, depending on the mass level.

The two variables, k and DelT, represent the thermal characteristics of the building—thermal resistance, mass, shading, external color, and the nocturnal ventilation rate.

DelT, the average elevation of the indoor maximum above the outdoor average should become smaller with higher resistance and mass, with lighter colors of the building's envelope, and with higher rates of nocturnal ventilation.

The k variable depends mainly on the effective thermal mass, also taking into account the effect of the surface area of the mass on its effectiveness in storing night coolness.

The data from the Pala monitoring study, and the study in Sede-Boqer, Israel, provided numerical values for these variables for the type of the monitored buildings. These values are shown in Table 3-1 and were used in calculating the predicted indoor maximum temperatures that are shown in Figures 3.14–3.16.

TABLE 3-1. EXPERIMENTALLY DERIVED VALUES OF THE K AND DELT VARIABLES

			The k Variable	
Low-Mass Buildings			k=0.8	
High-Mass Buildings			k=0.5	
	The DelT Variable (C)		*The DelT Variable (F)*	
Conditions	*Low-Mass*	*High-Mass*	*Low-Mass*	*High-Mass*
Windows Unshaded No Night Ventilation	12	8	22	14
Windows Shaded No Night Ventilation	10	6	18	11
Windows Shaded Night Ventilation Low Speed	9	3	16	6
Windows Shaded Night Ventilation Mid Speed	8	3	15	5
Windows Shaded Night Ventilation High Speed	8	2	14	4
High-Mass, Shaded, White color Natural Ventilation (Israeli study)		1		2

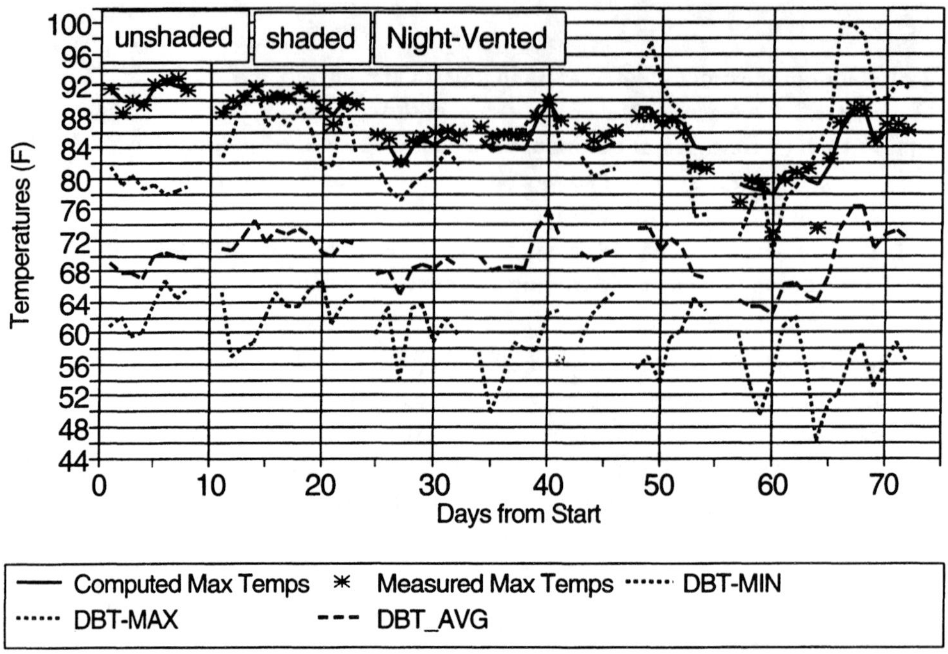

FIGURE 3.14 Measured and calculated indoor maximum temperatures of the low-mass building, together with the maxima, minima, and averages of the outdoor air temperature.

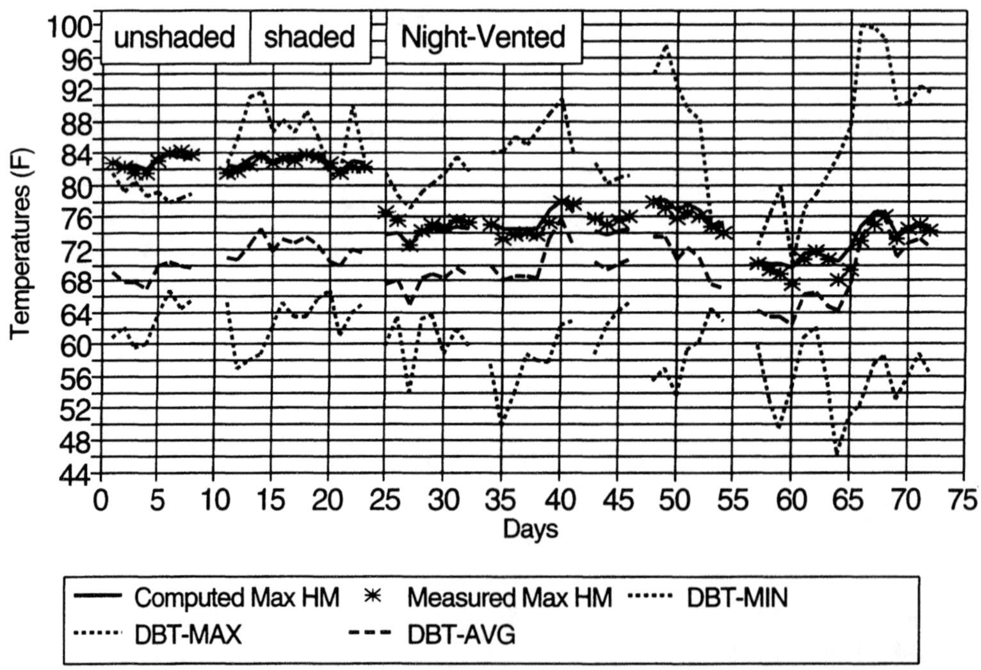

FIGURE 3.15 Measured and calculated indoor maximum temperatures of the high-mass building, together with the maxima, minima, and averages of the outdoor air temperature.

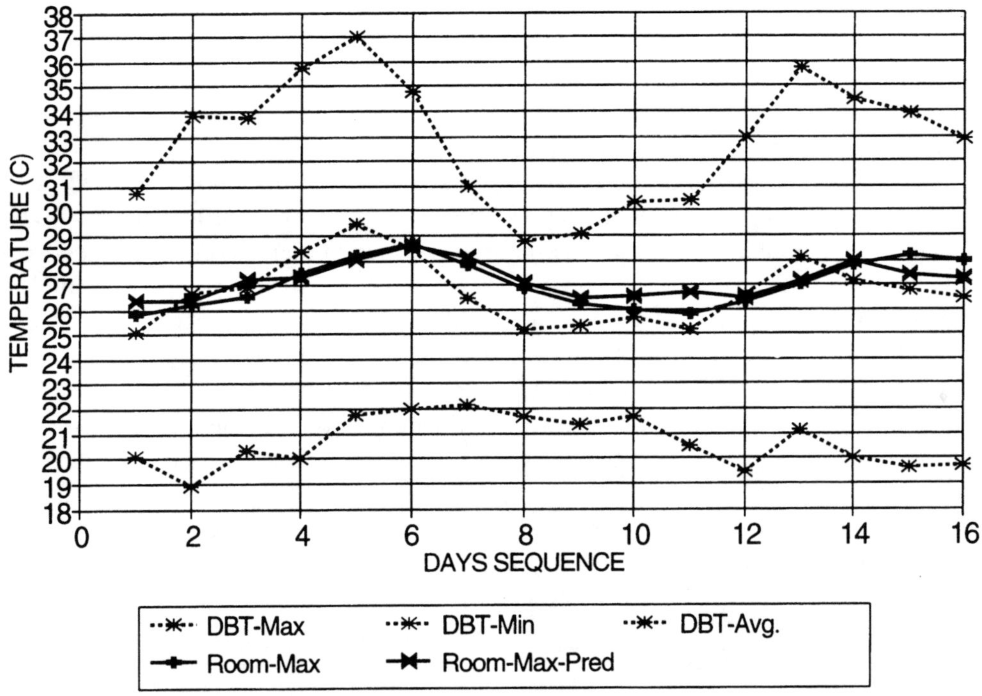

FIGURE 3.16 Measured and calculated indoor maximum temperatures of the high-mass room in Israel (same as in Figure 3.9).

Agreement Between Calculated and Measured Indoor Maximum Temperatures

Figures 3.14 and 3.15 show the measured and calculated indoor maximum temperatures of the low-mass and the high-mass buildings, respectively, together with the maxima, minima, and averages of the outdoor air temperature in the research project in California. It can be seen that, in spite of the very wide and sharp fluctuations in the outdoor climatic conditions, there is rather good agreement between the measured and the computed maximum temperatures. This demonstrates that the very simple formula can predict the indoor maximum not only under stable climate but also expresses the response of buildings to sharp changes in the climatic conditions.

Comparison With the Study in Israel

Predictions of the maximum temperature, by the formula developed above, are compared in Figure 3.16 to the temperatures measured in Israel (see Figures 3.8 and 3.9), assuming 0.5°C for the DelT variable. It can be seen that the formula developed on the basis of the experiments in California also predicts rather well the maximum temperatures measured in a different climate in a different building configuration.

From these experimental studies, it can be concluded that, with structural materials of high mass such as concrete and bricks in different forms (the common building materials in many countries), nocturnal natural ventilation can be an effective cooling system. It can lower daytime temperatures close to the level of the outdoor 24 hours' average temperature, especially in buildings with effective solar control. During sharp heat wave periods, such as the "Hamsin" in the Middle East or the "Santa Ana" winds in California, the indoor maximum can be lowered even below the outdoor average.

It should be noted that the 24 hours' average is usually lower than the mean of the maximum and minimum temperatures by about 1°K. In many countries and regions, the only climatic information available is the maximum, minimum, and the mean between them. Therefore, when the predictive formula is applied in such places, 1°K could be reduced from the mean to get an estimate of the 24 hours' daily average.

In arid and desert regions, with a diurnal temperature range in summer of about 15–20°K, expected indoor maximum temperature reduction can be up to about 10°K.

Although no daytime ventilation with outdoor hot air is desirable in buildings cooled by nocturnal ventilation, the indoor comfort limit can be extended by interior fans (e.g. ceiling fans) to about 30°C in developed countries and to about 32°C in hot developing countries. In this way the upper comfort limit is extended without heating the interior of the building by daytime ventilation.

Climatic Applicability of Nocturnal Ventilative Cooling

Taking into account the relationships between the ambient vapor pressure and the outdoor range in summer and between the indoor maximum temperature and the outdoor range, (described above), it is possible to specify, in a Building Bio-Climatic Chart, the outdoor maximum temperatures under which indoor comfort can be maintained in a well designed building as a function of the ambient vapor pressure. This is presented in Figure 3.17, together with an example of the simplified calculations that enable us to estimate the indoor temperatures.

In this example, it is assumed that the outdoor minimum, maximum, and mean temperatures are 18, 36, and 27°C, respectively. Thus the estimated 24 hours' outdoor average would be about 26°C. The indoor average temperature in a closed residential building would be elevated by solar radiation and internal heat generation to about 29°C. With an indoor swing of about 3°K, the maximum in the closed building would be about 30.5°C and the minimum about 27.5°C.

Nocturnal ventilation would lower the indoor maximum temperatures to about 27°C (one degree C above the outdoor average) and lower the minimum to about 22°C.

Figure 3.17 shows the climatic boundaries, in terms of the outdoor **maximum** daily temperatures, under which nocturnal convective cooling is applicable. The temperature limit is reduced with higher humidity, reflecting the corresponding decrease of the outdoor range with higher humidity. Increasing the indoor air speed by **internal** fans (not by ventilation) can extend the indoor comfort range without elevating the indoor temperature. A simple procedure for estimating the expected indoor maximum and minimum tempera-

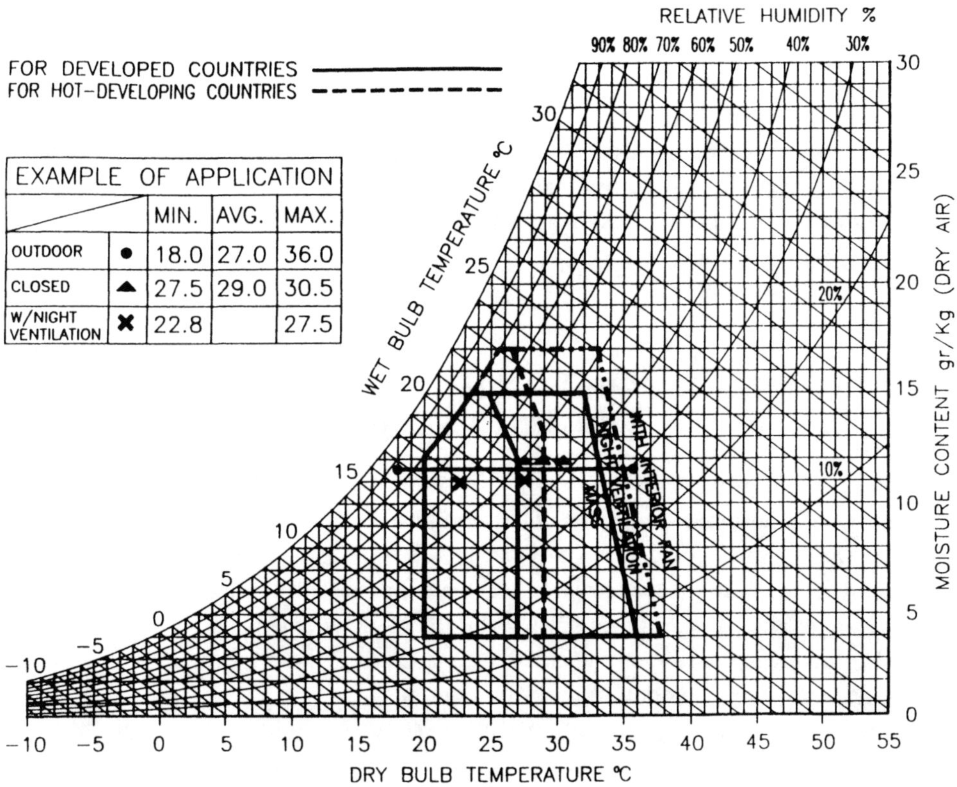

FIGURE 3.17 Building Bio-Climatic Chart showing the outdoor maximum temperatures under which indoor comfort can be maintained by nocturnal ventilation, together with an example of simplified calculations.

tures in high mass buildings that are either closed all the time or ventilated during the evening and night hours is presented later in this chapter. Figure 3.17 contains a table that shows the application of this procedure.

Nocturnal ventilative cooling is applicable mainly in arid and desert regions where the maximum temperature is below about 36°C. In desert regions with daytime temperatures above 36°C, night ventilation alone would not maintain the indoor daytime temperature at acceptable levels and other passive cooling systems, such as evaporative cooling (see Chapter 5), compression, or absorption air conditioning, should be used during the hot hours. But even in this case the application of nocturnal convective cooling can significantly reduce the length of time when the additional cooling systems will be needed.

As a rule of thumb it can be estimated that in arid and desert regions, with a summer diurnal temperature range of 15 to 20°K, the expected reduction of the indoor maximum temperature can be about 6 to 8°K below the outdoor maximum. On very hot days in desert regions, which usually have a larger diurnal range, the drop in indoor temperature during the time of the outdoor maximum may be up to about 10°K.

In arid regions, where the minimum is above about 20°C it is possible also to precool the ambient night air by evaporation. In this case the minimum wet-bulb temperature rather than the air (dry-bulb) temperature becomes the controlling factor in determining whether to cool the building by nocturnal ventilation.

In warm, humid regions, on the other hand, the typical small diurnal outdoor temperature range does not allow the daytime indoor temperature to be reduced significantly below the outdoor level. In this case comfort ventilation is usually the appropriate strategy for enhancing comfort, especially as it minimizes the elevation of the indoor vapor pressure resulting from such activities as breathing, sweating, cooking, and washing.

Implications of Different Storage Options of Ventilative Cooling

As mentioned above, there are several design options that can provide the mass that will serve as the nocturnal cold storage. They include: (1) using the structural mass of the building with a whole space ventilation; (2) using embedded air spaces (passages) within floors, ceilings and/or walls, through which outdoor air is circulated at night; and (3) using a specialized storage outside the cooled space, such as a rock bed or a water tank with embedded air tubes, cooled at night by outdoor air.

The characteristics, design implications, advantages, and limitations of these options are discussed below.

Cooling of the Structural Mass by Whole Space Ventilation

When a building whose structural mass is exposed to the interior space is ventilated at night the mass is cooled by convection from the inside, bypassing the thermal resistance of the envelope. The heat exchange between the mass and the interior space is through the surfaces of the mass that face that space. If fans, such as whole house exhaust fans, are used to provide the air flow in regions with low nighttime wind speeds the power requirements are minimal, compared with the other options.

During the daytime, the cooled mass, if adequately insulated from the outdoors, serves as a direct heat sink within the occupied space. It absorbs passively, by natural convection and radiation, the heat penetrating into and generated inside the building. As the mass is part of the structural system of the building it does not involve extra cost.

Design Implications of Whole Space Ventilation

The main design implications of nocturnal ventilative cooling by whole space ventilation involve the need to ensure sufficient mass, a large surface area of the mass exposed to the interior space, and thermal insulation external to the mass.

The capacity of any storage mass to serve as a heat sink during the following day is determined by the product of the heat capacity of the storage ($M\rho c$) times the range of the mass *average* temperature, from the minimum attainable at night to a maximum capable of keeping the indoor temperature below the upper limit of the comfort zone. Thus the total *nominal* amount of cold storage, $Q_{(store)}$, is given by:

$$Q_{(store)} = M.\rho.c.(T_m^{max} - T_m^{min})$$

where:

M = available volume of the storage mass, m^3
ρ = volumetric density of the mass, Kg/m^3
c = specific heat of the mass materials, about 220 Wh/(Kg.C)
T_m^{max} = average mass maximum temperature, °C
T_m^{min} = average mass minimum temperature, °C

The amount of available storage mass M, as well as the structural materials used in the building, are system design factors and can be chosen on the basis of technical and economic considerations. Only a certain fraction of the nominal mass is actually utilized effectively as a cold storage, especially when it is cooled at night by whole space ventilation.

The effectiveness of the structural mass to serve as nocturnal cold storage depends on the conductivity and heat capacity of the materials, the thickness of the mass elements and, especially on the surface area of the storage elements exposed to the indoor air. Quantitatively it can be expressed by the diurnal cooling capacity (DCC), a concept discussed elsewhere in this chapter.

The minimum average temperature to which the mass can be cooled at night depends on one hand on a meteorological factor, the outdoor minimum air temperature and, on the other hand, on the system design details: the mass elements and the rate and duration of the ventilation. The outdoor temperature, of course, is given by nature and cannot be modified by the design of the building or the energy system. However, in regions with suitable climate, the outdoor night air can be supercooled by evaporation (in conjunction with an evaporative cooling system) or by nocturnal radiation (in conjunction with a radiative cooling system).

Attainable Minimum Storage Temperature

When the structural mass surrounding the indoor space, including walls, floor, partitions, and ceiling, serves for night cold storage, that mass can be cooled by ventilating the indoor space at night. The factors that determine the attainable minimum mass temperature are of two types: on one hand physiological or functional factors, and on the other hand design details and the thermophysical properties of the storage material.

Physiological (comfort) or functional factors may limit the amount and duration of ventilation of an occupied building during the night. Comfort considerations may limit the airflow rate through the indoor space so that indoor air speeds are acceptable to the occupants. If nighttime indoor temperatures in a ventilated building are below the lower limit of the comfort zone, taking into account the higher indoor air speed, ventilation may produce discomfort or annoyance and eventually be stopped. In this case ventilative cooling of the storage mass may be terminated long before it can reach the minimum temperature attainable on the basis of climatic conditions.

When ventilation is natural—provided through open windows—privacy and security considerations may limit the time when windows remain open. In many cases windows will be closed when the room's occupants retire to sleep. In such cases convective cooling preferably should be based on mechanical exhaust ventilation with specialized and protected openings, even when climatic conditions permit convective cooling by all-night natural cross ventilation.

The system design details and thermophysical properties of the material that determine the attainable minimum temperature of the structural mass under a given minimum outdoor air temperature are the outdoor airflow rate through the indoor space, the surface area of the mass elements, the surface convective coefficient of the storage elements, the mass element thickness, and the thermal conductivity of the mass materials. In practice the indoor minimum temperature attainable by natural nocturnal ventilation may be limited by the reduced wind speed at night. In many regions almost still air conditions prevail after about 10 p.m., greatly reducing the rate of airflow through the building. In such regions mechanical nocturnal ventilation can allow this method to work. As the fan operation takes place during periods of minimum demand for electric power it is attractive to the power utilities and so power often may be supplied at reduced rates.

Useful Maximum Mass Temperature

The useful average maximum temperature, the temperature to which the storage mass can be heated during the day while the mass still absorbs heat while the indoor temperature remains below the upper limit of the comfort zone, depends on a physiological consideration—the upper limit of the comfort zone. It also depends on the difference between the surface temperature of the mass and the accompanying indoor air temperature. This temperature difference can be affected both by the design details of the building and by the cooling system.

It is during the time when the outdoor temperature drops sufficiently so that comfort can be provided by natural or mechanical ventilation with outdoor air that this indoor temperature is important. After that time, there is no further need to utilize the cold storage to ensure indoor thermal comfort. It should be remembered that in high-mass buildings maximum temperatures occur in the evenings. In many cases opening windows for cross ventilation would be desirable, from the comfort viewpoint, before the time of the indoor maximum.

The maximum temperature at which the storage mass can still be utilized depends on the indoor air upper limit for human comfort in unconditioned buildings. This limit, in turn, depends on the ambient vapor pressure and the indoor air speed. The vapor pressure depends on the local climate and thus sets a climatic limit for the applicability of the system.

Taking into account these two factors the upper temperature limit for nocturnal ventilative cooling may range under indoor still air conditions from about 25°C in semihumid regions (vapor pressure of 18 to 20 mmHg) to about 28°C in arid regions (vapor pressure below 15 mmHg). However, this upper limit of the *indoor-air* comfort zone can be extended upward by increasing the indoor air speed, *without* introducing the (warmer) outdoor air, with the help of fans. If a ceiling fan is used to increase indoor air speed to about 2 m/s, this limit is shifted upward to about 28°C in semihumid regions and 30°C in arid regions. Such an increase of the useful comfort range enables utilization of the cold-storing mass through a larger temperature swing, greatly enhancing the useful cold-storage capacity. This point is of particular interest in humid regions, where the useful range of the storage-mass temperature is very small to start with.

The maximum surface temperature of the storage mass under which this indoor-air temperature limit can be maintained is about 2°C below the upper comfort limit. Such a temperature difference is sufficient to maintain the required heat flow by long-wave radiation and natural convection, from the indoor air to the mass heat sink, when the storage elements surround most of the indoor space. To this end it is advisable to distribute the storage mass in as large an area as practical, using partitions, walls, ceilings, and floors.

When the indoor air speed is enhanced by fans, increasing the convective component of the surface coefficient, a smaller temperature difference (about 1°C) may be sufficient. Consequently the useful temperature range of the cold-storage mass and the amount of useful night cold storage with a given storage mass are increased.

Useful Temperature Range of Night Cold Storage

Summarizing the above discussions of attainable minimum and usable maximum temperatures of the night cold-storage mass, estimates of upper temperature limits are suggested and presented in Table 3.2. The table gives the maximum temperatures of the structural storage surfaces and the indoor air, with and without enhanced indoor air speed, for arid and semihumid regions.

From the table it can be seen that in semihumid regions, with vapor pressure of 17 to 19 mmHg, convective cooling can be contemplated only in conjunction with enhanced indoor air speed during the daytime hours. At still air conditions in a humid region, with usable maximum temperature of the mass surfaces of only 23°C, it is not possible to store any significant night cold energy. Enhanced indoor air speed of about 2 m/s raises the usable mass temperature to about 26°C, which is more attainable.

In humid regions, with vapor pressure above 22 mmHg, the outdoor minimum temperature is usually not low enough, and the diurnal range is too small, to permit any useful nocturnal cold storage for convective ventilation cooling. In such regions the appropriate approach may be ample daytime comfort ventilation—the indoor air temperature would remain close to the outdoor level but the higher air motion past the body would provide sufficient cooling by convection and sweat evaporation.

TABLE 3-2. MAXIMUM TEMPERATURE LIMITS (°C) OF INDOOR AIR AND STORAGE SURFACES IN ARID AND HUMID REGIONS

Climate	*Arid Still*	*V=2m/s*	*Semi-Humid Still*	*V=2m/s*
Indoor Air	28	30	25	27
Mass Surface	26	29	23	26

It should be noted that at some season most hot regions experience climatic conditions when night temperatures are suitable for convective cooling. A diurnal temperature range of about 10°C (18°F) is required to obtain useful lowering of the indoor daytime temperature by nocturnal cooling of the structural mass.

Enhancing Performance of Ventilative Cooling with Whole Space Ventilation

Several design factors can be considered for improving the utilization of the structural mass with whole space ventilation or for overcoming some of the comfort and functional problems associated with this cooling option. The following design factors are discussed below:

1. Increasing the surface area of the mass at the interface with the indoor space.
2. Choosing materials with high thermal conductivity and heat capacity for structural elements.
3. Increasing the heat-transfer rate (surface coefficient) by a higher air speed, produced by fans, next to the surfaces of the structural mass.

Increasing the Surface Area of the Thermal-Storage Mass

One of the major factors limiting the efficiency of the structural mass in storing thermal energy during diurnal cycles—using solar heating by direct gain in winter and ventilative cooling in summer—is a limited surface area of the mass, through which heat is exchanged between the indoor air and the materials composing the structural mass. The effect of the mass surface area is more critical in the case of nocturnal ventilative cooling than in solar heating by direct gain because the temperature differences that drive the heat flow are smaller.

Only convective heat transfer takes place while the mass is cooled during the night hours. In solar heating by direct gain, heat is transferred to the mass elements by both radiation and convection. The convective heat-transfer coefficient is only about one-half of the combined convective–radiant coefficient. As a result, a much larger surface area of thermal-storage elements is required for ventilative cooling than for passive solar heating.

There are two main design means for increasing the surface area of the heat-storing mass:

1. Spreading the mass by designing high-mass (load-bearing) exterior walls and internal partitions.
2. Profiling the indoor surfaces by architectural features such as grooves, and projections.

Optimal Materials for Nocturnal Cold Storage

As mentioned in the subsection "Attainable Minimum Storage Temperature," the thermophysical properties of materials that determine their effectiveness for storing night coolness are thermal conductivity, density, and heat capacity. This effectiveness is related to the ability of the storage element to lose and absorb heat in diurnal cycles—losing large quantities of heat during the night, when the building is ventilated, and absorbing heat during the daytime, when the temperature swings are smaller. To this end materials should have high conductivity and high density. The specific heat of all masonry materials is about the same—about 0.23 Wh/(kg°C) —so that the variables are density and conductivity. A quantitative procedure for computing the effective diurnal cold storage capacity is presented in the next section. From the point of view of storage the best material for structural storage elements is dense concrete.

Surface Coefficient in Convective Cooling

The main difficulty in modeling convective cooling with reasonable accuracy is in determining the surface convective coefficient as a function of the ventilation rate. Of course, a higher flow rate of outdoor air through the indoor space at night increases the convective coefficient and the cooling rate of the storage elements. However, the actual air speed along the various surfaces surrounding a space is very variable. Furthermore, the relationship between the airflow rate and the mass's cooling rate is rather complex, and any theoretical calculation, either analytic or by simulation, should be considered only as a rough estimate.

The reason for this uncertainty is that there is no clear and definable relationship between the flow rate through the building and the convective coefficient. The path and pattern of the airflow through the indoor space and the resulting air velocities along the various mass's surfaces depend in a complex way on the design details of the building. Relevant details include such factors as the location of openings within the space and the connection between the different rooms through which the air flows on its way from the inlet to the outlet openings. The airflow pattern also changes with changes in the wind direction relative to the inlet openings.

In unventilated buildings under still air conditions, the overall surface coefficient is rather small, about 8 W/m^2C. Even in buildings ventilated by natural or mechanical ventilation, the airflow often is concentrated in the central parts of the indoor space while the air in direct contact with the surfaces of the walls, ceilings, and floor flows at a much lower speed. As a result, the effect of the ventilation rate on the surface coefficient is often smaller than that expected from the overall airflow rate.

Another factor that complicates theoretical calculations is the significant variations in the air temperature that exist in practice within the ventilated space. In calculations, it is assumed that the mixing of the air across the flow front is perfect. The difference between the mass-element temperature and the indoor air temperature next to this element determines the rate of cooling under given air-speed conditions. The variations in indoor air temperature introduce another source of uncertainty into the modeling.

Experimentally, however, it is possible to get information about the expected attainable minimum temperature of the structural mass relative to a given outdoor minimum air temperature, as well as an approximate evaluation of heat extraction by night ventilation from a building with given design details.

By design detailing it is possible to direct the airflow within the building space in such a way as to maximize the flow next to the surfaces of the building elements, where thermal storage is concentrated. The pattern of the airflow is determined by such features as the geometry and details of the inlet openings and their location within the walls. With appropriate detailing it is possible to direct the main flow toward elements contemplated for thermal storage. These elements include ceilings and massive walls. In this way it is possible to significantly increase the surface coefficient and the rate of heat transfer between the heat-storing mass and the indoor air.

Night Airflow in Channels within the Structural Elements

As discussed in the subsection "Attainable Minimum Storage Temperature," ventilating a building through open windows during the night may involve problems of security and even discomfort—sometimes it may leave residents feeling too chilly. One way to overcome these difficulties and at the same time to increase the cooling rate of structural cold-storage elements is to blow outdoor air *through* channels embedded within the structural elements. Floors of reinforced concrete with embedded channels can be cast in situ, utilizing hollow concrete block. Also, prefabricated floor panels with embedded channels are available in many countries.

In fact there is already a passive system that stores solar energy in this way: the Barra system. This system was developed by Horazio Barra in Italy (Barra et al. 1987) as a passive solar heating system. (See Figure 3.18.) However, it can be used easily as a nocturnal

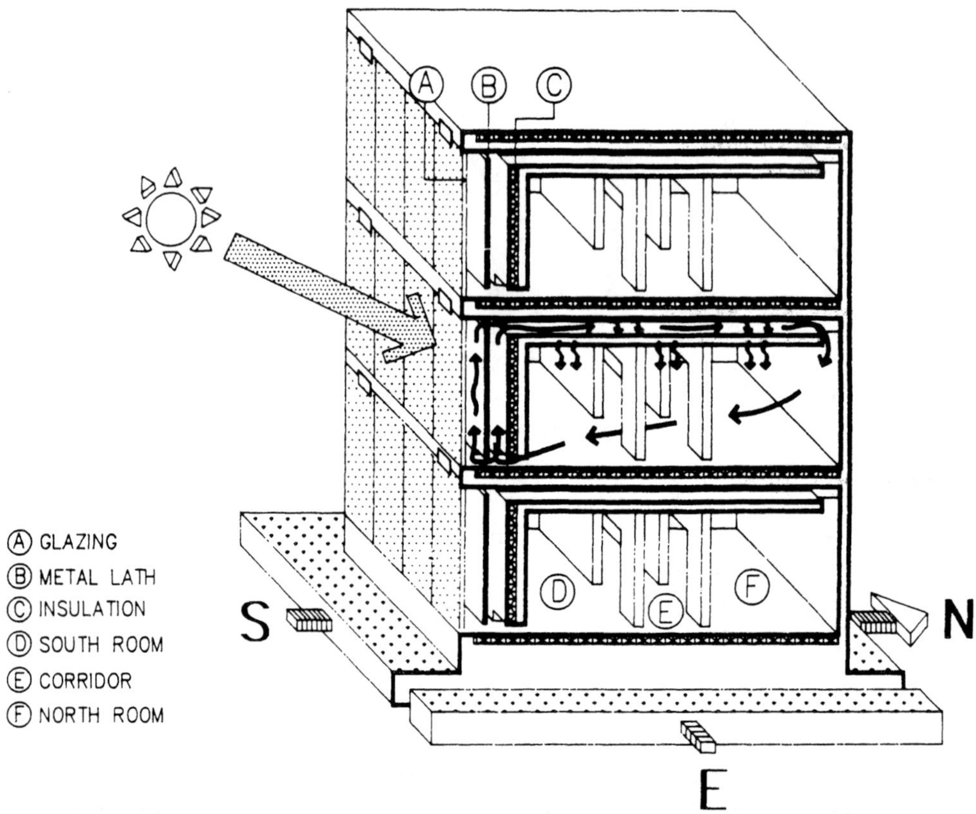

FIGURE 3.18 The Barra system: floors of reinforced concrete with embedded channels for air flow across the building.

ventilative cooling system as well. In the original design, when it is used as a heating system the southern wall is insulated and detailed as a thermosyphonic air heating solar collector. The hot air emerging from the insulated collecting wall flows horizontally within channels embedded inside a concrete ceiling, serving also for thermal storage. Part of the heat is stored in the concrete ceiling while the air, still warm, exits from the channels at the northern part of the building. The air thus warms the distant rooms first before it flows back through the building space to the inlets at the lower part of the sun-facing collecting wall. This assures more even temperature distribution throughout the whole house than is achievable with other passive solar systems.

This system can be modified and used as a cooling system also. To this end the ceiling channels should be open to the outdoors, at least on one side. At night a fan would blow the ambient air through the channels, into the interior space, and out. Thus cold night energy first will be stored up within the ceiling mass and then, at a somewhat elevated temperature, it will cool the interior space. As a result, ventilation can continue even when the outdoor night air is too cool for direct wholespace ventilation. And the functional and safety issues associated with windows that are kept open all through the night also would be eliminated.

The fact that the storage elements are completely interior, kept within the envelope of the insulated walls, ensures a high cold storage efficiency of the concrete ceiling. With insulated walls surrounding the living space, all the heat gain of the storage elements is from the interior space.

Using specialized channels to direct the night airflow inside the storing mass can improve heat exchange between the air and the storage mass in two ways:

1. The concentrated airflow within the channels inside the structure results in a higher speed and a higher surface coefficient for heat transfer from the structural mass to the air stream and a faster cooling of the mass.
2. A more intimate contact between the cold-storage mass and the cool night air is established, bypassing in part the thermal resistance for heat flow between the mass and the ventilation air.

During the daytime the cooled ceiling can absorb heat from the interior space passively, by natural convection and radiation, as is the case with whole space ventilation.

Night Cold Storage in Specialized Mass

As an alternative to passive storage of night coolness in the structural mass of a building it is possible to utilize a specialized storage mass, such as a rock bed, a water tank with embedded air tubes, or even the soil under a building. In such cases it is not possible to rely on nocturnal natural ventilation and daytime passive heat flow into the cooled mass, but it is necessary to use a blower (usually a centrifugal fan) to provide a reasonable flow rate.

The cooled mass can be utilized during the following day as a heat sink to absorb the heat penetrating or generated inside the building. Heat transfer from the building to the heat sink during the daytime hours can be provided by forced convection, with two modes:

1. Closed-circuit circulation of indoor air through the storage (circulation mode).
2. Outdoor air first forced through the cooled storage into the building and then exhausted from it (ventilation mode).

In high-mass buildings the use of specialized storage is always more expensive than structural storage, which, in fact, is obtained at no, or very little, extra cost beyond the cost of the structure itself. It also requires the consumption of energy to provide the airflow, both during the night, when the storage mass is cooled, and during the day, when the building is cooled.

On the other hand, the use of specialized cold storage has some advantages over structural storage:

1. It is possible to cool the storage mass all through the night without interference with the occupants' need for privacy and security.
2. When the night temperature is low, this becomes an advantage instead of a cause of discomfort.
3. With rock bed storage it is possible to bring the rocks' temperature at the end of the night closer to the outdoor minimum.
4. With soil storage and earth tubes it is possible to provide a relatively larger storage capacity than is practical with structural storage.

These advantages have to be weighted, of course, against the extra cost, maintenance, and consumption of energy inherent in the hybrid system of specialized storage.

Diurnal Cooling Capacity and Its Computation

The performance of buildings cooled by nocturnal ventilation with structural cold storage depends greatly on the *effective* heat storage capacity of the building. A low thermal storage capacity limits the ability of the building to absorb heat with a relatively small temperature elevation. During the night hours the indoor temperature in ventilated buildings lacking sufficient mass may reach levels too low for comfort, so occupants are likely to close the windows. As a result, a building lacking sufficient mass would heat up fast during the daytime hours and reach higher maximum temperatures than a high-mass building ventilated at night.

The specific heat (c) of all masonry materials, including concrete, bricks, and stone, is about the same—0.23 Wh/(kg°C). Therefore, the nominal heat capacity NHC per unit area of the storage element made of masonry materials is proportional to the density, ρ, in kilograms per cubic meter and the thickness, L, in meters:

$$NHC = 0.23 \times \rho \times L$$

The amount of mass by itself is not sufficient to determine the ability of the building to store night coolness and then to utilize it as a heat sink during the following day. Because of the low rate of convective heat transfer between the mass and the air flowing through the building and the limited conductivity of the storage material, only a limited thickness of the structural storage elements effectively participates in the night cold-storage process. This storage effectiveness depends on the thermophysical properties of the material, its thermal conductivity, k, and heat capacity, ρc, as well as on the thickness of the storage element.

At night the flow of cooler air cools successive layers of the storage elements, while the inner layers remain at a higher temperature than the near surface layer. During the daytime hours, when heat is absorbed from the interior space, the exposed layers are heated first and reach higher temperatures than the inner layers. Thus, beyond a given depth, the mass is not effective in serving as a heat sink.

A high thermal conductivity of the material increases the depth within the storage element, which effectively participates in night cold storage. The composite property that determines the effectiveness of the heat-storage elements is the "diurnal heat capacity" (dhc). Elements with higher dhc values are more effective in storing night coolness than elements with lower values of dhc.

The original concept of the dhc was developed by Balcomb et al. (1980) to allow calculations of the Solar Saving Fraction (SSF) in the Solar Load Ratio (SLR) model for Direct Gain passive solar buildings, taking into account different storage materials and thickness of the storage elements. The original SLR model was applicable only to 9 building configurations: combinations of 3 thickness levels (only concrete), 2 ratios of mass to glazing areas, with or without standard night insulation. According to the original dhc formulation its value is peaked at some thickness, L_{max}, of the storage element and then it drops down with further thickness.

Givoni (1983) has developed a generalized predictive model for Direct Gain, enabling the evaluation of the SSF for any design configuration of a building and using any material for thermal storage. Givoni (1987) has analyzed the effect of heat capacity in Direct Gain buildings, reaching the conclusion that, taking into account longer term storage, (beyond 24 hours), there is no drop in the effective storage capacity with depth beyond L_{max}. He has developed simple, linearized algorithms for calculating the effective heat capacity (ehc) for any material and thickness of the storage elements. A distinction was made between the ehc in "solar" and "nonsolar" rooms and a formula was developed for the effective heat capacity of a building as a whole (EHC).

Elaine Best of the Los Alamos National Laboratory (LANL) has compared the SSF calculated by the Givoni (1987) model and detailed annual hourly simulations for two cities with very different climates, Boston and Albuquerque. She has found very good agreement between the two methods. This agreement can be seen in Figure 3.19 (Givoni 1987—private communication with Best).

Wray and Best (1987) have used Givoni's (1987) work as a starting point to develop the dhc analytic model further.

The procedure suggested below for calculating the cold storage capacity (CSC) is based in part on Givoni's (1987) work and in part on Wray and Best's (1987) work.

Calculation of the Cold Storage Capacity (CSC)

The modification that must be introduced in quantitative evaluation of the dhc when this concept is applied to ventilative cooling results from the difference in the coupling of the energy source and the storage element. This point will be discussed after presentation of

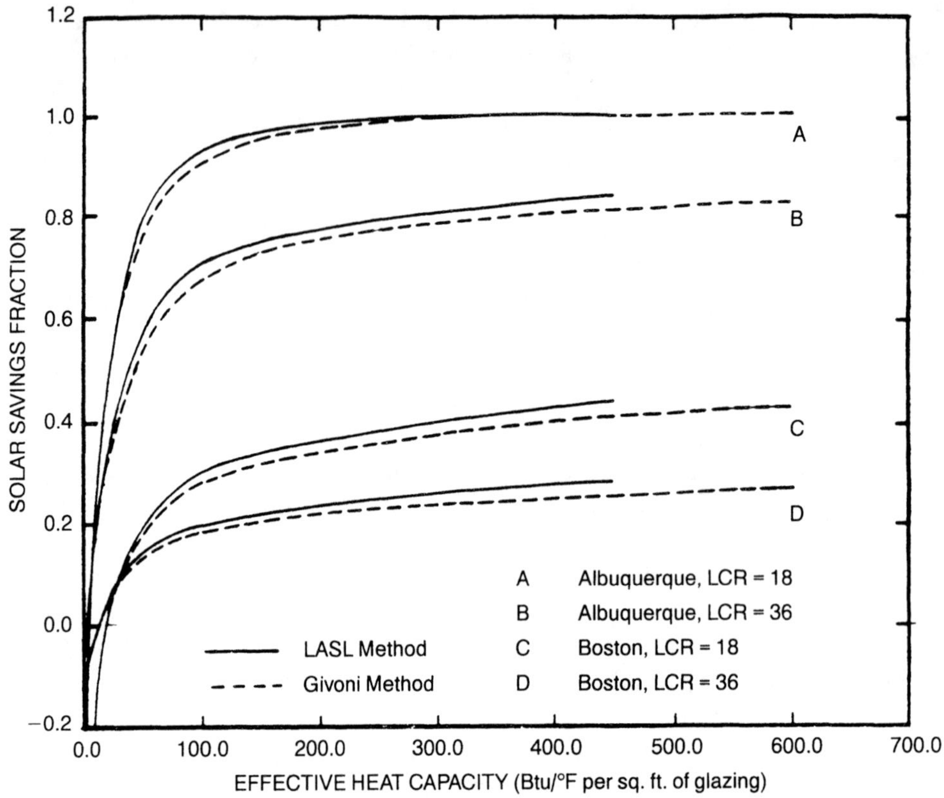

FIGURE 3.19 Comparison of calculations of the Solar Saving Fraction (SSF), in Boston and in Albuquerque, by LANL simulation system and by Givoni's model (after Elaine Best).

Wray and Best's original formulas for calculating the dhc in the case of Direct Gain solar buildings, with some changes in the notations to accommodate differences in the two sources:

Let: X = thickness
k = conductivity
ρ = density
c = specific heat
P = Pi = 3.1415 . . .

1 The thickness, X, of the mass element is expressed in Wray and Best (1987) in a dimensionless form: t:
$t = X/(24k/P\rho c)^{0.5}$
2 The asymptotic (maximum) value of dhc (max) is given by:
$dhc_{max} = (24k\rho c/2P)^{0.5}$
3 The actual value of the dhc for a storage element with a given thickness is:
$dhc = f(t)dhc_{max}$
where f(t) is:
$f(t)=[(\cosh 2t-\cos 2t)/(\cosh 2t+\cos 2t)]$

Modification of the dhc for Ventilative Cooling

When a storage element is located in a room in which solar energy penetrates directly (a "solar" room), its surface receives energy mainly in the form of radiation. The radiant flux can be direct solar impingement; reflected solar radiation from other, illuminated surfaces; or long-wave radiation emitted by the irradiated surfaces. Therefore, in such rooms the environmental temperature of the space is close to, or even slightly higher than, the indoor air temperature. The diurnal heat capacity of the storage element can be defined as the ratio of the heat flux into the element to the range of its surface temperature, or dQ/dT.

The formulas of the dhc presented above refer to this situation. However, when warm air from a solar room is transferred by convection to a nonsolar room, the range of the indoor air temperature is always larger than that of the surfaces of thermal-storage elements located in such rooms. This is because heat flows during the daytime from the indoor air into the storage elements, and the flow is reversed during the night hours. The result is that the range of the environmental temperature dT, which serves as the basis for calculating the dhc (dQ/dT), is smaller in the convective zone than in a solar zone.

Consequently, the dhc of storage elements located in a convective zone, as is the case with storage elements exposed to the outdoor air flowing through the building, is smaller than the dhc of such storage elements that are in solar rooms. The dhc in convective zones actually decreases to about 40 percent of its value in solar rooms (Givoni 1987). Therefore, the cold storage capacity, CSC, of storage elements, in the case of convective cooling, is given by:

$$\mathbf{CSC = 0.4 \times dhc}$$

Diurnal Cooling Capacity of the Building as a Whole

For each storage element with a given surface area, Ai, the total diurnal cooling capacity (dcc) is given by

$$\mathbf{dcc = Ai * CSC}$$

The diurnal cooling capacity of a building cooled by night ventilation (DCC) is given by the sum of the dcc of all its storage elements:

$$\mathbf{DCC = \Sigma\ (Ai * CSC)}$$

This DCC should be related to the heat gain coefficient of the building (BHGC) in which it serves as a heat sink. Therefore, the criterion suggested for evaluating the effectiveness of storage in serving as a heat sink can be defined as the diurnal cooling capacity of the building (DCCB), given by the ratio between the DCC and the building's heat-gain coefficient BHGC:

$$\mathbf{DCCB = DCC/BHGC}$$

It can be assumed that as this ratio rises, the building will have a lower rate of temperature elevation during the daytime hours, when it is not ventilated. If this assumption is validated experimentally, then the thermal performance of buildings cooled by nocturnal convection could be predicted on the basis of the DCCB.

REFERENCES

Aynsley, R. M., W. Melbourne, and B. J. Vickery. 1977. *Architectural Aerodynamics*. London: Applied Science Publishers.

Balcomb, J. D., D. Barley, R. D. McFarland, J. Perry, W. O. Wray, and S. Noll. *Passive Solar Design Handbook*. Vol. II. U.S. Department of Energy, NTIS, 1980.

Chandra, S, P. Fairey, and M. Houston. *A Handbook for Designing Ventilated Buildings*. 1984. Cape Canaveral: Florida Solar Energy Center.

Chandra, S., M. Houston, P. Fairey, and A. A. Kerestecioglu. 1983. "Wing Walls to Improve Natural Ventilation: Full Scale Results and Design Strategies." *Proceedings of the Eighth National Passive Solar Conference*, Glorieta, New Mexico, September, 1983.

Davies, M. G. "The Thermal Admittance of Layered Wells." 1978. *Building Science*, vol. 8, pp. 207–220.

Givoni, B. *Basic Study of Ventilation Problems in Housing in Hot Countries*. 1962. Building Research Station Report. Technion-Israel Institute of Technology.

——— *Man, Climate and Architecture*. 2nd ed. 1976. London; Applied Science Publishers; paperpack, New York; Van Nostrand Reinhold Company, 1981.

——— *Ventilation Problems in Hot Countries*. 1968. Building Research Station Report, Technion-Israel Institute of Technology.

Givoni, B. 1983a. "Review of Passive Heating and Cooling Research." PLEA '93, Oxford: Pergamon Press, pp. 339–352.

——— 1983b. "A Generalized Predictive Model for Direct Gain." Research Report. Institute for Desert Research, Ben Gurion University of the Negev.

——— *Building and Urban Design Guidelines for Different Climates*. 1994. New York: Van Nostrand Reinhold.

Holleman, T. R. "Air Flow Through Conventional Window Openings." 1951. Research Report No. 33, Texas Engineering Experiment Station, College Station.

Humphreys, M. A. 1975. "Field Studies of Thermal Comfort Compared and Applied." Building Research Establishment, Current Paper. CP 76/75, UK: Garston, Watford.

McIntyre, D. A. 1978. "Preferred Air Speeds for Comfort in Warm Conditions." *ASHRAE Transactions* V–84, pp. 264–277.

Nicol, J. F. 1974. "An Analysis of Some Observations of Thermal Comfort in Roorkee, India and Baghdad, Iraq." Annals of Human Biology, V–1, pp. 411–426.

Tanabe, S. I. 1988. "Thermal Comfort Requirement in Japan." Tokyo: Waseda University.

Webb, C. G. 1959. "An Analysis of Some Observations of Thermal Comfort in an Equatorial Climate." British J. Industrial Medicine, V–16, pp. 297–310.

Wise, A. F. E. "Wind Effect Due to Groups of Buildings." *Phil. Trans. Royal Society London* A269 pp. 469–485.

Wise, A. F. E., D. E. Sexton, and M. S. T. Lillywhite. 1965. "Studies of Air Flow Round Buildings." *Architects Journal* V–141, pp. 1185–1189.

Wray, W. O., and E. D. Best. 1987. "Thermal Storage in Direct Gain Buildings: The Effective Heat Capacity and a Generalized Solar Load Ratio Correlation." *Passive Solar Journal* V–4, No 1, pp. 41–61.

Wu, H. 1990. "Moisture Uptake and Energy Efficiency Associated with the Use of Evaporative Coolers." Research Report, Envir. Testing Lab., Tempe: Arizona State U.

RELATED BIBLIOGRAPHY

Akins, R. E., J. A. Peterka, and J. E. Cermak. 1979. "Average Pressure Coefficients for Rectangular Buildings," Proceedings of the Fifth International Conference on Wind Engineering, Fort Collins, Colorado. New York: Pergamon Press.

Alberts, W. 1982. "Modeling the Wind in the Town Planning Process." *Energy and Buildings*, vol. 4, no. 1, pp. 71–76.

ASHRAE Handbook and Product Directory—Fundamentals Volume. 1981. Atlanta: Amer. Soc. Heating, Refrig. and Air-Conditioning Engs., Inc. 1986. Chap. 14, "Air Flow in Buildings."

Aynsley, R. M. "Wind Generated Natural Ventilation of Housing for Thermal Comfort in Hot, Humid Climates." 1979. Fifth International Wind Engineering Conference.

——— 1980. "Tropical Housing Comfort by Natural Air Flow." *Building Research and Practice*, pp. 242–252.

Baer, S. 1980. "Concentrating Skylights, Insulation and Ventilation." in Tom Wilson (ed.), *Home Remedies—A Guidebook for Residential Retrofit.* Philadelphia: Mid-Atlantic Solar Energy Association.

——— 1980. "Cooling with Nighttime Air." *Alternate Sources of Energy*, vol. 41, pp. 22–26.

——— 1981. *Cooling with Night Air.* Albuquerque: Zomeworks Corporation.

——— 1981. "Pressure Coefficients to Evaluate Airflow Pattern in Wind Towers." *Proceedings of the International Passive and Hybrid Cooling Conference*, Miami Beach, Fla.

Bahadori, M. N. "Passive Cooling Systems in Iranian Architecture," *Scientific American*, vol. 238, no. 2, February 1978.

Balcomb, J. D. 1983. *Heat Storage and Distribution inside Passive Solar Buildings*, Los Alamos National Laboratory Report LA-9694-MS.

Barnaby, C. S., D. H. Hall, and E. Dean. 1980. "Structural Mass Cooling in a Commercial Building Using Hollow Core Concrete Plank." *Proceedings of the Fifth National Passive Solar Conference*, Amherst, Mass., pp. 747–751.

Caudhill, W. W. 1952. *Geometry of Classrooms as Related to Natural Lighting and Natural Ventilation*, Research Report No. 36, Texas Eng. Exper. Station, College Station.

Caudhill, W. W., S. E. Crites, and E. G. Smith. 1951. *Some General Considerations in the Natural Ventilation of Buildings*, Research Report No. 22, Texas Engineering Experiment Station, College Station.

Cermak, J. E. 1976. "Aerodynamics of Buildings." *Annual Review of Fluid Mechanics*, vol. 8, p. 75.

Cermak, J. E., M. Poreh, J. A. Peterka, and S. S. Ayad. *Wind Tunnel Investigations of Natural Ventilation*, ASCE Convention and Exhibit Preprint 82–519, New Orleans, Oct. 25–29, 1982.

Chand, I. "Effect of the Distribution of Fenestration Area on the Quantum of Natural Ventilation of Buildings," *Architectural Science Review*, vol. 13, no. 4, December 1970, pp. 130–133.

Chandra, S., P. Fairey, and M. Houston. 1983. "A Handbook for Designing Ventilated Buildings." Cape Canaveral, Florida: Florida Solar Energy Center.

Chandra, S., and A. A. Kerestecioglu. "Heat Transfer in Naturally Ventilated Rooms: Data from Full-Scale Measurements, FSEC-PF-45-83," ASHRAE Semi-Annual Meeting, Atlanta, January 1984.

Chandra, S., K. Ruberg, and A. Kerestecioglu. "Outdoor Testing of Small Scale Naturally Ventilated Models," *Building and Environment*, Vol. 18, no. 1/2, 1983, pp. 45–53.

Dick, J. B. "The Fundamentals of Natural Ventilation of Houses," *Journal, Institution of Heating and Ventilating Engineers*, June 1950, pp. 123–132.

Evans, B. H. 1957. *Natural Air Flow Around Buildings*, Research Report No. 59, Texas Engineering Experiment Station, Texas A & M College System.

Fanger, O. P. 1972. *Thermal Comfort.* New York: McGraw-Hill.

Givoni, B. 1981. "Convective Ventilation Cooling," DOE Passive Cooling Workshop, Philadelphia.

Kammerud, R., E. Ceballos, B. Curtis, W. Place, and B. Anderson. "Ventilation Cooling of Residential Buildings," ASHRAE Semiannual Meeting, Atlanta, January 1984.

Kusuda, T. Savings in Electric Cooling Energy by the Use of a Whole House Fan, NBS Technical Note 1138, May 1981.

Lee, B. E., M. Hussain, and B. Soliman. "Predicting Natural Ventilation Forces upon Low-Rise Buildings, *ASHRAE Journal*, February 1980.

Milne, M., and B. Givoni. 1979. "Architectural Design Based on Climate." In Donald Watson, ed. *Energy Conservation through Building Design.* New York: McGraw-Hill.

O'Hare, Michael, and Richard E. Kronauer. "Fence Designs to Keep Wind From Being a Nuisance," *Architectural Record*, July 1969, pp. 151–156.

Olgyay, V. 1963. *Design with Climate: Bioclimatic Approach to Architectural Regionalism.* Princeton, N. J.: Princeton University Press.

Schubert, R. P., and B. Kennedy. "The Testing of Full Scale Ventilator Cap Types to Determine Their Effect on Natural Ventilation," *Proceedings of the Annual Meeting of AS/ISES*, May 1981, pp. 901–905.

Sobin, H. J. "Window Design for Passive Ventilative Cooling: An Experimental Model-Scale Study." *Proceedings of the International Passive/Hybrid Cooling Conference*, Miami Beach, Fla., November 1981, pp. 191–195.

Soliman, B. F., and F. R. Frick. *Effect of Building Grouping on Wind Induced Natural Ventilation*, University of Sheffield, England.

* *

4

RADIANT COOLING

Of all the natural energies available for cooling by passive or semipassive systems outgoing long-wave radiation has received the widest research attention in recent years. Extensive investigations dealing with the potential for radiant cooling, attempts at developing specialized materials aimed at maximizing cooling derivable from this source, and mathematical modeling of the physical processes involved have been undertaken by a number of researchers in several countries.

As a result of these studies the amount of information on radiant cooling vastly exceeds that available on other passive cooling sources. This chapter summarizes the available information, discusses its applicability under different climatic and building type conditions, and evaluates it in view of the personal research experience of the author.

INTRODUCTION: OVERVIEW OF RADIANT COOLING OPTIONS

Radiant Cooling Processes

Any surface emits radiation with a spectrum of wavelengths, which depends on its temperature. At the ordinary temperatures found on earth the emitted radiation is in the long-wave range. When a given surface faces other surfaces at similar temperatures (for example, the ground around the building or the walls of other buildings) the net gain or loss is rather small. However, when a given surface such as the roof of a building, is exposed to the sky, the situation is different. The downward flux of atmospheric long-wave radiation is weaker than the radiation emitted upward by ordinary surfaces. The result is a net long-wave radiant heat loss and a cooling of surfaces exposed to the sky.

Emission of long-wave radiation takes place continuously, day and night. However, during the daytime the surfaces radiating in the long-wave part of the spectrum are exposed to solar radiation. The solar radiation absorbed at the surface (which depends on its solar absorptivity; that is, its color) produces heating, which in most cases outweighs the cooling effect produced by the emission of long-wave radiation. Therefore, net radiant cooling can be obtained only during the night hours. For this reason radiant cooling is often referred to as nocturnal radiation.

The long-wave radiation emitted by the building is continuous over the range of 5 to 30 microns μ, with peak radiation at about 10 u ("gray-body radiation"). The spectrum of the radiation emitted by the atmosphere, however, may or may not follow a similar pattern. Its spectrum pattern depends on the moisture content of the air (measured, among other ways, by the vapor pressure or the dew-point temperature) and, especially, on cloud conditions (See the section on heat transfer processes for details).

The net radiant heat loss from a radiator placed on the roof of a building is the balance between the energy flux emitted by the radiator and the incoming atmospheric radiation absorbed by the radiator. It can be about 70 W/m^2, or even more, from a radiator with an "ordinary" surface under clear night sky in a desert, and it decreases with cloudiness and the humidity of the ambient air. It drops to about 40 to 50 W/m^2 under clear sky in humid areas and is negligible under cloudy sky. The net radiant loss provides the potential for radiant cooling.

In calculating the net radiant heat loss two optional concepts can be used: the "effective sky temperature" and the "sky emissivity." The effective sky temperature is defined as the temperature of a black body that radiates toward the ground with a continuous spectrum at the same energy flux as the measured atmospheric radiation under given climatic conditions. The sky emissivity assumes that the same radiant flux is emitted from an atmosphere with the same temperature as that of the air near the ground but with a lower emissivity. Both the effective sky temperature and the sky emissivity depend mainly on cloudiness conditions and ambient humidity.

Not all the radiant net loss can always be utilized as a cooling energy for a building. A nocturnal radiator at a temperature below the ambient air gains heat by convection. Some cold energy is also lost during the process of transfer from the radiator to the building. In this case the effective cooling of the building is less than the radiant net loss. However, when the radiating surface is warmer than the ambient air (for example, when a building has a concrete roof) all of the radiant potential is utilized and, in addition, convection to the ambient air also becomes a cooling source.

Several design options can be considered for utilizing the physical process of radiant heat loss as a cooling source for buildings. They involve three basic types of nocturnal radiators: (a) a high-mass roof with movable insulation serving as a combined radiator/cold storage element; (b) a lightweight, usually metallic radiator that cools ambient air below its initial temperature; and (c) unglazed water-type solar collectors. Each one of these radiators can be developed into several radiant cooling systems.

A high-mass roof can be a structural concrete roof or a steel deck with water-filled bags over it. Movable insulation panels expose the roof mass to the sky during the night and insulate it during the daytime.

A lightweight radiator with an air space and ambient air flow underneath can cool the flowing air by several degrees. The cool air then is directed into the building to provide instantaneous cooling during the night and to cool the interior mass of the building by convection, thus creating a cold storage for the following day.

Unglazed solar collectors can be used during the night to cool water. The cooled water can be circulated in pipes embedded within a concrete roof (or walls) and thus preparing cool mass as a cold energy reservoir (heat sink) for absorbing the heat penetrating into the building during the daytime. Alternatively the circulated water can cool a pond over the roof that, together with the roof mass, forms a combined heat sink.

More detailed discussions of the implementation details of these options is presented below.

High-Mass Roofs with Operable Insulation

If a roof of high thermal mass is exposed to the sky during the night the mass is cooled down by long-wave radiation, and often also by convection to the cooler outdoor night air, rendering the roof's mass into a cold energy storage. When the cooled mass is thermally coupled with the interior space and insulated from the ambient environment during the daytime hours, it absorbs heat from the space below. Therefore, in order to serve as a cooling system, the roof has to be insulated from the sun and the hotter ambient air during the daytime hours. This will minimize its heat gain from the outdoor environment. Therefore, the system has to include movable insulation above the roof that exposes it during the night but insulates it during the daytime.

With the provision of operable insulation a high-mass conductive roof can act as an efficient combined radiator/storage cooling element. The thermal mass can be provided in different form. A high-mass structural dense concrete roof, when equipped with movable insulation, is a simple implementation of this concept. Figure 4.1 is a sketch of this concept.

Another option is a building with a flat structural steeldeck roof with water-filled plastic bags and movable insulating panels over it. This is the "Skytherm" system developed by Harold Hay. This is the only radiant cooling (and solar heating) system that has been commercialized and actually applied in a number of buildings. (See below for a detailed discussion).

The thermal mass of either type cools down at night at a lower rate than the ambient air. During some or even most of the night, its surface has a higher temperature than the air. Consequently it is cooled also by convection, in addition to the radiant heat loss. A higher wind speed, therefore, increases the cooling rate of high-mass roofs. The nightly average cooling rate of the storage mass can be above 120 W/m^2 in a desert area that is windy at night (for example, Palm Springs in California or the Arava region in Israel).

Such roofs thus serve as a combined nocturnal radiator and cold-storage mass, passively cooling the space below by conduction through the ceiling. With a suitable design it is possible to cool the storage mass to a temperature well below the daily average of the outdoor air temperature. Experimental data on the performance of such high-mass radiant cooling systems is presented later in this chapter.

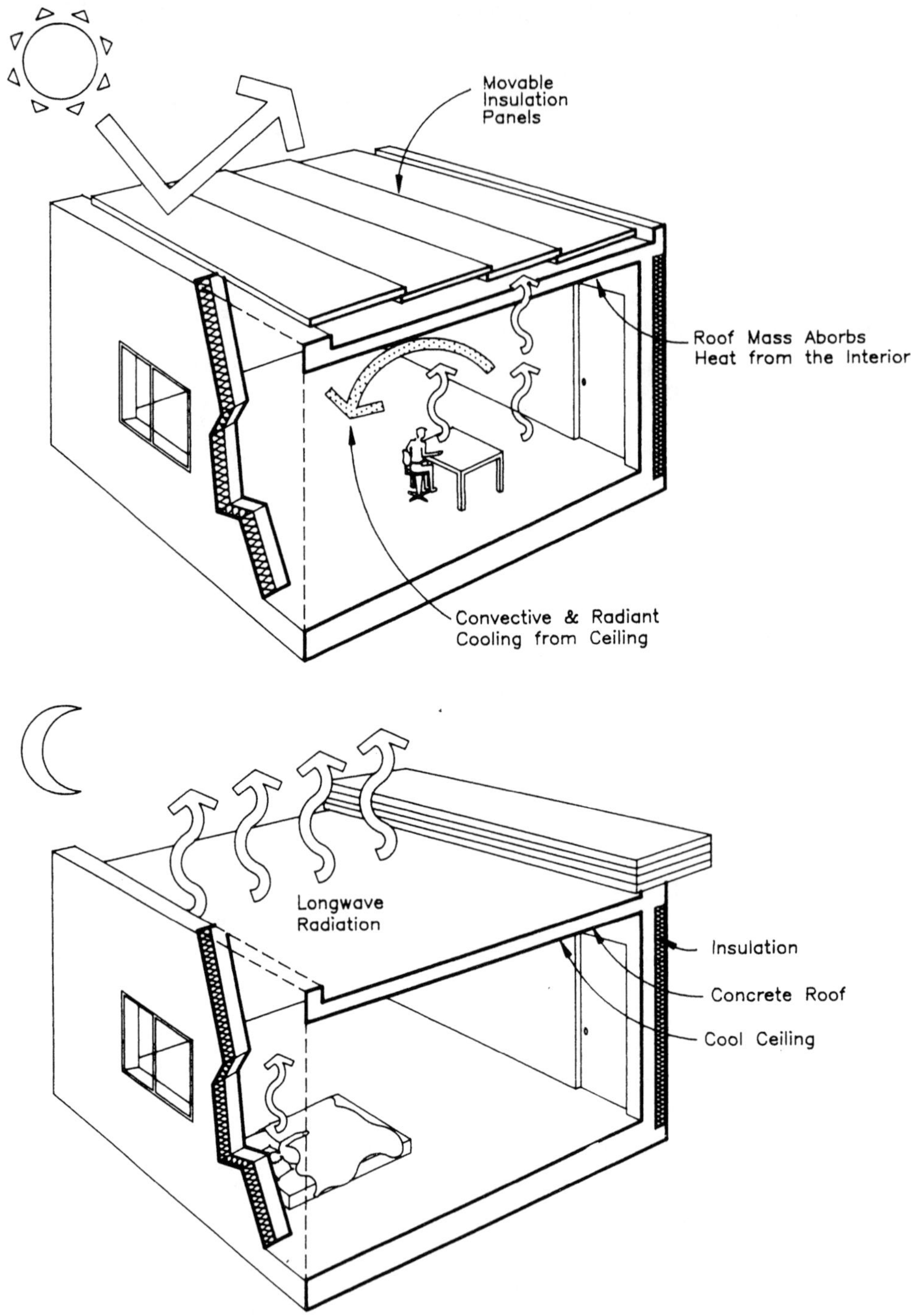

FIGURE 4.1 Concrete roof with operable insulation as a nocturnal radiator.

Lightweight Nocturnal Radiators

A thin plate above the roof, insulated underneath, does not store much solar energy, nor does it gain heat from the space below the roof. The emission of long-wave radiation lowers the temperature of such a lightweight radiator below the ambient air level. Consequently, convective heat gain from the air counteracts the radiant heat loss until an equilibrium is reached between heat gain and loss, and the radiator's temperature is stabilized at some level below the ambient air. The convective heat gain depends on wind speed near the emitting surface—as wind speed increases the temperature of the emitting surface becomes higher, drawing closer to that of the ambient air.

With conventionally insulated roofs the net heat loss at the external surface of the roof cannot be utilized directly for cooling the building, because of the thermal resistance of the roof. To utilize the cooling effect of nocturnal radiation, the cold energy produced by the radiating surface has to be transferred into the building's interior. When ambient air is sucked at night by a fan through the "channel" between the cooled radiator and the insulation layer underneath, it is cooled by the radiator (convective cooling). The cooled air can be directed into the space to be cooled, thus cooling the inhabitants in the space and the interior mass. Nocturnal radiant cooling thus can be stored in the interior structural mass of the building, with the cooled mass acting as a heat sink during the following day. Alternatively, the cold energy can be stored in a specialized mass—for example, in a gravel bed under the building. Figure 4.2 illustrates a building with a metallic nocturnal radiator and cold storage in the interior structural mass of the building.

In some buildings that are occupied mainly during the evening and at night hours—for example, some residential buildings—the air cooled by the long-wave radiation under a specialized radiator is utilized directly to cool the indoor space, and so it is not necessary to store the cold energy. Such radiant cooling is applicable in such low-mass buildings as those with stud wall wood construction. On the other hand, in buildings occupied mainly during the daytime, the cold produced at night has to cool some storage mass that subsequently can be utilized during the following day as a heat sink. Thus, daytime utilization of radiant cooling is possible only in high-mass buildings, or in low-mass buildings with supplemental specialized storage mass.

Unglazed Solar Collectors as Nocturnal Radiators

It is possible to install unglazed solar collectors over the roof and to circulate water through them at night during the summer. The cooled water can be utilized in one of the following ways:

- **A** The cooled water can be circulated through the roof material—the concrete. In this way the roof serves as the thermal storage element and the ceiling as a radiant cooling panel above the space to be cooled. (See Figure 4.3.)
- **B** The cooled water can be circulated through a water pond over the roof and below the collectors (a system developed and tested by Y. Etzion, 1989). The pond then serves as the main thermal storage element.

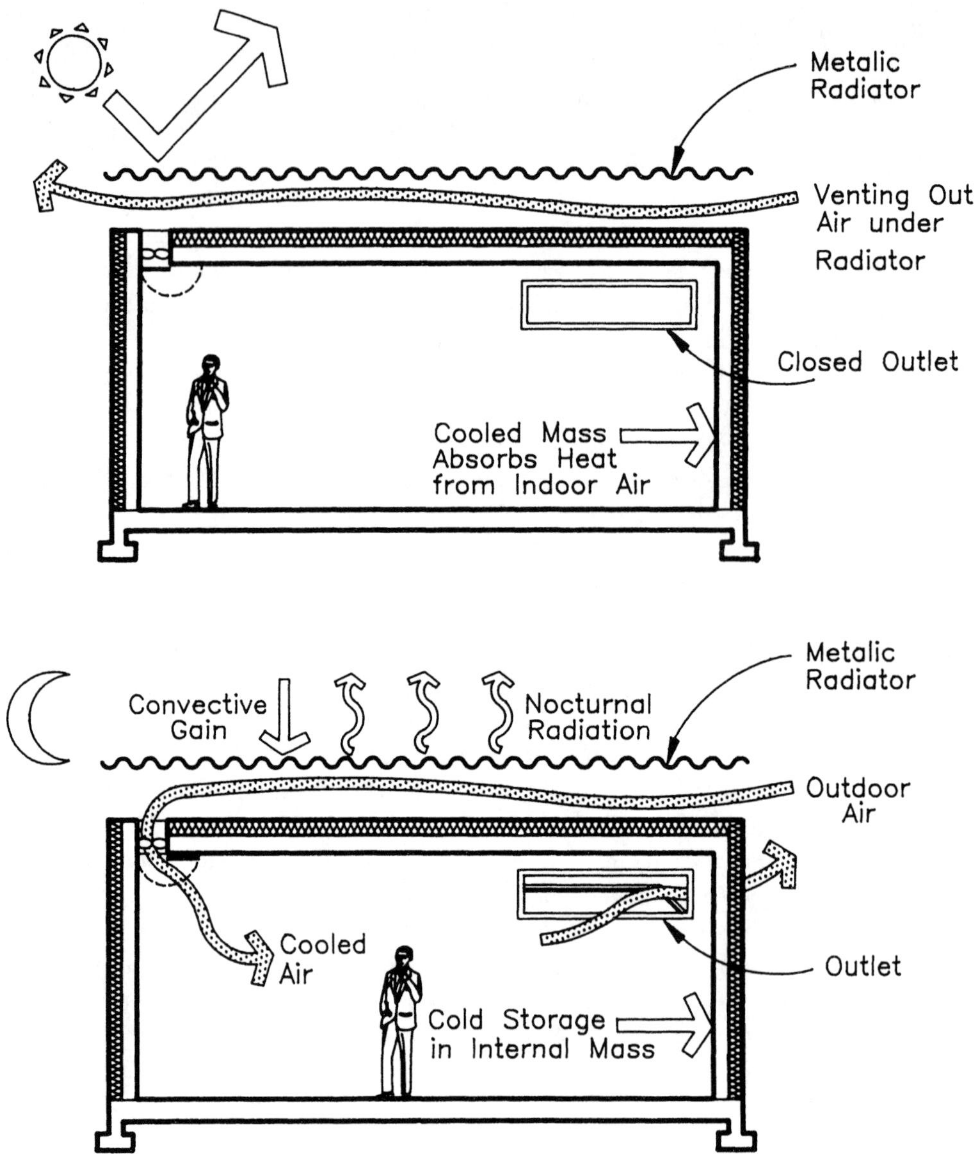

FIGURE 4.2 Cooling processes of a lightweight nocturnal radiator.

In winter it is possible to use the low-efficiency unglazed collectors to heat water during the daytime.

Design Issues in Radiant Cooling Systems

Actual applicability and choice of type and design details of a radiant cooling system depend on the type of roof (for example, flat or pitched, load-bearing or nonload-bearing), the type of the building (single story or multistory, lightweight or high-mass) and the

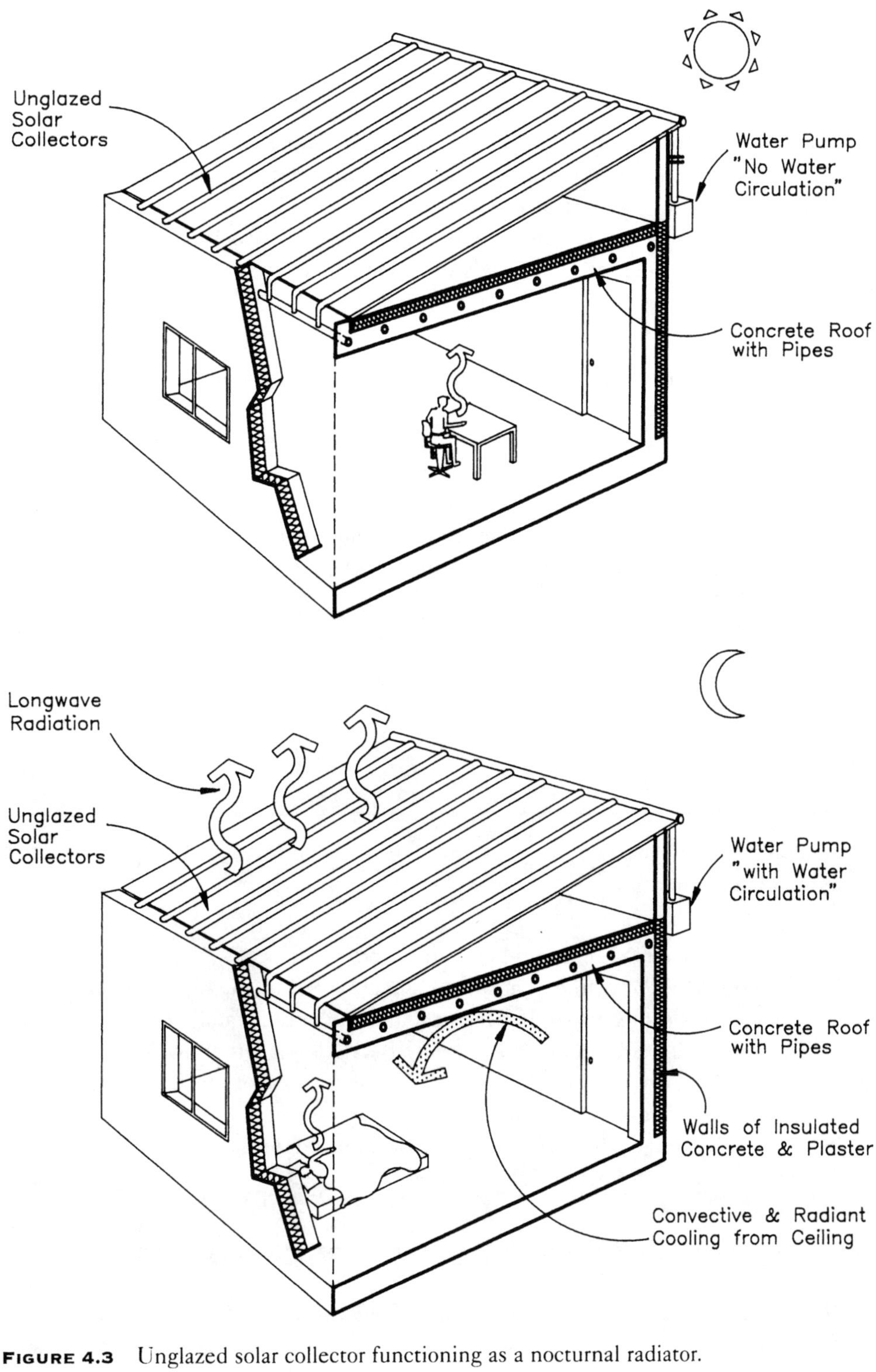

FIGURE 4.3 Unglazed solar collector functioning as a nocturnal radiator.

climatic conditions. This choice also may be influenced by the desired period of indoor cooling. For example, residents may need the building cooled only during the evening and night hours, or cold storage for the following day may be necessary. Another issue is the desired mode of heat transfer between the storage medium and the indoor space—conductive or convective—and the resulting indoor air speeds and temperature swings. This last point affects the comfort sensations of the occupants because it determines the relative roles of radiative and convective heat exchange between the human body and the indoor environment.

Several issues should be considered when choosing a cooling system that uses nocturnal long-wave radiation. The issues should also be considered when design details are decided.

1. The applicability of radiant cooling under the specific climatic conditions of the site should be checked.
2. The applicability of radiant cooling for the specific building type in question should be checked.
3. The specific system should be chosen. Subissues involve deciding whether to use the roof itself as a radiator/storage/transfer unit, whether to apply a specialized lightweight metallic radiator, or whether to use unglazed solar collectors as nocturnal radiators.

Heat Transfer Processes In Radiant Cooling

All systems for radiant cooling of buildings involve five processes of heat transfer that affect the performance of the system:

1. Long-wave radiation emitted by the radiator.
2. Radiation emitted by the atmosphere and absorbed by the radiating surface.
3. Convective heat exchange between the radiator and the ambient air.
4. Cold energy transfer from the radiator to the building, either by conduction (with concrete roof radiators) or by forced air flow in the case of specialized lightweight (metallic) radiators.
5. Convective heat exchange between the radiator and the heat transfer medium, the air (in the case of a lightweight radiator).

The system's performance depends in part on the climatic conditions and in part on its design details.

The balance between the two first processes yields the net radiant heat loss, which represents the climatic potential for radiant cooling. The balance between the net radiant loss and the convective exchange with the ambient air yields the stagnation temperature of the radiator. The heat exchange between the radiator and the air flowing underneath, and the air flow rate, determine the temperature of the air exiting from the radiator and the cooling energy delivered to the building.

The Radiation Emitted by a Long-wave Radiator

The intensity of long-wave radiation emitted by a specialized radiator, or by the roof itself when it serves as the radiating surface, depends on only two factors: the (absolute) temperature of the radiating surface, T_r and its emissivity, E_r:

$$R_{emit} = S * E_r * T_r^4$$

when S is the Stefan-Boltzman constant;

$$S = 0.567 * 10^{-8}; \ (w/m^2K^4).$$

The emissivity of a given radiating surface is a physical property and represents the potential of the surface to emit radiation relative to a perfect "black" surface. Most common materials (except metals) have an emissivity of about 0.9. When a metallic material serves as a nocturnal radiator it should be painted with a paint that has an emissivity of about 0.9. The paint layer then serves as the actual radiating surface.

The temperature of a radiator, and especially its relation to ambient air temperature, depends on its mass and on the way in which it is utilized. Therefore it is one of the system design factors affecting its performance. A high-mass radiator (for example, a concrete roof or water bags above a metallic deck) is at a temperature above the ambient air for most of the night. Consequently, its radiant heat loss is augmented by convective heat loss to the ambient air and its total heat dissipation is further increased. The temperature of an insulated lightweight radiator (metallic or asbestos cement) is below that of the ambient air and the radiant loss is counterbalanced by convective heat gain. Thus the radiator type is very significant in determining total energy loss from the radiator (combined net radiant loss and convective heat exchange). The thermal contact between the radiator and the building to be cooled can also affect the temperature of the radiator and its overall efficiency.

This section summarizes the research leading to the prediction of net radiant loss and the stagnation temperature, and on techniques to lower them. A mathematical model calculating the performance of a radiant cooling system with a metallic radiator, the exit air temperature, and the cooling energy delivered to the building is described in Appendix 1.

Atmospheric Radiation

Any surface exposed to the sky receives long-wave radiation, which is emitted by the atmosphere. As is known, the atmosphere consists mainly of nitrogen (about 78%) and oxygen (about 21%), with only minute quantities of the other constituents, mainly water vapor, CO_2, and dust. Oxygen and nitrogen are mostly transparent to this part of the radiation spectrum. However, water vapor emits and absorbs radiation at wavelengths around 6.6 and 18 μ while CO_2 emits and absorbs radiation around 15 μ. Clouds, in particular, greatly increase the incoming atmospheric radiation. The radiation emitted by the atmosphere downward toward the earth is defined as "atmospheric back radiation."

Atmospheric back radiation depends mainly on cloud conditions and to some extent also on the water vapor content of the ambient air. Under clear sky conditions and low humidity the atmospheric radiation is concentrated in distinct wavelength bands—the 3 to 8 μ and the 13 to 20 μ bands. Between the 8 and 13μ atmospheric back radiation is weak, and the 8 to 13 μ band therefore is called the "atmospheric window." As the water-vapor content of the air increases, the intensity of the atmospheric back radiation in the 8 to 13 μ band also increases, resulting in stronger total atmospheric back radiation towards the earth.

Clouds, especially low clouds, emit radiation through the whole long-wave spectrum, so that under an overcast sky the phenomenon of the sky window practically disappears. Thus, under cloudy sky conditions atmospheric radiation reaches its maximum.

There are two ways by which the atmospheric back radiation can be expressed quantitatively. The first is to assume that the sky emits radiation as a black body, with a given effective sky temperature, T_s. Atmospheric emissivity is then taken as a unit (E_s = 1.0). The other way is to assume that the atmosphere has the same temperature as the air near the ground but its emissivity is less than unity: $E_s<1.0$, like radiation emitted from a gray body. The lower sky emissivity is used to express quantitatively the reduced intensity of the atmospheric radiant flux.

Climatic Potential for Radiant Cooling

Extensive experimental studies to evaluate the climatic potential of radiant cooling were performed by several investigators, in particular at the Lawrence Berkeley Laboratory (LBL) in Berkeley, California, and at Trinity University, San Antonio, Texas. A detailed summary of these studies is presented in Martin (1989). A shorter summary of some of these studies is presented below.

The formulae suggested by the Trinity research team for calculating the sky emissivity, E_s, and the effective sky temperature as functions of the Dew Point temperature, T_{dp}, the ambient temperature, T_a, and cloudiness, are:

$$E_s = 0.787+0.0028*T_{dp}$$

The clear sky temperature is then given in absolute temperatures by:

$$T_{clear\ sky}\ (K) = (E_s)^{1/4} * T_a$$

Cloudy sky temperature is given by:

$$T_{cloudy} = (C_a)^{1/4} * T_{clear\ sky}$$

The incoming radiation, R_{sky}, is then given by:

$$R_{sky} = \sigma^*(T_{sky})^{1/4}$$

where σ is the Stefan-Boltzman constant.

Clark and Allen (1978) and Berdhal and Martin (1984) have measured the emissivity of clear sky in different locations in the United States. Berdhal and Martin have developed the following formula expressing the emissivity of clear sky, E_{clear}, as a function of the ambient dew point temperature, T_{dp}, (Martin 1989):

$$E_{clear} = 0.711 + 0.56\,(T_{dp}/100) + 0.73(T_{dp}/100)^2$$

Clark and Allen (1978) have developed another formula for estimating the sky emissivity, E_s, as a function of the dew point temperature:

$$E_s = 0.787 + 0.764 \ln\left((T_{dp}+273)/273\right)$$

Clark (1981) has developed the following formula, expressing the effect of cloudiness (n) on the atmospheric radiation, R_{sky}:

The effect of cloudiness is expressed by a factor Ca:

$$Ca = 1+0.0224n-0.0035n^2+0.00028n^3$$

The sky radiation is then given by:

$$R_{sky} = R_{clear}\,(1+0.0224n+0.0035n^2+0.00028n^3)$$

Berdhal and Fromberg (1982) have developed a model of clear sky emissivity on the basis of measurements in Gaithersburg, Maryland, St. Louis, Missouri, and Tucson, Arizona.

Clark and Berdhal (1980) have plotted maps of the distribution of the net cooling rate obtainable from nocturnal radiation across the United States for different temperatures of the radiator.

Convective Heat Exchange with Ambient Air

In practically all cases of radiant cooling there is also a convective component of heat exchange interacting with the radiative processes. For exposed long-wave radiators the convective exchange is a function of the convective coefficient, which depends on the wind speed near the radiator. It is proportional to the temperature difference between the radiator and the ambient air.

The convective coefficient, however, is one of the most difficult factors to estimate accurately. It depends on wind speed next to the radiating surface and on whether the airflow above the surface is laminar or turbulent. In reality any value assumed for air speed should be considered only as a rough estimate. The only known wind parameters relevant to the building are usually the wind speed and direction at some meteorological station, which at best is close, but usually may be at some distance or even remote, from the building in question.

Meteorological stations are usually located in open areas, where the wind speed is higher than in a built up area. Furthermore, the wind fluctuates constantly, both in direction and in speed, so that the relationship between wind conditions at the measuring point and those next to the building with the radiator are changing constantly. In addition, such design details of the roof itself as parapets can greatly affect the air speed next to the radiator under a given wind speed measured near the building. Thus, there is no way to know with reasonable accuracy the nature and speed of the wind next to the radiating surface, and it should be realized that any value assigned to the convection coefficient in the equations given by various models is only a rough estimate.

In those radiant cooling systems where it can be expected that the temperature of the radiating surface for most of the night will be above ambient air temperature (for example, when a massive concrete roof or water bags over the roof serve directly as the radiators), a high wind speed increases the cooling rate by increasing the convective heat loss. On the other hand, in those radiant cooling systems in which the radiator temperature is planned to be below that of the ambient air (that is, when a lightweight air-cooling radiator is used) infrared transparent windscreens can be applied to minimize convective heat gain from the warmer air.

The general formula for convective heat exchange between the radiator and the ambient air is:

$$Q_c = h_c * (T_r - T_a)$$

where h_c is the convection coefficient. The problem is in evaluating the value of h_c as a function of the wind conditions. Various authors have suggested different formulae for evaluating this coefficient (among them Clark and Berdahl (1981), Duffie and Beckmann (1974), and Hansen and Yellott (1978). Givoni (Mostrel and Givoni 1982), based on the analysis of various experimental data involving exposed and wind-screened radiators, has suggested the following formulae:

For an exposed radiator:

$$h_c = 0.6 + 3.5 * (V)^{0.5}$$

For a radiator covered by a single Polyethylene layer:

$$h_c = 0.5 + 1.2 * (V)^{0.5}$$

For a double-covered radiator:

$$h_c = 0.3 + 0.8 * (V)^{0.5}$$

The Stagnation Temperature of Lightweight Radiators

Formulae predicting the stagnation temperature of exposed and screened radiators as a function of the meteorological (temperature, humidity and wind speed) and screening conditions of the radiator, has been developed by Mostrel and Givoni (1982) by balancing the net radiant loss and the convective gain of a stagnant radiator. This formula does not deal with cloudiness. The predictions by this formula have been compared with experimental data.

When a radiator is wind-screened by polyethylene layers both the absorption of sky radiation, R_{abs}, and the emittance of radiation to the sky, R_{emit}, are modified by the long-wave transmission of the windscreens, trans, and the emissivity of the radiator, E_r:

$$R_{abs} = E_r * trans * R_{sky}$$
$$R_{emit} = E_r * trans * \sigma * (T_r)^4$$

where σ is the Stefan-Boltzmann constant, $0.567*10^{-8}$; in W/m^2K^4.

The values of trans are assumed to be 70% for a single layer and 60% for a double layer.

The net radiant loss can thus be evaluated, under different cloudiness conditions, by the formula (Mostrel and Givoni 1982):

$$R_{net} = C_n * trans * E_r * 0.554 * (T_r - T_{sky})$$

where C_n is the Clark and Allen (1978) formula for the cloudiness coefficient:

$$C_n = 1 - 0.056 * n$$

and n is the cloud cover index.

By balancing the net radiant loss and the convective gain, ($R_{net} = Q_c$), the following linearized formula (Mostrel and Givoni 1982) predicts the stagnation temperature, T_{stag}, of a metallic radiator as a function of the air and sky temperatures, cloudiness, and the windscreen transmittance:

$$T_{stag} = \frac{0.544 * C_n * trans * E_r * T_{clear\ sky} + h_c * T_a}{0.544 * C_n * trans * E_r + h_c}$$

A detailed description of the experimental studies in Sede Boqer and the development of the stagnation formula is presented in Givoni (1982) and in Mostrel and Givoni (1982).

Figure 4.4, a to d, shows the correspondence between the computed and measured diurnal patterns of the stagnation temperature in four different experiments. It can be seen that as long as dew has not formed on the polyethylene there is good correspondence between the predicted and the measured data. However, the formula does not take into account the phenomenon of dew formation. Therefore, the formula does not predict the disappearance of the effect of the wind screening on the radiator's temperature.

The Utilized Radiant Energy

Using radiant heat loss as a cooling source for a building depends on two factors: (a) the net radiant energy loss from the radiator and (b) the convective heat gain of the radiator from the air flowing under it.

Net energy loss from a radiator is the balance between three components of heat exchange: the radiation emitted by the surface, the atmospheric radiation absorbed by the radiator (which together determine the net radiant loss), and the convective heat exchange between the radiator and the ambient air above it (a function of the radiator's temperature, its design details, and the wind speed). The direction of the convective heat exchange—

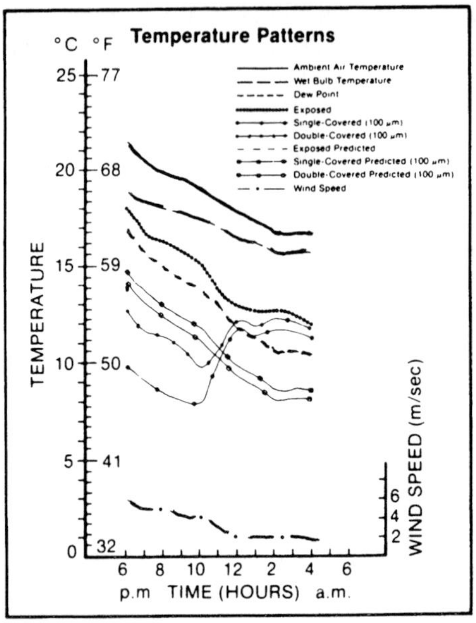

a) : On September 18-19, 1981, a humid, cloudy night, wind speed affected the difference between the double- and single-covered windscreens. The double-covered radiator's predicted temperature was 6°C (*10.8°F*) higher than that of the single-covered's predicted temperature, until around midnight.

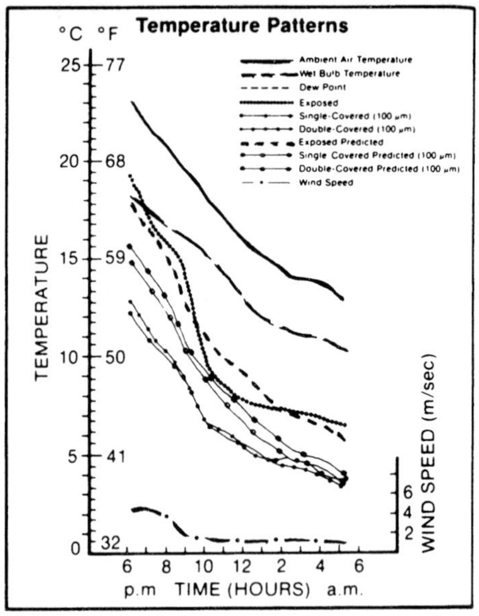

c) : The predicted and the measured temperature patterns for the dry, clear night of October 13-14, 1981, are very similar.

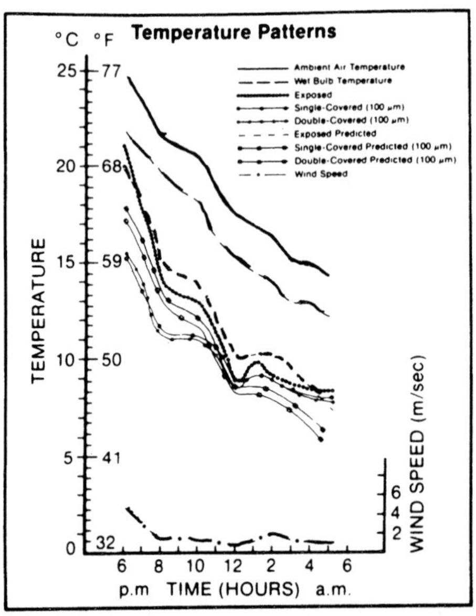

b) : On the dry, cloudy night of October 17-18, 1981, the difference between the predicted and measured temperatures of single- and double-covered radiators was not significant.

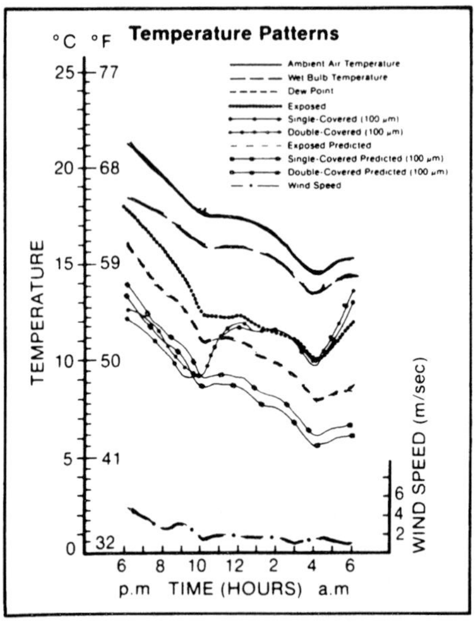

d) : Humid, clear nights, such as the evening of October 11-12, 1981, affect predicted and measured results of windscreened radiators. After dew forms, the measured patterns exhibit a sharp rise not anticipated in the predicted patterns.

FIGURE 4.4 Correspondence between measured stagnation temperatures of metallic radiators and temperatures predicted by Givoni's formula.

whether it provides or extracts heat from the radiator—depends, of course, on the temperature relationship between the radiator and the ambient air.

A mathematical model calculating the utilized radiant heat loss has been developed by Ingersoll and Givoni and is described in the appendix to this chapter.

Techniques to Increase the Net Heat Loss

Two techniques can be considered, separately or jointly, for increasing the rate of the net heat loss from radiators:

- **A** Applying a windscreen, transparent to long-wave radiation, to reduce convective heat gain from the ambient air.
- **B** Applying a selective surface to the radiator, which emits mainly in the spectrum of the sky window (between 8 and 13 μ) and is reflective in the rest of the long-wave spectrum and preferably also in the short wave (solar) spectrum.

Convective Gain Suppression by Wind-Screens

At night the temperature of lightweight radiators drops well below the ambient air temperature. The radiator then gains heat from the ambient air by convection. A higher wind speed increases the heat gain and reduces the temperature drop of the radiator.

Therefore, when it is desired to lower a radiator's temperature well below the ambient level it would be desirable to minimize convection without interfering too much with the emission of long-wave radiation. This is always the case when a specialized lightweight nocturnal radiator is used.

An inexpensive material that is about 75% transparent to long-wave radiation (infrared-transparent) is polyethylene film without ultraviolet inhibitors. If a thin film of polyethylene is stretched firmly over a radiator, supported by either air pressure or tension so that it does not flutter in the wind, a stagnant cold air layer is formed over the radiator. The stagnant air layer and the film serve as transparent insulation, minimizing convective heat gain and lowering the temperature of the radiator below the temperature drop of an exposed radiator.

Radiative properties of polyethylene film (without ultraviolet inhibitors) were measured by Clark and Blanpied (1979). They studied the film's infrared transmissivity, its reflectivity, and its emissivity. Typical weighted averages (for the different angles of incident) are:

Transmissivity = 0.75
Reflectivity = 0.10
Emissivity = 0.15

Even a single layer of polyethylene reduces emitted radiation by about 25 percent because of its imperfect transmissivity. However, convective heat gain, when the radiator's temperature is below that of the ambient air, is reduced to a much greater extent, especially under windy conditions. Thus, the net effect is to increase the net heat loss and the temperature depression of the radiator below ambient air temperature.

At the Institute for Desert Research in Sede Boqer, Israel, extensive measurements of the stagnation temperature of metallic radiators, both exposed and screened by polyethylene, were carried on in a lengthy experimental series that lasted several months and were performed under different meteorological and wind screen conditions (Givoni 1982-b, Mostrel and Givoni 1982).

The radiators consisted of white-painted 45×45 cm thin iron plates that were insulated by 10 cm thick expanded polystyrene on their sides and bottom. An air space of 1 cm was left between the iron plate and the bottom insulation. The wind screens consisted of either one or two layers of polyethylene film, the first stretched about 3 cm above the radiating plate and the second (when applied) 2 cm above the first. Detailed information on this study is reported in Givoni (1982) and Mostrel and Givoni (1982).

Figure 4.5a-d (Givoni 1982) shows separate temperature patterns of metallic radiators with and without a polyethylene wind-screen during four nights with different humidity and cloudiness conditions, from 6 p.m. through 6 a.m. The figures also contain the measured wind speed.

Figure 4.5a shows the temperature patterns of three radiators during a clear night with low humidity (October 13–14, 1981). The first radiator was exposed, the second had one layer of polyethylene, and the third had two layers of wind screen. The figure also shows the ambient wind speed, which was about 4 m/s from 6 to 7 p.m., then dropped gradually to about 1.5.m/s at 9 p.m. This was a clear night with very low humidity.

It can be seen that the difference between the single and double layers of polyethylene was very small. Apparently, the reduced convective heat gain with the double wind screen is balanced by a reduction in the radiant loss. It can also be seen that, as expected, during most of the night the temperature drop of the screened radiators is larger than that of the exposed radiator (about 10 and 7°C, respectively). No condensation was observed on the polyethylene.

It can also be seen that the screened radiators were less sensitive to wind speed. With the higher wind speed the temperature drop of the screened radiators was more than twice that of the exposed one but with the lower wind speed the additional temperature drop brought about by the wind screens was only about one third of the drop of the exposed radiator.

Figure 4.5b shows the temperature patterns during a medium-humid night, with higher wind speed than in Figure 4.5a, and with passing clouds (August 28–29, 1981).

The temperature of the screened radiator again remained lower than that of the exposed one throughout the night. When it became cloudy the temperatures of both the exposed and the screened radiators rose sharply.

However, the patterns observed with the screened radiators during these two nights were not typical of others in the Sede Boqer study. On most nights in this desert climate the air temperature dropped down while the absolute humidity rose slightly. (Apparently this pattern is a result of the stability of the air near the ground.) Condensation started to form on the polyethylene sheets at about midnight or even earlier, and then the difference between exposed and protected radiators gradually disappeared, as can be seen in Figures 4.5c and 4.5d. Figure 4.5d also shows a sharp rise in radiator temperature as a cloud passed by between 3 and 4 a.m.

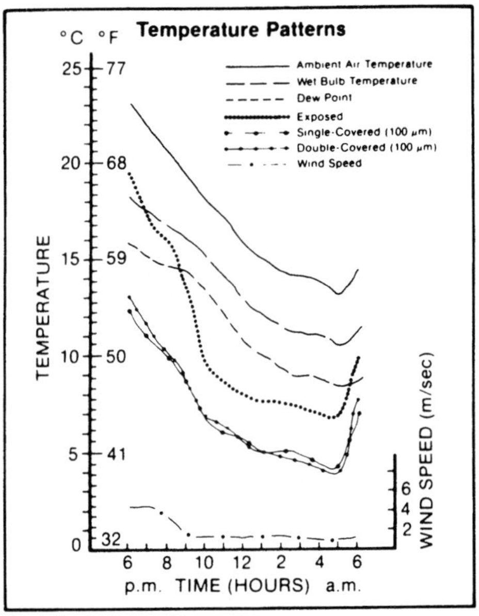

a) : The graph of the night of October 13-14, 1981, compares the temperature patterns of exposed, single-covered, and double-covered radiators on a dry, clear night. The temperatures of the windscreened radiators were lower than the exposed radiator's temperature.

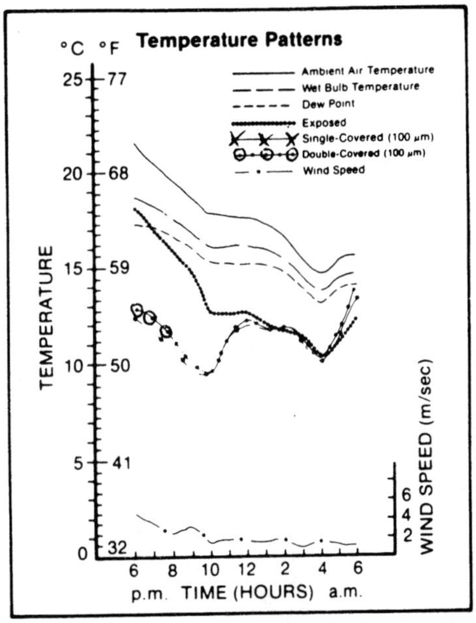

c) : The time lag between dew formation and when the windscreened radiators' and the exposed radiator's temperatures are equal is the same for both single- and double-covered radiators. The night of October 11-12, 1981, also illustrates that dew forms on both of the covered radiators at about the same time

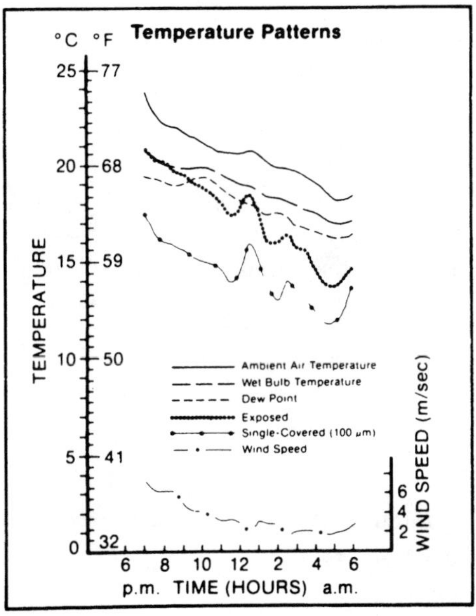

b) : A graph of the dry, cloudy night of August 28-29, 1981, compares the effect of cloudiness on an exposed and a covered radiator. The temperature of the windscreened radiator again was lower than that of the exposed one.

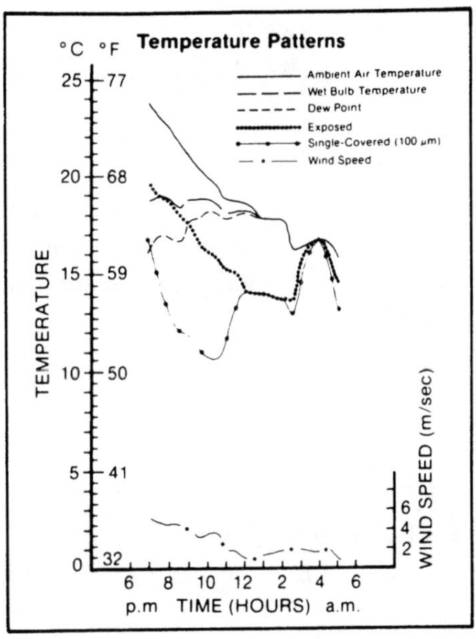

d) : The night of July 12-13, 1981, was humid and cloudy. This graph shows the combined effects of dew formation and cloudiness on an exposed and a covered radiator. After a time lag of one and one-half hours, the temperature of the windscreened radiator reached that of the exposed radiator.

FIGURE 4.5 Temperatures of exposed and wind-screened metallic radiators. Effects of clouds and dew formation.

Selective Long-Wave Radiators

To maximize net long-wave radiant heat loss the radiative surface should reflect away incoming atmospheric radiation from water, CO_2, and dust particles in the air while strongly emitting at the long-wave spectrum of the atmospheric window. An ideal long-wave radiator should have high absorptivity (and, therefore, high emissivity) in the 8 to 13 μ spectral band and high reflectivity (and, therefore, low absorptivity) above and below this band. In particular, it would be desirable to have a surface with high reflectivity also in the solar spectrum (0.4 to 3.0 μ) so that radiant cooling will continue during the daytime. Theoretical discussion of the potential for radiant cooling that can be obtained by a selective surface was presented by Catalanotti et al (1975). They have also built an experimental model utilizing as a radiator a sheet of Tedlar (polyvinyl fluoride) with a thin film of evaporated aluminum. A windscreen of thin polyethylene film has minimized the convective gain. The experimenters have reported a stagnation temperature depression of about 12°C. Daytime radiant cooling, with the radiator shaded from the direct solar radiation, was reported to yield temperature depressions of up to 15°C.

This report has prompted widespread experimental research in several countries, aimed at reproducing similar results. To the best knowledge of the author, all these "validation" studies have not demonstrated a significant effect of the selective surfaces that have been tested. Some of these studies are summarized below.

Experimental comparisons of radiant cooling obtainable by a gray-body emitter and a selective emitter have been made by Mitchell in Australia, Sakkal, Martin, and Berdahl of the Lawrence Berkeley Laboratory, Miller and Bradley of the Desert Research Institute in Nevada, and Landro and McCormick in Australia. The main results of these studies are summarized below.

Mitchell

Mitchell (1976) reported on extensive research done on selective surfaces of nocturnal radiators at the Commonwealth Scientific and Industrial Research Organization in Australia. In particular, the following surfaces were compared directly, under the same climatic conditions:

— Anodized aluminum
— Aluminized Tedlar
— Black paint

In spectral measurements the anodized aluminum has demonstrated selectivity. However, at night, the depression of the three surfaces were about the same, from 6 to 8°C. Thus the spectral selectivity of the anodized aluminum did not result in a higher total radiant heat loss.

Sakkal et al.

The experimental study done by Sakkal et al. in Berkeley consisted of exposing two horizontal plates covered with a polyethylene windscreen to a clear night sky and allowing them to cool until an equilibrium temperature was achieved. The gray-body emitter was an aluminum plate 0.8 mm thick and painted white. The selective surface consisted of an aluminum layer evaporated onto a 0.1-mm Tedlar sheet that was adhered to the aluminum radiator.

During the experimental comparison the ambient air temperature was 12.5°C and dew-point temperature was 8°C. The white radiator was cooled to an equilibrium temperature of 3.5°C, while the selective emitter reached 3.3°C. Thus, the stagnation temperature depression with both surfaces was about 9°C. When the temperature depression of the radiator below ambient-air temperature was less than 5°C, the ordinary surface actually cooled faster than the selective one. Only at temperature depressions of more than 5°C did the selective surface attain a somewhat lower temperature than the ordinary surface. The actual difference between them was negligible.

From the results of this experiment it seems that the added efficiency of this selective emitter, as compared with an ordinary painted aluminum plate, was obviously negligible. Another observation was the condensation of dew on the supper surface of the polyethylene windscreen.

Miller and Bradley

In the study done by Miller and Bradley at the Nevada Desert Research Institute the performance of ordinary surfaces (painted black and white) was compared with that of a selective surface made of anodized aluminum. At night, with a still ambient air temperature of 5°C, the selective surface reached a stagnation temperature of –12°C, the white surface and the black surface reached about –10°C.

Kruskopf

At the Lawrence Berkeley Laboratory, Kruskopf et al compared the performance of aluminized polyvinyl fluoride with that of a surface painted white. Very little difference was observed when surface temperatures were about 6°C below ambient air temperature. The comparison was even less favorable at higher radiator temperature.

Landro and McCormic

Landro and McCormic (1980) have compared the radiant heat loss of ordinary (gray) and selective radiators, as a function of the radiator's surface temperature, ambient temperature, and humidity. The selective surface was aluminized PVF. Both radiators were paired with polyethylene wind screens.

Figure 4.6 (after Landro & McCormic) shows the measured net cooling power (w/m^2) under clear sky, as a function of the temperature difference between the ambient air and the radiators. The stagnation temperature drops are 10 and 14°C for the gray and the selective radiators, respectively. The difference in the cooling power decreases along with the temperature difference between the ambient air and the radiator. At temperature difference smaller than 3°C the performance of the gray surface is better than that of the selective one.

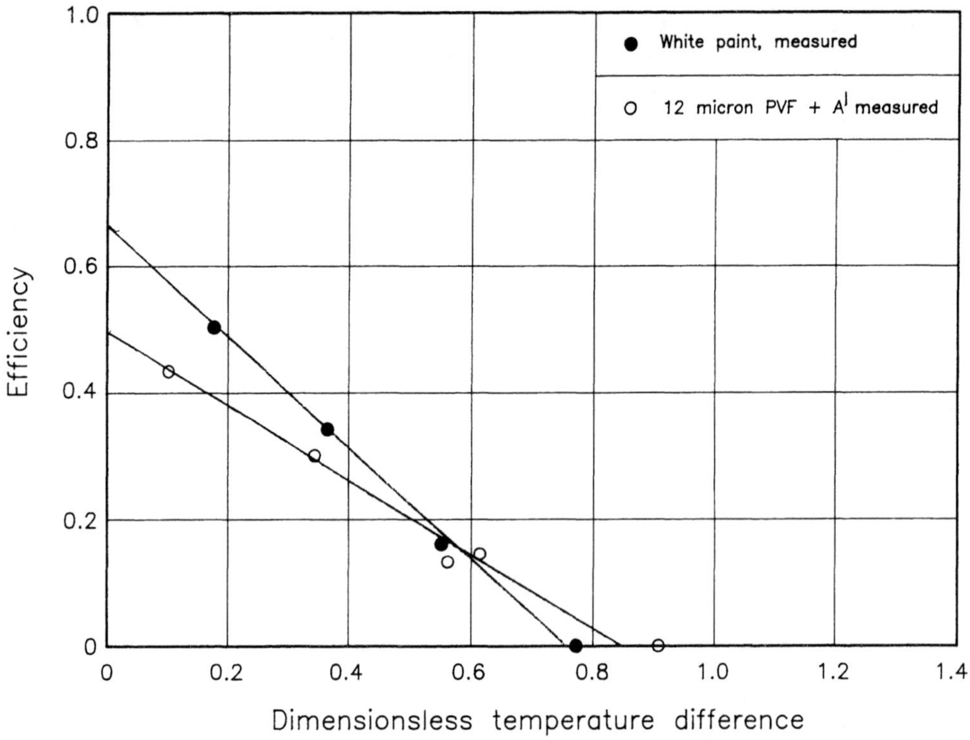

FIGURE 4.6 Comparison between ordinary and long-wave selective surfaces (after Landro and McCormic (1980).

Discussion of Selective Surfaces

The results of all the studies mentioned demonstrate that, with the selective surfaces tested, no distinct advantage in the radiant cooling potential was observed in the selective surfaces over radiators with ordinary surfaces. Taking into account the fact that the small effects of selectivity that have been observed in these studies were seen under stagnation conditions, without any utilization of cooling, the practicality of using selective surfaces (always more expensive than ordinary surfaces) is questionable, at least given present technology. When the roof itself serves as a combined radiator and cold-storage element, and the radiating surface temperature is closer to ambient air temperature, a selective radiator might even be less effective than an ordinary surface.

Another point that should not be overlooked is the probable effect of condensation on the performance of selective surfaces. Even if a lower temperature were attained by such surfaces, it might increase the likelihood of condensation over the selective surface itself. Once the selective surface has been covered by a water film, it would lose its selective radiative properties. This point deserves experimental examination.

As research on the development of new materials for selective surfaces is going on in several places, it is possible that eventually materials with a performance closer to the ideal will be available.

Concepts and Systems Utilizing Radiant Cooling

As mentioned in the introduction, several options are possible for utilizing nocturnal radiation as a cooling source for buildings. This section is, on the one hand, a summary of the state of the art in radiant cooling, based on a literature review, and, on the other hand, the author's personal evaluation of the different systems. This evaluation is, of course, subjective and reflects the judgment of the author, based on his experimental and theoretical research. Several ideas about options for utilizing radiant cooling have not been applied, and in some cases not even tested in a laboratory setup. These ideas also are described and discussed in this section.

The "Skytherm" System and its Variants

The first roof system utilizing nocturnal radiation for cooling a building, and the only one commercially available, is the "Skytherm," developed by Hay (1978). In this system the structural roof consists of a horizontal metal deck. Thermal mass is provided by large plastic bags, filled with water, that are placed above the metal. Horizontal insulating panels, moved by a motor, insulate the water bags during the daytime and expose them to the sky at night, when they slide aside and stack over a porch or a garage. The metal ceiling serves as a cooling panel for the space below.

Figure 4.7 shows a scheme of the Skytherm system.

During the winter this process is reversed, and the system acts as a passive solar heating system. Because of the good thermal coupling between the water and the indoor space and the high heat capacity of water, this system offers an efficient cooling mechanism. A number of buildings utilizing the Skytherm cooling system have already been built. Several researchers have developed and tested various modifications of Hay's original design, attempting to overcome some technical difficulties that were experienced with the horizontal panels.

The following buildings with the Skytherm system, which have been tested for their cooling performance, will be discussed below:

- — The Phoenix prototype
- — The Atascadero building
- — The New Mexico State University building
- — A building with the modified roof pond in Pala.

The report delivered by Marlatt et al (1984), of the Energy Technology Engineering Center, to the US Department of Energy, describes the various roof pond systems and provides information on their performance. Most of the information in the various research reports deals with the heating performance in winter, but some of them also cover the cooling performance in summer. Details of some of the various studies covering summer cooling, based on the Marlatt report and the chapter of E. Clark (1987), are presented below. Only the discussion of the Pala roof pond is based on the original research report on the Pala project (Clinton—no date).

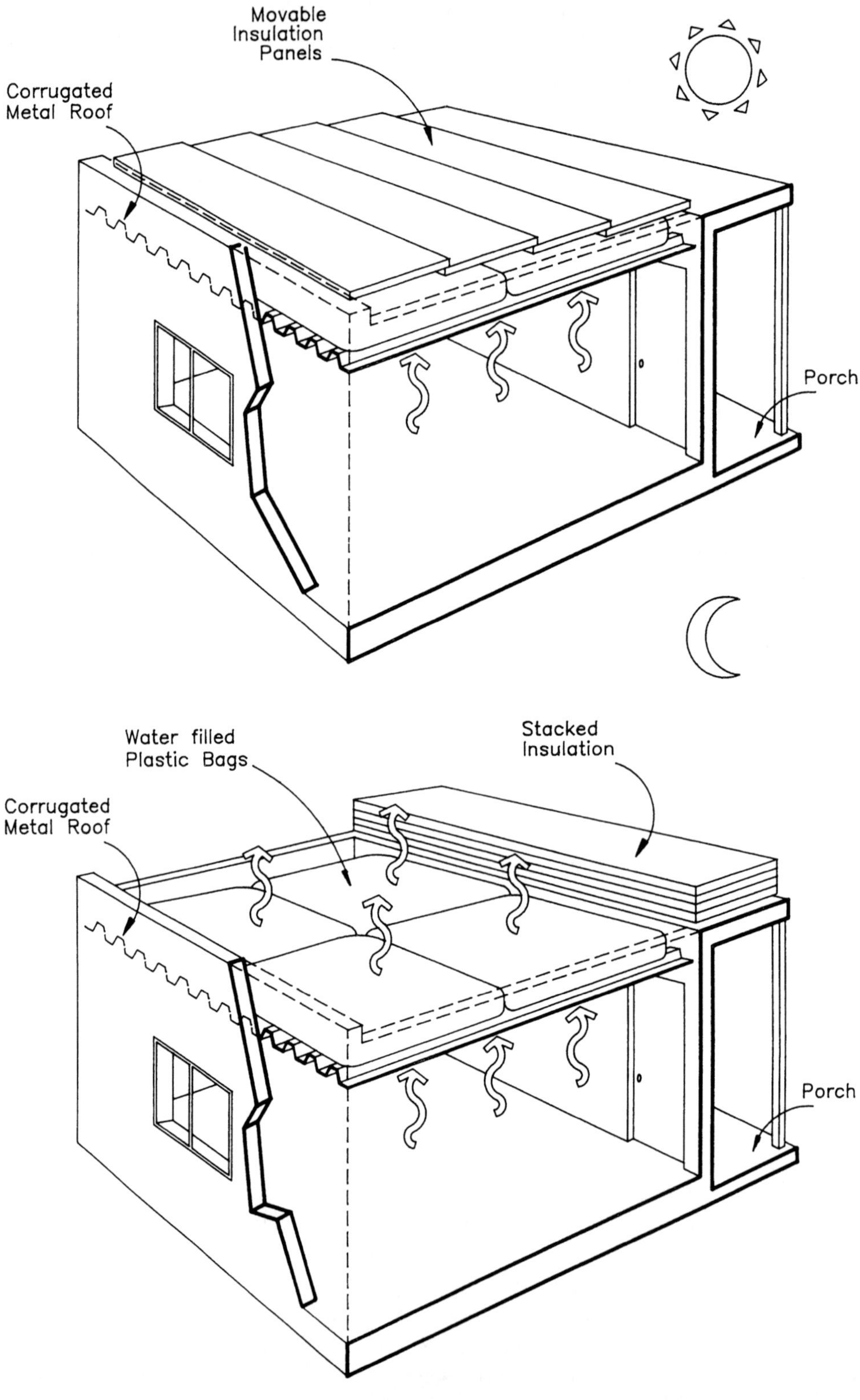

FIGURE 4.7 The "Skytherm" System.

The Prototype in Phoenix, Arizona

The Phoenix prototype was a 10 by 12 ft (3×3.6 m.) room constructed with vermiculite-filled concrete block walls, insulated by 1.5-in. rigid polyurethane. The roof was built of corrugated steel sheets. A polyethylene liner formed three water-proved ponds. The insulation over the pond consisted of 1-½-in.-thick polyurethane panels, operable manually by nylon clothlines with help of pulleys. When retracted, the panels were stacked atop the carport. In summer the pond was exposed during the night and insulated during the day. In winter the schedule was reversed.

As reported by Marlatt et al. (1984), in the summer the water was exposed on the roof (that is, it was not contained in bags). Thus the cooling was mainly by evaporation, not radiation. The interior temperature was maintained in a range between 68 and 82°F. During the three summer months, with maximum temperatures up to 110°F and high humidity, a fan-coil unit circulated the pond's water through the unit.

The House in Atascadero, California

The Atascadero house, built and originally owned by Hay, was the first full size building utilizing the Skytherm. It is a three-bedroom single-story building made of hollow lightweight concrete and insulated wood-frame walls. The structural roof consists of corrugated steel deck plates. Thermal storage is in large water-filled plastic bags. Movable custom-made insulating panels, operated automatically by a motor, and stacked over the carport when retracted, cover the water bags.

The performance of the Atascadero house was monitored and evaluated by the California Polytechnic State University (1975). Most of the information in the Marlatt Report deals with the heating performance of the building in winter, but some information is also provided on the summer cooling performance.

The interior temperature was kept within the range from 20 to 23.5°C throughout the summer by the roof pond. A typical performance of the Atascadero building during a summer day, as reported in Marlatt (1984, Figure 6-11) is reproduced in Figure 4.8. In evaluating this performance it should be noted that in summer the outdoor average temperature in Atascadero is within the comfort range. So, in practice, just the thermal mass of the water and the walls, even if insulated by fixed insulation, could maintain a comfortable indoor temperature most of the time under the climatic conditions of the site.

New Mexico State University Building (NMSU), Las Cruces

The building built and tested by the NMSU is a three-bedroom single-story house. Walls are of load-bearing reinforced concrete, 15 cm thick, and the floor, too, is concrete. Las Cruces has a more severe summer climate than Atascadero; thus, it could provide more realistic conditions for evaluating the cooling performance of the Skytherm system.

The Marlatt (1984) Report cites data from two NMSU Master Theses measurements, taken during the summers of 1979 and 1981, that indicate excellent cooling performance. However, it seems to this author, upon inspection of the data, that the cooling performance data is quite questionable. For instance, Figure 6-22 in the Marlatt Report shows indoor daily average temperatures *lower* than the daily average water temperature, when the

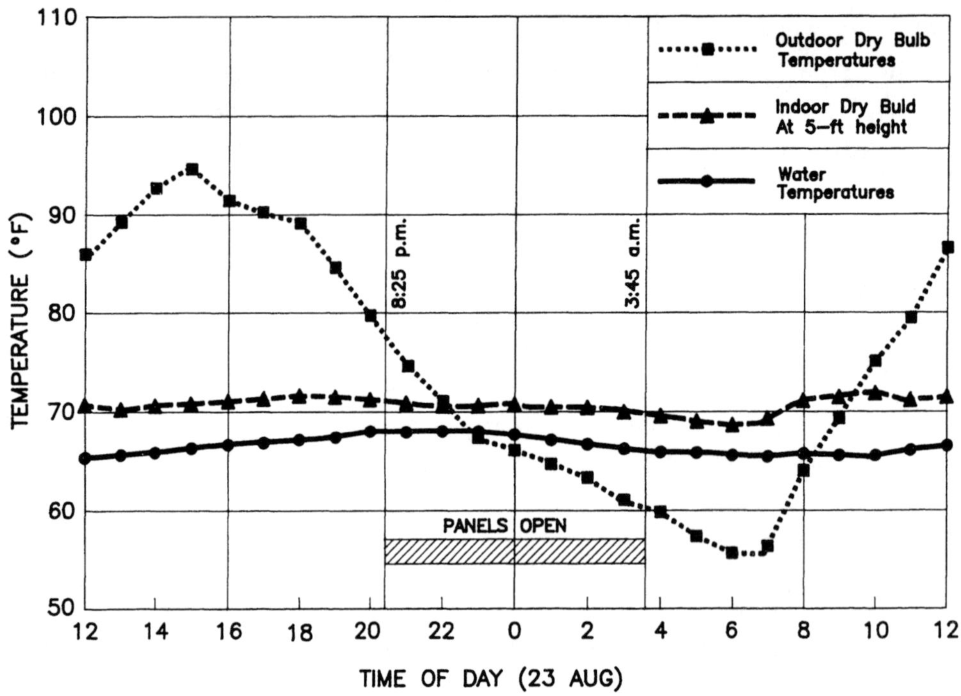

FIGURE 4.8 Cooling performance of the "Skytherm" in Atascadero, California.

outdoor temperature was well above the indoor. Another figure in the Marlatt report (Figure 6-21) shows a three-day diurnal temperature pattern with an afternoon indoor temperature *lower* than the water temperature, even when ambient temperature was above 100°F (38°C). There is no way that a passive buildings can have indoor daytime temperatures lower than its cooling source.

The "Pala" Roof Pond Variant

The Pala Passive Solar Facility, owned by the San Diego Gas and Electric Company, has eight test buildings, but in the framework of this book only two buildings, which are relevant to radiant cooling, will be discussed: a "conventional" building serving as a reference and a building with a roof-pond and operable insulation different than that of the Skytherm system. Each building has a total floor area of about 25 m^2, equally divided into two rooms by an internal wall. The ceiling height is 2.4 m. The roof-pond building has a flat steel deck structural roof, supporting large water filled plastic bags, similar to the "Skytherm" system. However, the operable insulation panels were of a different design. Instead of the panels stacking horizontally they moved accordion-style and hinged together with standard door hinges. They slid on tracks mounted vertically on top of the parapet of the roof, as shown in Figure 4.9.

Figures 4.10a and 4.10b show the diurnal temperature patterns of the conventional and roof-pond buildings in Pala, together with the ambient temperature, during two days in

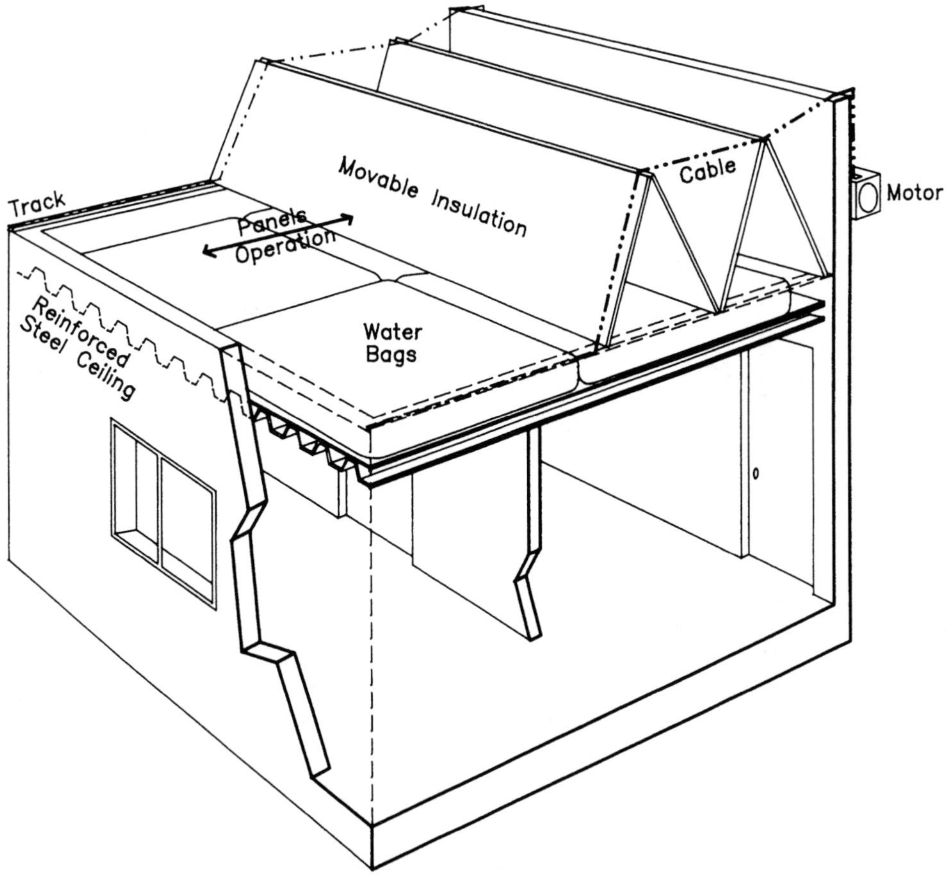

FIGURE 4.9 The Pala version of the "Skytherm" system.

July and two days in October, 1982. The indoor temperature of the roof-pond building was below that of the conventional building throughout that time, especially with respect to the maxima. In July the maximum temperatures of the roof-pond building were about 11°F (6°C) below the maxima of the conventional building, and in October the difference was about 16°F (9°C).

The difference in performance between July and October was attributed by Clinton to the fact that nights are longer in the fall, and so the water in the bags had more time to cool.

However, these figures bring out a very important point concerning the interaction between passive cooling and the building design appropriate for a given climate.

The relationship between outdoor and indoor temperatures was rather similar in July and in October. Solar gain through the small unshaded windows and solar absorption in the building's envelope have raised the indoor temperatures so that in July the average temperature of both buildings was above the outdoor average. The temperature of the conventional building was almost always above the outdoor level and its maximum was about the same as outdoors. But the temperature of the roof pond building was too high

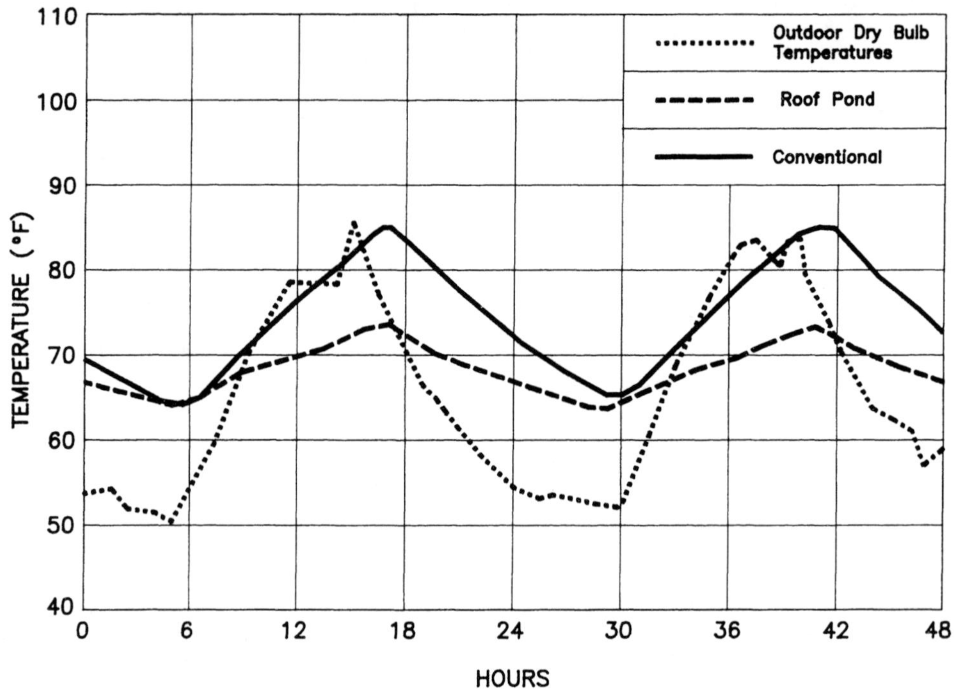

FIGURE 4.10A Cooling performance of the Pala version of the "Skytherm" system in July.

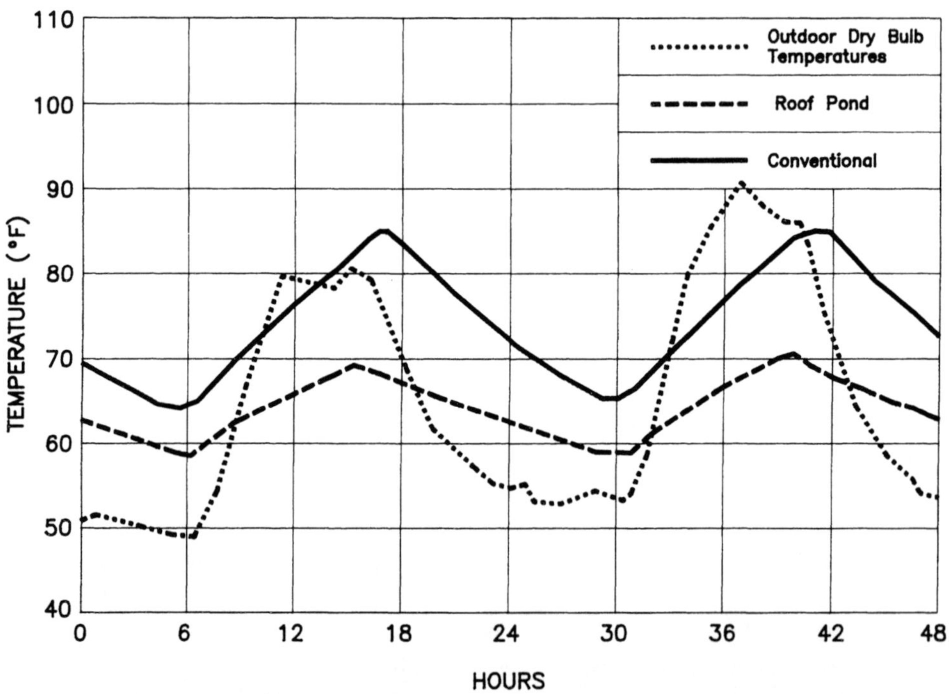

FIGURE 4.10B Cooling performance of the Pala version of the "Skytherm" system in October.

relative to the outdoor conditions, indicating that the total direct and indirect solar heat gain was more than the cooling provided by the roof pond, although its area was the same as the floor's area.

Summary of the Cooling Experience with the Skytherm System

Unfortunately, none of the reported data allows the author to evaluate the cooling performance of the water-filled plastic bags when exposed at night and insulated during the day. The Phoenix prototype, as reported in the Marlatt (1984) report, was tested in the summer with the water exposed rather than kept within the plastic bags, and thus the cooling process was predominantly by evaporation, not by nocturnal radiation. The Las Cruces cooling data is questionable, and in Atascadero and Pala the summers are quite mild, with average temperatures within the comfort range. Although a very extensive experimental evaluation of the Skytherm system took place at Trinity University in San Antonio, Texas, the reported information, both in the Marlatt Report and the chapters dealing with radiant cooling in the Cook's book *Passive Cooling* (Martin 1989 and Clark 1989), do not provide any experimental data on its cooling performance.

The main problem encountered with the various variations of the Skytherm system seems to concern the movable insulation. On the basis of research reports done by people who have tested the system [Marlatt et al 1984, Clinton (no date), Clark 1989], it is the opinion of the author that the availability of a simple, inexpensive, convenient, and trouble-free system of movable insulation is still in question.

Clark (1989), in summarizing the experience gained with the different buildings that have applied the Skytherm system, notes:

> *As a one-of-a-time kind installation, the conventional horizontally rolling panels have also been expensive and mechanically unreliable.*

Clinton notes:

> *Major problems cited with roof pond buildings in the past have been their questionable cost-effectiveness for residential buildings . . . and uncertain reliability of operation of the movable insulation. . . .*

A personal note: While visiting the California Polytechnic State Institute in the summer of 1974 Professor Haggard took me to the building to demonstrate its operation. When he started to retract the insulating panels the motor stopped midway. Professor Haggard commented that this was not an uncommon occurrence. I have heard other comments about problems with the movable panels from other people who were researching this system.

As was illustrated in the performance of the Pala building, application of a roof pond in a hot region does not relieve a building designer from the need to provide appropriate solar protection and sufficient thermal resistance and heat capacity to the building. The same applies, of course, to the other passive cooling systems as well.

Concrete Roofs as Nocturnal Radiators

In many countries horizontal concrete roofs are the standard practice. If such roofs could be equipped with one form or another of operable insulation that could cover the roof during the day and expose it to the sky during the night, the roofs would function as effective nocturnal radiators and cold storage elements in one, very common, building element.

The performance of this concept was tested by Givoni (1981) at the Institute for Desert Research of Ben Gurion University in Sede Boqer, Israel, with small lightweight test cells (55×55×50 cm) with walls of Polystyrene sandwiched between plywood sheets with a 10 cm thick concrete roof. The roof was covered during the daytime hours by a polystyrene plate, 5 cm thick, and it was exposed during the nights. The experimental cell was compared with a "control" of a similar cell with a superinsulated roof.

The average results are summarized in the following table:

	Minimum	*Maximum*	*Average*
Outdoor temperature	14.5	27.0	20.7
Indoor—experimental	14.0	21.3	17.2
Roof surface (exp)	12.0	19.3	15.2
Indoor—control	19.4	25.6	22.5

Thus, the concrete nocturnal radiator has lowered the indoor maximum temperature by 5.7°C. (46% of the outdoor range) and the average by 3.5°C below the outdoors and by 4.3 and 5.4°C, respectively, below the temperature of a similar cell with well insulated roof.

Etzion and Dover (1989), at the same Institute, later tested this concept with larger test cells. The cell's dimensions were 136×136×108 cm, with walls constructed of 6 cm thick polystyrene sandwiched between two sheets of plywood, painted white. The roof was 10 cm concrete, with a movable insulation panel of 10 cm polystyrene. The roof insulation was removed during the nights and installed back during the daytime hours. The higher height to floor area ratio increased the relative heat gain through the walls while the better daytime insulation reduced the heat gain through the roof during the daytime, relative to the smaller cells in the Givoni experiment.

Figure 4.11 shows diurnal patterns of the outdoor and the interior temperatures during a sequence of four days in July, 1987.

With averages of the outdoor maxima of 38, minima of 19, and average of 29°C, the respective indoor air temperatures were 29.5, 17.5 and 24°C. Thus, the concrete nocturnal radiator with the higher daytime insulation lowered the indoor maximum temperature by 8.5°C (45% of the outdoor range) and the indoor average by 5°C, below the outdoor's respective levels.

Figure 4.12 shows the performance of the test cell with concrete nocturnal radiator over a period of seven days in September, 1987. Again, the indoor temperatures were maintained well below the outdoor, and the indoor maximum was lowered below the outdoor's level by about 50% of the outdoor range.

It is of interest to note that in the these three independent studies the indoor maximum temperatures were lowered below the outdoor's maxima by a similar fraction of the outdoor's range (45–50%).

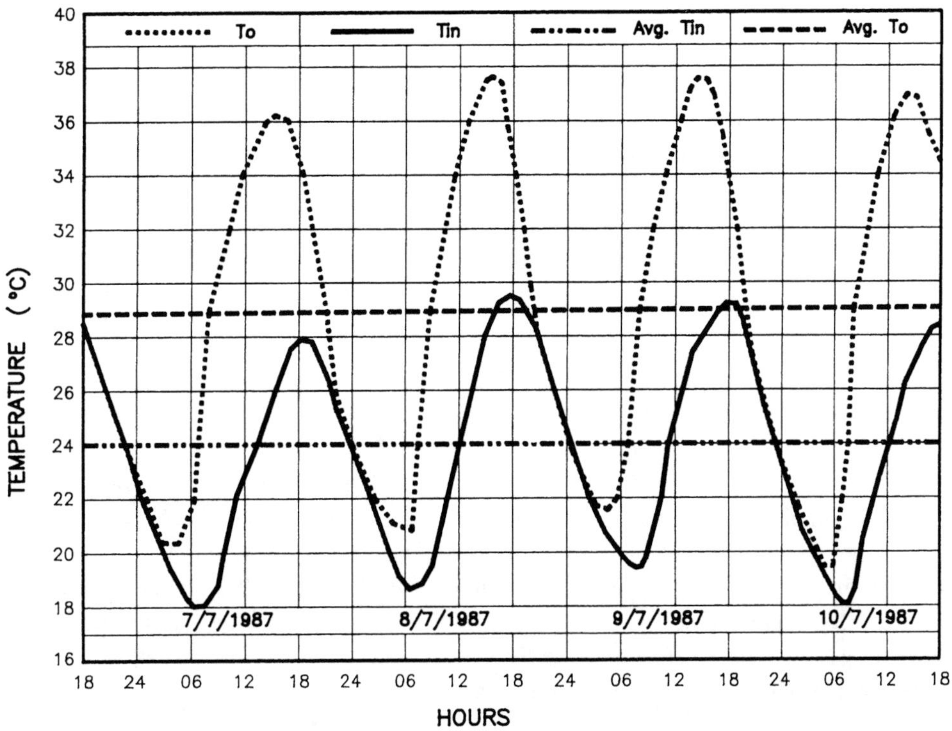

FIGURE 4.11 Cooling performance of concrete roof with operable insulation, acting as a nocturnal radiator, during 4 days in July 1987 (after Etzion and Dover 1989).

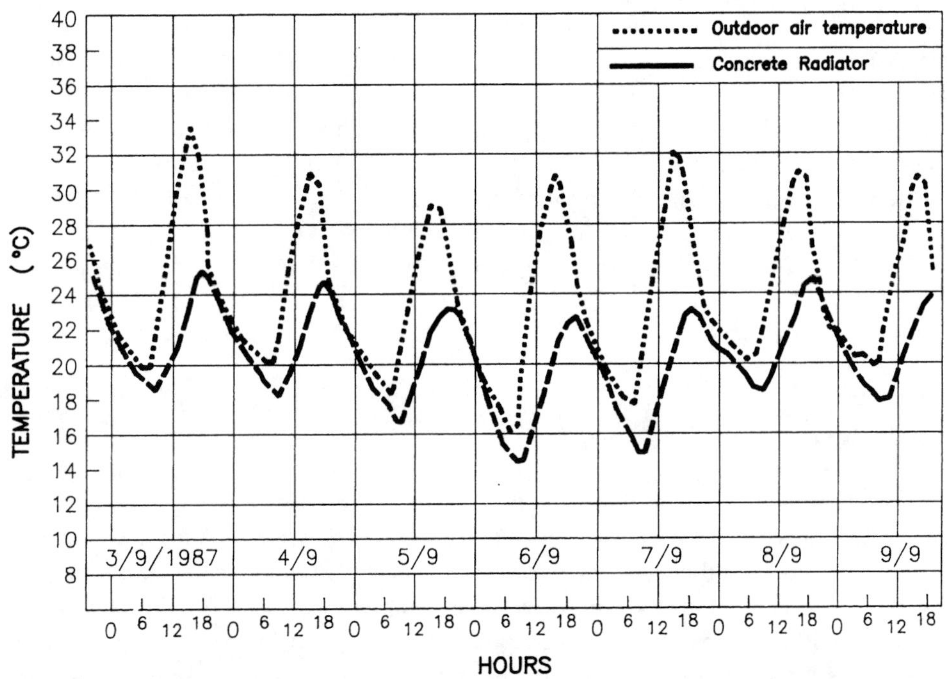

FIGURE 4.12 Cooling performance of concrete roof with operable insulation, acting as a nocturnal radiator, during 7 days in September 1987 (after Etzion and Dover 1989).

The Prospects of Utilizing Concrete Roofs as Nocturnal Radiators

Although it has been demonstrated that concrete roofs, which are very common in many countries, can provide effective radiant cooling, this can be achieved only with the application of movable insulation. At present, in the opinion of the author, practical and inexpensive movable insulation is not available, as was discussed in the section dealing with the Skytherm system.

To the extent that simple movable insulation will be developed, concrete roofs could provide simple, inexpensive and effective radiant cooling in many hot regions of the world.

Metallic Lightweight Radiators

Using the high-mass roofs as nocturnal radiators makes it necessary to install operable insulation, which technically is a problematic item. However, instead of cooling the mass of the storage medium itself, it is possible to cool a specialized long-wave radiator, positioned above an insulated roof, to temperatures below those attainable when the storage mass is cooled directly. The radiator preferably would be a metallic one, in order to minimize the thermal resistance between the upper surface where the radiant loss takes place and the lower surface where the cooling of the flowing air occurs. However, asbestos/cement radiators also have been tested by the author (see details below). The radiator should be insulated underneath, with a gap between the radiator and the insulation, to prevent heat flow to the radiator from the roof below.

When radiant heat loss from a lightweight radiator is not utilized by air flow under the radiator (stagnation conditions), there is a balance between the net radiant heat loss and the heat gain by convection from the ambient air. This balance results in a temperature depression of the radiative surface below the ambient air, when heat loss by radiation equals heat gain by convection. Under these conditions the temperature difference between the ambient air and the radiating surface is highest (if stable conditions in wind speed and cloudiness are assumed). The temperature thus attained is termed the "stagnation temperature," T_{rs}, and the temperature drop of the stagnant radiator below the ambient air is termed the "stagnation temperature drop," dT_{rs}.

The stagnation temperature of lightweight nocturnal radiators was measured by Givoni in two climatic regions in Israel: in Haifa (a coastal city with a humid climate) and in Sede Boqer (an arid climate).

The Haifa Study: Effect of Radiator Color

In Haifa (Givoni 1976) the stagnation temperature was measured with exposed radiators made of asbestos cement sheets without a polyethylene wind screen. Almost every night a lot of dew formed over the radiators. In spite of the heat gain from the condensation of the dew, the surface temperature, during clear nights, dropped significantly below the ambient air. Figure 4.13 shows patterns of stagnation temperatures of gray and white horizontal panels of insulated asbestos cement, measured in Haifa in midsummer 1964 for three clear days.

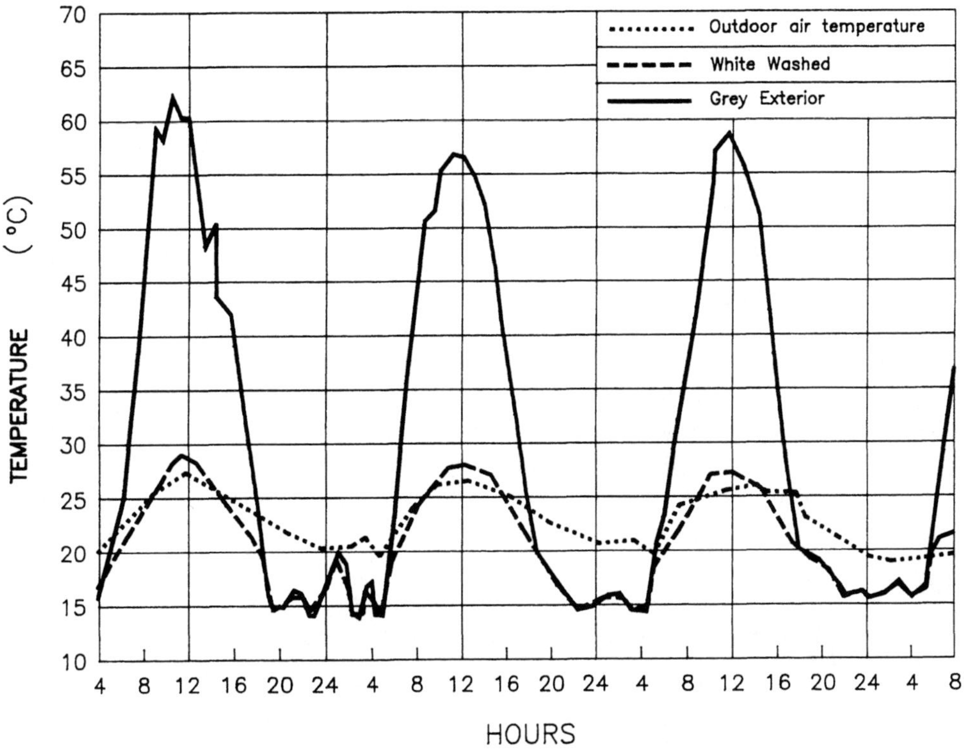

FIGURE 4.13 Surface temperatures of gray and white lightweight radiators.

It can be seen that the nighttime temperatures of the two surfaces were practically identical and even in this humid climate both have dropped by about 5 to 6°K below the ambient air. This figure demonstrates two points of practical significance:

- **A** As long as the sky is clear, radiant cooling has an effective potential even in humid regions.
- **B** The color of the radiator does not affect its performance for nocturnal radiant cooling but during the daytime it would result in very different temperatures.

Consequently, an exposed dark radiator (over a well-insulated roof!) can be utilized in winter during the daytime as a low-cost low-efficiency air heating system and in summer as a radiant cooling system. In both cases thermal storage could be provided by the interior structural mass. This study has demonstrated that even in a humid region an exposed radiator is cooled during clear nights by 4 to 6°K below the ambient dry bulb temperature.

The Sede Boqer Study: Effects of Polyethylene Wind-Screens

The experiments in the arid climate of Sede Boqer, dealing with the effect of polyethylene wind screens on the stagnation temperatures of metallic radiators, were discussed in detail above. The temperature drop under stagnation conditions was about 5 to 8°C for exposed radiators and about 9 to 11°C for the screened radiators. However, when dew was precipitated over the polyethylene the effect of the wind screening practically disappeared. It was also demonstrated that there is no significant difference between single or double layers of polyethylene.

Effect of Radiator's Utilization on its Temperature

When outdoor air flows under the radiator while insulation prevents heat flow to the airstream from the roof below, its temperature drops toward the stagnation level. The practical cooling of the air is always less than the stagnation temperature depression.

Higher air speed under the radiator and higher flow rates per unit area of the radiator increase the heat transfer from the flowing air to the radiator. Thus more heat is actually dissipated by the radiant loss. However, this thermal improvement is accompanied by a higher electrical consumption by the fan moving the air.

A mathematical model calculating the temperature of the air flowing under a metallic radiator is described in the appendix to this chapter.

Screened Asbestos-Cement Sheets as Nocturnal Radiators

Givoni and Hoffman (1970) have studied the possibility of using corrugated asbestos-cement sheets, which are a common roof material in many developing countries, as nocturnal radiators. They have measured the indoor temperatures of a room cooled *passively* by an asbestos-cement roof, covered with polyethylene wind screen, functioning as a nocturnal radiator. The research took place in the hot desert town of Eilat in Israel. The room built specifically for this study was 6 × 3 m with walls and roof built of 20 cm concrete and insulated externally by 5 cm polystyrene and a white corrugated asbestos-cement sheet. The roof sloped down from a height of 2.8 m at the north wall to 2 m at the south wall. It was shorter than the length of the room by 1 meter before the north and the south walls, creating gaps in which hinged operable insulated panels were installed. A polyethylene sheet was stretched 10 cm above the asbestos cement, so that when the roof panels were opened at night the indoor air could exit through the higher northern gap, flowing down the sloped roof by gravity under the polyethylene and above the asbestos sheets, which, in turn, were cooled by radiant heat loss to the sky. The flowing air re-entered the room through the open southern gap (Figure 4.14).

At night, when the roof openings were open and the windows closed, the indoor air rose at the higher roof opening, entered the space between the radiator and the wind screen, and, after being cooled as it flowed down, re-entered the room through the lower roof opening. During the daytime, the insulation panels closed the roof openings and any air flow between the room and the radiator was prevented.

From the thermal aspect the performance of this passive radiant system was impressive, especially during the first month of the study, June 1979, when the humidity was very low and the polyethylene still new and dust free.

Figure 4.15 shows indoor and outdoor temperatures during a sequence of four days, June 18 to 21st, 1970. During the nights the indoor temperature was lowered by the thermosyphonic flow (without any ventilation), reaching a minimum about 3 to 4°C above the outdoor minimum. The indoor average temperature was well below the outdoor

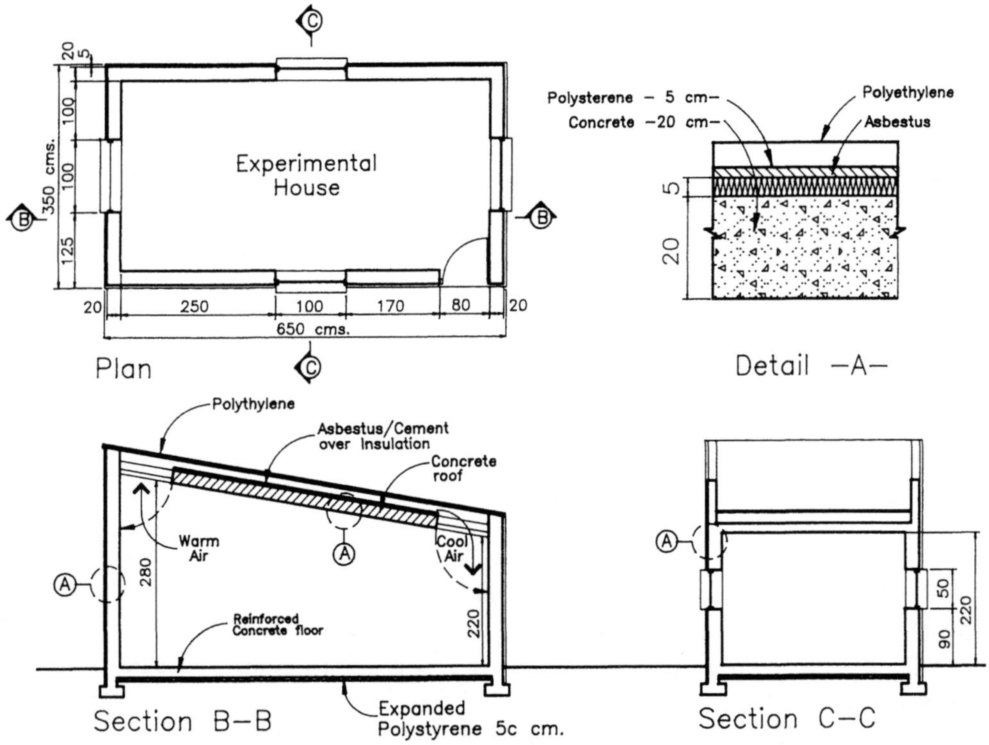

FIGURE 4.14 Plans of the test room cooled by an Asbestos-cement radiator in Eilat, Israel.

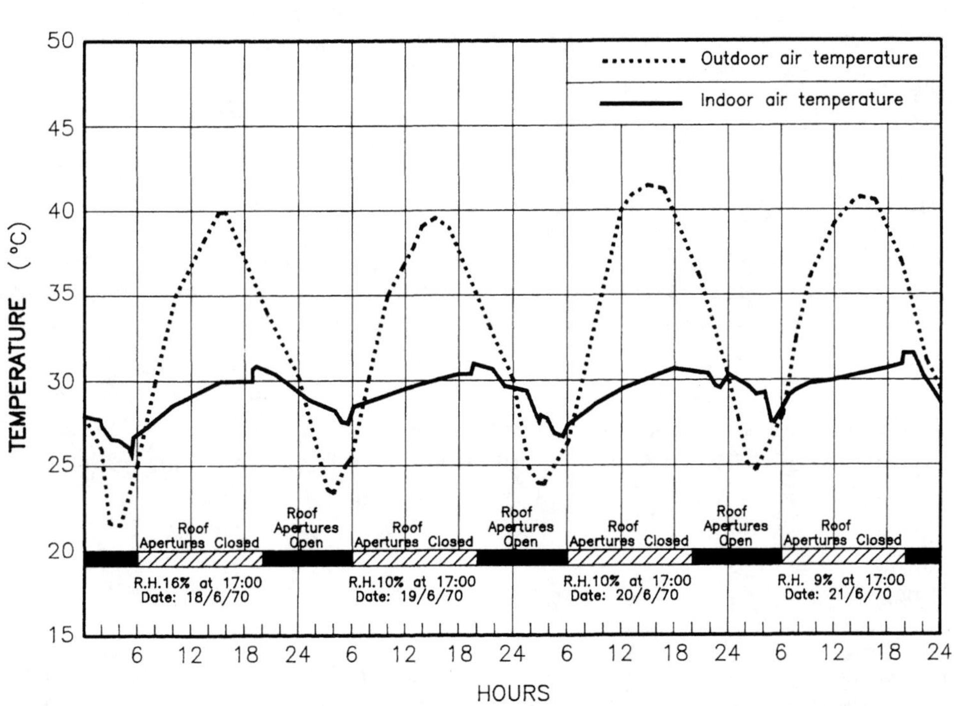

FIGURE 4.15 Indoor temperatures of the room cooled by a nocturnal Radiator in Eilat, Israel.

average and the maximum temperature was lowered below the outdoor's maximum by 8 to 10°C (about ⅔ of the outdoor range).

With the closure of the roof openings in the morning there was a sharp rise, of about 1 to 2 degrees, in the air temperature, resulting from equalization of the temperatures of the indoor air and the surfaces of the concrete elements around it. During the daytime hours the rise of the indoor air temperature, after this initial jump, was very moderate, about 3°C for the whole day, indicating the effectiveness of the high mass of the building in storing the nocturnal cold energy.

It can also be seen that only at about 3 a.m. did the outdoor temperature become lower than the indoor. In fact, when the windows were opened together with the roof openings, there was a sharp rise in the indoor temperature, although the minimum temperatures were somewhat lower.

The temperature of the air under the polyethylene was significantly cooler at night than the indoor air, indicating that increasing the flow rate by a fan would improve the cooling performance.

It should be noted, however, that the performance of the system did deteriorate with time and this deterioration was already noticeable in July. In part this deterioration was attributed to dust accumulation on the polyethylene, which reduced its transparency to long-wave radiation, and in part to a higher humidity level.

This study thus has brought attention to practical problems involved in the use of the polyethylene wind screen. Dust accumulates on the screen so that the screen has to be cleaned frequently. Even for a single room the cleaning process was cumbersome. It became clear that with a full-size building cleaning the wind screen is not a practical proposition. Even with cleaning, the thermal performance of the system, as well as the polyethylene material, gradually deteriorated. The polyethylene had to be restretched periodically and by August the indoor average temperature was about the same as the outdoor.

Unglazed Solar Collectors Used as Nocturnal Radiators

The idea of utilizing unglazed solar collectors as nocturnal radiators for summer cooling and as solar collectors for winter heating was suggested by McCormick and Landro (1980). The system they have suggested consists of a radiator/collector and a rock pile thermal store, with ducting and fans for the air flow control. At night, the cooled air can be drawn directly into the rooms or directed to the rock pile store. During the day, when cooling is required, the air is circulated through the store. No information on any testing of the system proposed by McCormick and Landro is known to the author.

A Roof Pond Cooled by Unglazed Solar Collectors

Etzion and Erell, at the Institute for Desert Research of Ben Gurion University in Israel, have used unglazed water heating solar collectors (copper tubes attached to a copper plate) as nocturnal radiators (Etzion and Erell 1991, Erell and Etzion 1991). Their test chamber consisted of a 2m × 2m, external dimensions (1.6 × 1.6m internal) × 2m high (internal) structure, with concrete floor and walls of 20 cm hollow concrete blocks, externally insulated with polystyrene 5 cm thick, and plastered. The roof was concrete, 10 cm thick, with a water pond over it, serving as thermal storage, with insulation panels floating over the water. The collectors, with total area of 5.5 m^2, were coupled with a roof pond. Thus, in this test cell, the collectors' area was much larger than the interior floor area.

In summer, the pond water was circulated during the night hours through the unglazed collectors, cooling the pond and the concrete roof beneath it. In winter the water circulation took place during the daytime, thus heating the water. Heat flow between the pond and the room below was by conduction through the roof. The system was tested for both cooling in summer and heating in winter, but only the cooling performance will be discussed here.

Figure 4.16 (Etzion & Erell 1991) shows diurnal patterns of outdoor air, indoor air, ceiling, and water temperatures during three days in August 1988. Outdoor minimum temperatures were about 19°C and the maxima was about 33.5°C. The daily average indoor temperature was about 23.5°C, which was about 2.7°C below the outdoor average. The average maximum temperature was about 24.2°C.

The temperature of the water circulating through the unglazed collectors was higher than the outdoor between about 10 p.m. and 6 a.m. Thus, convective cooling did accompany the radiant heat loss most of the night. The measured average total (radiant and

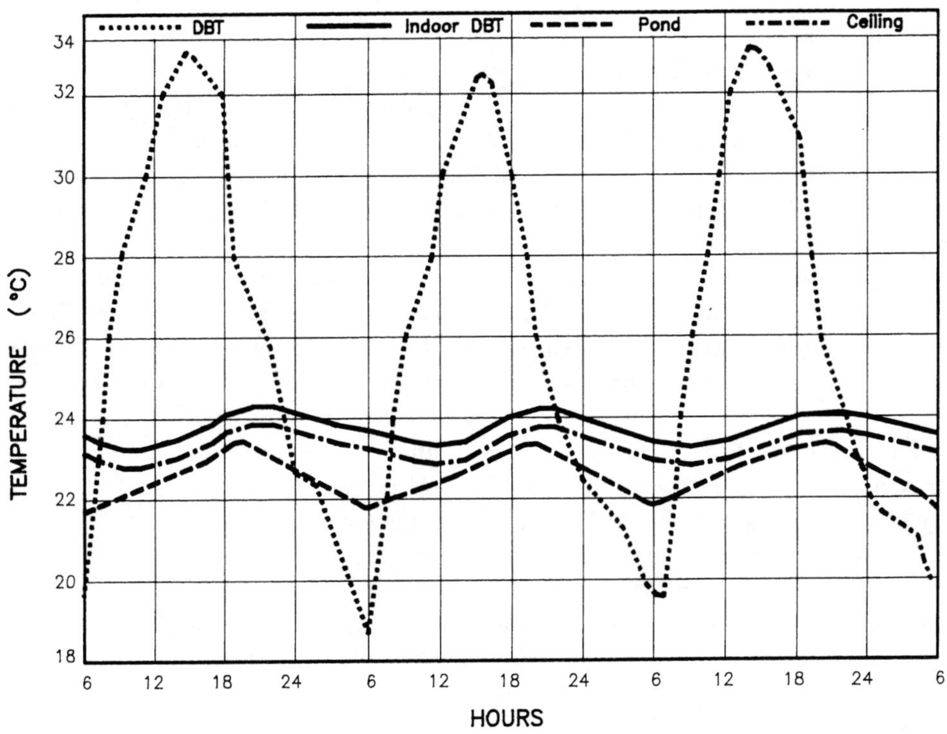

FIGURE 4.16 Indoor air, ceiling, and pond temperatures in a test cell cooled by unglazed solar collectors acting as nocturnal radiators (after Etzion and Erell 1991).

convective) cooling rate of the water, while flowing through the collectors, was about 55 watts/m2 of collector area. This rate is lower than that expected in a desert area from a radiator at a temperature higher than the ambient temperature. It apparently was due to resistance to heat flow from the tubes to the cooler radiating areas in between, but the large collector to floor area ratio apparently has provided a total cooling rate sufficient to achieve such a remarkable measured temperature reduction.

The Roof Radiation Trap

Based on the lessons from the Eilat building showing the impracticality of polyethylene wind screening, as well as on the desire to incorporate passive solar heating in the system for the winter, Givoni (1977) has proposed a passive solar heating and nocturnal radiant cooling system called a "Roof Radiation Trap" for buildings with concrete roofs. The radiation trap consists of an inclined fixed insulation layer rising to the south, above the concrete roof. Painted corrugated metal sheets (the nocturnal radiators) are installed above the insulation, forming a channel for outdoor air flow under the radiator during the night. The southern gap between the concrete roof and the insulation is glazed and equipped with hinged insulation panel, to serve in winter as a solar radiation trap. (See Figure 4.17).

In summer, when the radiation trap is used for cooling, penetration of solar radiation is prevented by the closed insulation panel. At night the painted metal layer is cooled by outgoing radiation and the air under the corrugations is blown either into the building or through a gravel bed, which provides thermal storage.

In winter, in the heating mode, the hinged panel in the southern gap is opened during the day, allowing solar penetration through the glazing and its absorption in the black painted concrete roof. The panel is closed during the night, insulating the glazing. The concrete roof serves as a radiant heating panel for the space below. Thus, in winter, the concrete roof acts as a horizontal Trombe wall.

The performance of the radiator, under stagnation conditions and when air flows under it, was tested by Givoni, at the Institute for Desert Research of Ben Gurion University in Sede Boqer, Israel, with a small lightweight model (internal dimension: 50×100 cm). It was built of 50 mm polystyrene plates sandwiched between plywood sheets. The roof was inclined, with an average height of 50 cm, and was covered by a corrugated aluminum, painted light gray, serving as the nocturnal radiator, with an air space of 3 cm between the radiator and the roof of the model. Outdoor air was sucked during the night by a small exhaust fan, under the radiator, through the model, and outward.

Figure 4.18 (top and bottom) shows diurnal temperature patterns in two days in July, 1982, under two conditions—stagnation and nocturnal operation of the fan. Figure 4.18a shows a diurnal temperature pattern of the DBT and the radiator under stagnant conditions. Even before sunset, after about 4 p.m. local time, the radiator's temperature started to drop below the ambient temperature as the long-wave radiant loss started to exceed the convective gain.

The temperature difference increased until, at about 9 p.m., a stable temperature drop of about 7 to 8°C was established. After midnight the temperature drop did fluctuate,

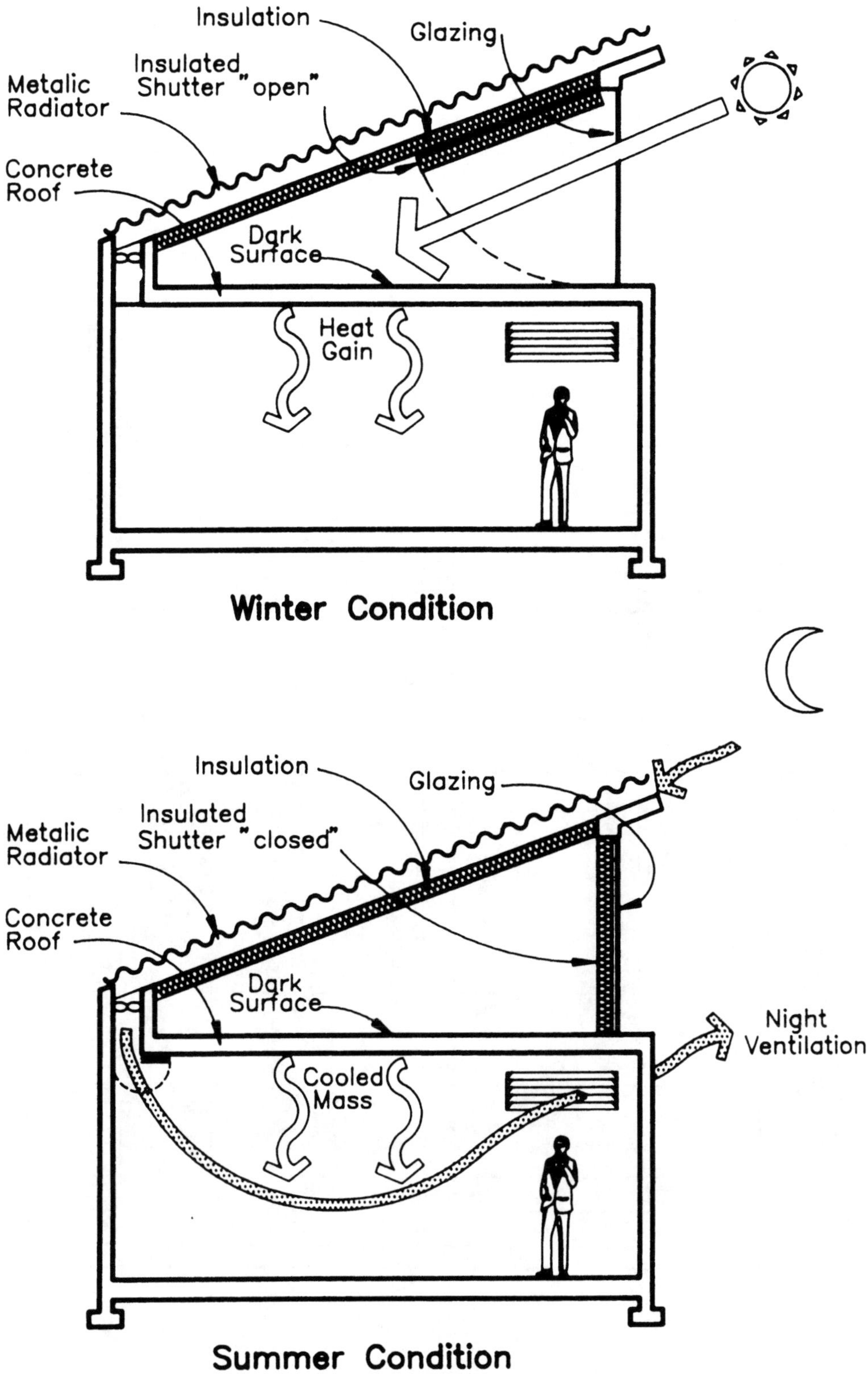

FIGURE 4.17 The "Roof Radiation Trap."

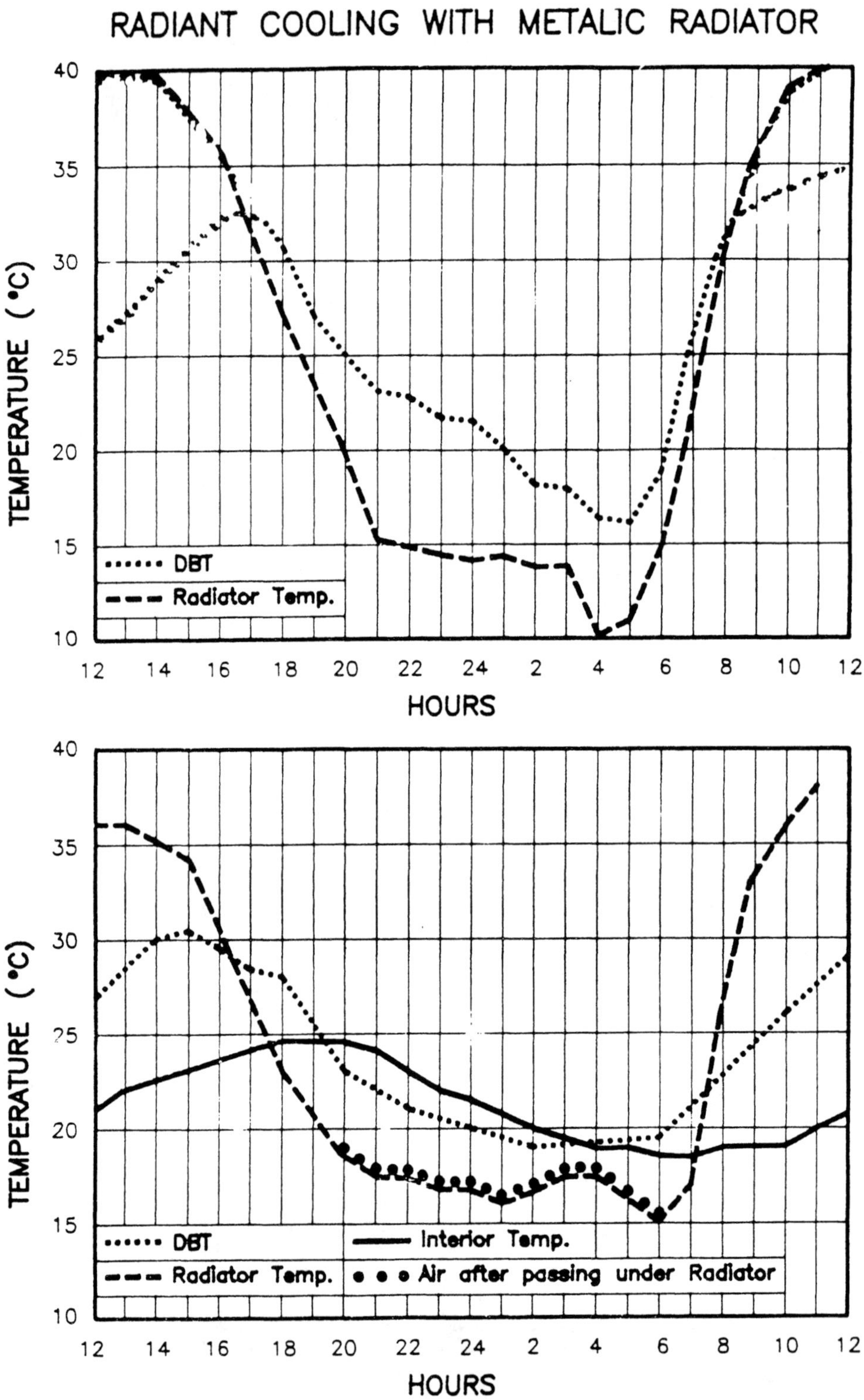

FIGURE 4.18 (Top): Stagnation temperatures of the Roof Radiation Trap. (Bottom): Temperature of air flowing under the nocturnal radiator.

apparently due to changes in the wind speed and/or cloudiness, which were not measured and observed in these experiments.

Figure 4.18b shows the model's diurnal temperature pattern when the fan operates between 8 p.m. and 6 a.m. (DBT, the radiator and the air emerging from the radiator).

From 4 p.m. till 8 p.m., before the start of the fan operation, the radiator's temperature drop was similar to the pattern in the stagnation experiment. However, with the start of the forced air flow at 8 p.m., the radiator's rate of cooling decreased. The temperature drop of the radiator, after initiation of the flow, was about 3 to 4°C until 1 p.m., when, probably, passing clouds markedly reduced the temperature drop. The sucked air was cooled by the radiator. The fan was rather small, producing a low flow rate. As a result, at the exit the air reached a temperature very close to the average temperature of the radiator. Thus, the air flowing under the radiator was cooled by about one half of the stagnation temperature depression.

In reality the exit air temperature drop is expected to be between ⅓ and ⅔ of the stagnation depression, depending mainly on the length of the radiator and the flow rate.

A mathematical model for calculating the exit air temperature, developed by Ingersoll and Givoni, is described for the first time in the appendix to this chapter.

Untested Concepts for Radiant Cooling

This subsection describes and discusses *untested* ideas on the application of radiant cooling. The goal of the discussion is to interest researchers in testing their performance and applicability.

Metallic and Asbestos-Cement Roofs in Developing Countries

In many developing countries corrugated metal or asbestos-cement roofs are very common. During the nights the low-mass roof cools down rather quickly, acting in effect as an effective nocturnal radiator located directly above the living space. The indoor **night** conditions in such buildings are often more comfortable than in buildings with high-mass roofs. However, during the daytime hours the indoor climate in buildings with such roofs is often uncomfortably hot, as the uninsulated lightweight roofs are heated to much higher temperatures than a massive concrete roof would be.

Installing centrally hinged **interior** insulating parallel plates under the roof can greatly reduce the daytime heating without interfering too much with the cooling effect of such roofs during the nights. When the plates are in a horizontal position (closed), during the daytime, they form a continuous insulation layer under the roof, minimizing the heat flow into the interior space. During the nights the plates should be turned into a vertical position, enabling radiant and convective heat flow from the interior space to the ceiling, which is cooled by the long-wave radiation to the sky. (See Figure 4.19.)

Interior insulation plates are not exposed to the wind and the rain and thus can be simpler in construction, lighter, and much less expensive than external insulation panels. The changes in their position, vertical or horizontal, can be controlled from the interior manually, through such means as a rope.

A major potential hazard with interior insulation that is made of expanded plastic materials is fire. A possible design made of noncombustible operable interior insulation

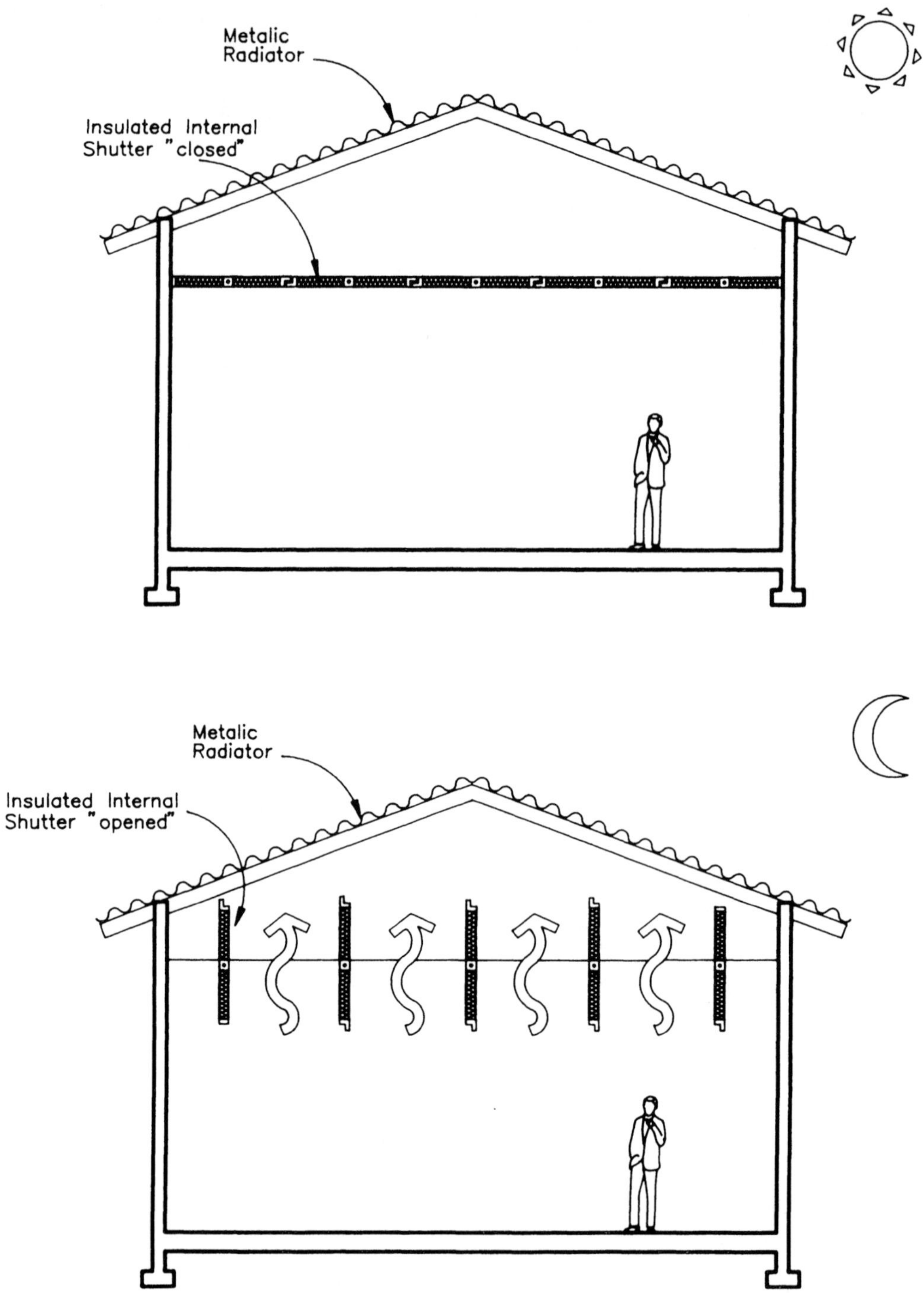

FIGURE 4.19 A scheme for utilizing metallic roofs in developing countries as nocturnal radiators with daytime insulation from the roof.

is a wood frame lined with aluminum foil. Such panels, when in horizontal (closed) position during the day, act as an effective radiant barrier between the hot roof and the interior space. Such simple devices can, with appropriate instructions, be "manufactured" and installed by the residents themselves and they also can be retroffited easily in existing buildings.

During the nights in hot humid regions, indoor water vapor may condense on the interior surface of the metallic roof. This is most likely when the radiator is metallic, but may also happen when thin asbestos-cement sheets are used. Design details that would facilitate the collection and drainage of the condensate can actually lower the interior humidity level and enhance the comfort of the occupants.

It seems to the author that research on the actual performance, and design development of such devices, would be very useful.

Enhanced Nocturnal Radiant Cooling with Desiccant Dehumidification In Humid Regions

A novel concept for combining nocturnal radiant cooling with desiccant dehumidification, providing a potentially cooling system in humid regions, has been developed at the Florida Solar Energy Center (FSEC) by Philip Fairey and Robin Vieira (1984). When humid air passes over a thin desiccant bed, such as silica gel, water vapor from the air is absorbed by the desiccant material. The process of water absorption is accompanied by a release of sensible heat, raising the temperature of both the desiccant bed and the dehumidified air. Thus latent heat is converted into sensible heat.

The FSEC concept is to dehumidify the indoor air and the structural mass, which is made of desiccant impregnated hollow concrete blocks, in a well-insulated house during the night hours. This will be achieved by providing closed-loop air circulation between the building's mass and a dry desiccant bed. The desiccant bed itself, located under an inclined dark metal roof (a nocturnal radiator), will be cooled at night by heat transfer to the roof, which in turn is cooled by convection and nocturnal radiation. The temperature of the radiator will be elevated by the absorption of vapor in the desiccant bed underneath and consequently the radiant heat process will be augmented. The desiccant bed under the dark metal roof will be regenerated by thermosyphonic flow of solar heated air.

During the daytime hours the heat and vapor penetrating into and generated inside the building will be absorbed, according to the proposal, by the cool, dry mass. This will be achieved by closed-loop air circulation during the daytime hours between the interior space and the desiccant-impregnated hollow concrete blocks. The issue of temperature elevation of the concrete blocks, as a result of the interior vapor absorption, has not been addressed in the FSEC proposal.

As discussed earlier, nocturnal radiant heat loss is greatly enhanced when the temperature of the radiating surface is above ambient air temperature. The combination of nocturnal radiant and convective heat loss can keep the desiccant material, if it is thermally coupled with the radiator, at a temperature substantially lower than it would be if the dehumidification process were not accompanied by simultaneous heat loss to the environment. In turn, the relatively low temperature of the desiccant, which is kept throughout the night hours, ensures its continued high water-absorption capacity.

Summary: Applicability of Radiant Cooling

The applicability of radiant cooling as a major mode of heat dissipation from a building depends on the building type and, especially, on the structure of the roof. The thermal performance of radiant cooling, and hence the practicality of its application, depends greatly on the climate.

Radiant cooling takes place only at night. For buildings that are occupied during the evening hours, in regions lacking night winds and with ambient temperatures that are too high from the comfort aspect, the method's benefits are instantaneous, as buildings are ventilated with air at a lower temperature. However, in order to benefit from nocturnal radiant cooling during the following daytime hours the cold generated by the radiator has to be stored in a thermal-storage mass. By ventilating the building with the cooled air the cooling process is utilized twice, improving the occupant's comfort directly and in the process also cooling down the structural mass of the building. The cooled mass serves, during the following day, as a passive sink for heat that is generated inside the building or that penetrates through the building's envelope.

Climatic Applicability of Radiant Cooling

From the information presented in this chapter, it is clear that the best prospect for using radiant heat loss as a significant mode of cooling is in arid regions that have clear skies and low wind speed at night in the summer. In such regions the potential for radiant cooling is maximized for all the different systems. With specialized air cooling radiators, the ambient air can be cooled at night in arid regions by about 3 to 4°C by a nocturnal radiator. However, even in humid regions with predominantly clear skies and still air during the night hours radiant cooling can be considered as a practical option. In such regions ambient night air could be cooled by about 2 to 3°C by nocturnal radiation, with the additional benefit of some dehumidification of the air introduced into the building, resulting from condensation over the bottom surface of the radiator.

On the other hand, in regions with predominantly cloudy skies and/or high wind speed during the nights, the net radiant heat loss is usually too small. Usually there is a temperature drop of only about 1°C below ambient air temperature, which seems to be too small to justify cooling systems based on radiant heat loss.

The situation is different when the radiant heat loss is supplemented by convection, as is the case when a high-mass roof (concrete or water bags over a metal deck), coupled with operable insulation, serves the radiator. With high-mass radiators the temperature of the radiating surface is above the ambient temperature most of the night. Radiant heat loss is thus augmented by the higher temperature of the radiator, in addition to the convective cooling of the thermal storage that is an integral component of the system. A similar situation also exists when the radiator consists of unglazed solar collectors through which water is circulated at night and cools the storage element, which is either a roof-pond or a concrete roof with embedded water pipes.

High-mass radiators, coupled by direct conduction with the interior space and insulated during the daytime, are applicable even in places where the cooling effect delivered to the building by specialized metallic radiators is too small, as in warm-humid regions.

Applicability of Radiant Cooling for Different Building Types

When examining building types it should be noted that, because of the low intensity of radiant heat loss, a large area of a radiator is required to provide significant cooling for a building. As the roof is the most reasonable location for a nocturnal radiator, radiant cooling is applicable almost exclusively to single-story buildings or to the top floor of a multistory building.

The type of roof also determines, from the structural aspect, what type of radiant cooling would be applicable to a given building. Specifically, only roofs with the load bearing capacity of about 200 to 300 Kg/m^2 can support a roof pond. Concrete roof thickness of about 25 cm or more (or equivalent mass) is needed to provide reasonable diurnal thermal storage for a roof that is exposed during the night and insulated during the daytime.

Buildings with Nonload-Bearing Roofs

Buildings with nonload-bearing roofs (such as wood-structure roofs) can use radiant cooling only with a specialized lightweight radiator. As such systems do not have an integral thermal storage, their most direct application would be when the cooled air can be used directly to enhance comfort. They can be used, for example, in buildings that are generally occupied—and so must be cooled—during the evenings and nights. Such buildings, when located in regions that are too warm during the evenings, are good candidates for radiant cooling with specialized air cooling radiators. They can be ventilated during the evening hours with air that has been cooled below the ambient temperature by the radiator. In particular, good candidates for direct use of radiatively cooled air are public buildings that are used in the evenings and that require a high ventilation rate because of high occupancy—for example, theaters and lecture halls.

Increasing the rate of airflow under the radiator to provide the high ventilation rates required in public buildings results in a smaller temperature drop but higher efficiency of cooling. Lowering the temperature of the outdoor air in the evening, even by about 2°C, with a high rate of airflow, can be quite useful in many cases.

Lightweight buildings, such as the houses with stud-wall construction common in the United States, cannot store the cold generated at night by a specialized radiator to lower the indoor daytime temperature. Therefore, such buildings are not good candidates for radiant cooling. However, a well-insulated building that has a lightweight roof and high-mass walls (made, for example, of concrete or bricks) can utilize the residual effect of night flushing with radiatively cooled air. The indoor temperature of such buildings would be lower during the daytime hours than the temperature of a similar building cooled at night by "ordinary" ventilation—that is, with ambient air.

APPENDIX: *The Ingersoll/Givoni Model of Radiant Cooling*

Mathematical modeling of radiant cooling involves the following issues:

1. Evaluating the **climatic potential** for the net radiant heat loss as a function of the ambient air temperature, humidity, and cloudiness.

2 Computing the "stagnation" temperature, incorporating the effect of the ambient wind speed.
3 Computing the temperature of the cold transfer medium and the actual cooling energy transferred to the building, as a function of the design details of the radiator and the air flow rate.

Extensive theoretical and experimental studies on the effect of the meteorological conditions on the potential for radiant cooling and the resulting net radiant loss have been conducted by investigators at the Lawrence Berkeley Laboratory in Berkeley, California, and at Trinity University in San Antonio, Texas. Their findings were summarized on page 91–92. However, these studies did not address the stagnation temperature achieved by a nocturnal radiator, the temperature of air cooled by the radiator, and the cold energy transferred to the building. A more detailed summary of this previous research can be found in Martin (1989).

The subject of the temperature of a stagnant radiator, with or without a wind screen, was addressed by Givoni (1982) and by Mostrel and Givoni (1982). The formulae for predicting the stagnation temperature of exposed and of screened radiators were presented on page 93.

A comprehensive mathematical model of a radiant cooling system utilizing a metallic radiator (without a wind-screen) was developed by Ingersoll and Givoni (1992) and is described below.

The model calculates the temperature of the air exiting from a nocturnal radiator, based on the different components of heat exchange between the radiator and the outdoor environment and between the radiator and the air flowing under it. It takes into account not only the effect of the ambient climatic factors (air temperature, humidity, wind speed, and cloudiness) on the net heat loss by the radiator but also the effect of the design details of the radiator and the air flow rate on heat transfer from the flowing air to the radiator. This model calculates not only the potential of radiant cooling but also the stagnation temperature of the radiator and the temperature of air cooled while flowing under it. The "operational" formulae of the model and the sequence of calculations are given below.

Formulae and Sequence of Calculations

INPUT DATA:

Radiator:

Length, L, (m)
Width, W, (m)
Spacing between radiator and insulation, D, (m)
Emissivity, E_R, (#)
Flow rate, F, (m^3/hr)

Climate:

Air temperature, T_A, (oC)
Wind speed, V, (m/s)

Dew Point, T_{DP}, ($_{0}C$)
Cloudiness, n, (#)

Calculations (assuming the "constant" properties of air).

Preliminary calculations:

1. Sky emissivity (after Clark and Allen 1978):
   $E_S = 0.787 + 0.764 \ln ((T_{DP}+273)/273)$

2. Cloudiness coefficient (after Clark 1981):
   $C_a = 1 + 0.0224\, n - 0.0035\, n^2 + 0.00028\, n^3$

3. Outdoor surface coefficient:
   $h_A = 0.6 + 6.64\, (V)^{0.5}$

4. Air flow velocity:
   $U_F = F/(3600(W.D))$; (m/s)

5. Reynolds number:
   $Re = 138430\, (D.U_F)$

6. Radiation coefficient:
   $h_R = \sigma(T_A+273)^3*[1+(E_S.C_A)^{.25}+(E_S.C_a)^{0.5}+(E_S.C_A)^{0.75}$
   $\sigma = 5.67 \times 10^{-0.8}$

7. Internal surface coefficient:
   $h_F = 2.7826\, ((U_F^{0.8})/(D^{0.2}))$

Performance calculations:

8. The stagnation temperature, T_{RS}
   $T_{RS} = [(T_A+273)*[h_A+E_R(E_SC_A)^{0.25}*h_F]\,/(h_A+h_R)\,]-273.$

9. Exit air temperature; $T_F(L)$:
   $T_F(L)=T_{RS}+(T_A-T_{RS})*\exp\{-(0.82*10^{-3}h_F LW)/[1+(h_F/(h_A+E_R h_R))*F/3600]\}$

10. The cooling energy delivered to the building, Q:
    $Q = 1220 * F * (T_A - T_{F(L)})/3600$; (W)

Numerical Examples of the Use of the Model

As a numerical example a radiator with the following design characteristics is chosen:

Length L = 15 m; Width W = 6 m; and spacing D = 5 cm. The exterior surface of the radiator is painted, resulting in surface emissivity E_R = 0.9. Three air flow rates are considered: 6000, 3000 and 1500 m^3/h.

The corresponding Re numbers and average air speeds are:

F = 6000 m^3/h : Re = 38760, U_F = 5.6 m/s
F = 3000 m^3/h : Re = 19380, U_F = 2.8 m/s
F = 1500 m^3/h : Re = 9690, U_F = 1.4 m/s

From equation 7 the corresponding internal coefficients of convective heat transfer (h_F) are calculated for different flow rates (F):

F = 6000 m^3/h : h_F = 20.100 $W/m^{2\circ}C$
F = 3000 m^3/h : h_F = 11.545 $W/m^{2\circ}C$
F = 1500 m^3/h : h_F = 6.631 $W/m^{2\circ}C$

The following weather parameters are assumed:

Ambient Air Temperature: 27°C
Dew Point Temperature: 13°C
Wind Speed: 0.45 or 2.25 m/s.
Cloud cover: n = 3 or n = 0

Using eq. (3) we obtain:
h_A = 10.561 $W/m^{2\circ}C$ at 2.25 m/s,
h_A = 5.055 $W/m^{2\circ}C$ at 0.45 m/s.

From eq. (1) we find:
E_S = 0.823

From eq. (2) we obtain:
C_A = 1 for n = 0; and C_A = 1.043 for n = 3.

Using eq. (6) and for T_A = 27°C we have:
h_r = 5.79 W/m^2C

nearly independent of T_A, E_S and C_N.

The stagnation temperature T_{RS} of the radiator is calculated from eq. (8) for different wind speeds (V) and cloudiness conditions (n):

V=0.45 m/s; n=0: T_RS=292.7°K = 19.8°C; dT=7.3°K
V=0.45 m/s; n=3: T_RS=294.3°K = 21.2°C; dT=5.7°K
V=2.25 m/s; n=0: T_RS=295.3°K = 22.2°C; dT=4.7°K
V=2.25 m/s; n=3: T_RS=296.3°K = 23.2°C; dT=3.7°K

The corresponding exit air temperature, T_F, is calculated from eq. (9). Results are given in Table 4-1 and the calculated total cooling power delivered by the system (eq. 10) in Table 4-2.

System Power Requirements

Fan power is required in such a radiant cooling system to draw outside air through the radiator channel and into the building. The selection of fans is a function of the particular radiant system as it relates to a specific building. The interested reader may consult the sources of technical information on this subject, including the Handbook of the American Society of Heating, Refrigerating and Air-Conditioning Engineers (ASHRAE) (1989). For most typical applications fan power would usually be on the order of about one hundred up to about 500 watts. Typical COP (Coefficient of Performance) of a radiant cooling system in a region with clear sky conditions (with stagnation temperature drop of about 6 to 8°C) is of the order of seven to ten.

TABLE 4.1. RADIATOR EXIT AIR TEMPERATURE, T_F, AND AIR TEMPERATURE DEPRESSION, T, BELOW AMBIENT

Wind Speed (m/s)	*Cloud cover n*	*Air flow Rate(m³/h)* 6000		3000		1500		*Stagnation Temp. Drop $T_{RS}(C)$*
		$T_f(K)$,	$T(C)$;	$T_f(K)$,	$T(C)$;	$T_f(K)$,	$T(C)$	
0.45	0	298.0	2.0	297.1	2.9	296.2	3.8	7.3
0.45	3	298.5	1.5	297.8	2.2	297.0	3.0	5.7
2.25	0	298.5	1.5	298.5	2.1	297.3	2.7	4.7
2.25	3	298.8	1.2	298.3	1.7	297.9	2.1	3.7

TABLE 4.2. POTENTIAL RATE OF COOLTH ENERGY (POWER) THAT MAY BE TRANSFERRED TO THE BUILDING WITH A NOCTURNAL RADIATOR AND IN EXCESS OF THAT DUE TO NIGHT VENTILATION

Wind Speed (m/s)	*Cloud cover #*	*Air Flow Rate (m³h)* 6000		3000		1500	
		kW	*W/m²*	*kW*	*W/m²*	*kW*	*W/m²*
0.45	0	4.22	46.9	3.06	34.0	2.0	22.3
0.45	3	3.16	35.2	2.32	25.8	1.6	17.6
2.25	0	3.16	35.2	2.21	24.6	1.4	15.8
2.25	3	2.53	28.1	1.79	19.9	1.1	12.3

Comments on Modelling Radiant Cooling

Any mathematical modelling of radiant cooling involves assigning values to the external convective coefficient (a function of the wind speed next to the radiator) and the sky cloudiness conditions. The actual assigned values of these parameters can, in reality, be just rough estimates. For this reason it is the personal opinion of the author that actual measurements of the basic performance parameter of a nocturnal radiator, the "stagnation temperature," should be evaluated experimentally in any location where radiant cooling is being considered.

More Comments on the Use of Polyethylene Wind-Screening

Research in Eilat has demonstrated that the dust accumulation over the polyethylene has significantly reduced its performance as a long-wave transparent wind screen. The study in Sede Boqer revealed a phenomenon that had not yet been taken into account in estimating the effect of a polyethylene windscreen: the dependence of the effect of the polyethylene wind screening on dew formation, even in arid regions. When an exposed radiator is cooled, during clear nights, by about 5 to 8°K below the ambient dry bulb temperature, it may also drop below the dew point temperature. However, dew formation over an exposed radiator does not affect its performance because water has a very high emissivity.

The situation is different when the radiator is covered by polyethylene. The temperatures of radiators with a wind screen, during clear dry nights, are much lower than those of the exposed ones, and are more likely to drop even below the ambient dew point temperature. Dew is likely to form over the polyethylene at night whenever the temperature of the film itself drops below the ambient dew point temperature. The polyethylene film itself, being mostly transparent to infrared radiation, has low emissivity and thus would not cool significantly by radiant heat loss. However, when the radiator, under the windscreen, is cooled sufficiently below ambient dew point temperature, a layer of stagnant **cold** air is formed between the radiator and the windscreen.

The underside of the polyethylene film is then cooled by conduction, and its temperature may also drop below the ambient dew-point temperature. When the film's temperature drops below this level, dew may begin to condense over the film. Once condensation starts over the polyethylene, the process is self-accelerating: water absorbs and emits long-wave radiation, and therefore the film attains high emissivity. When dust particles settle over the film they enhance the initiation of dew formation.

The water film over the polyethylene thus becomes the **radiating surface,** and because of its very low heat capacity (due to its thinness) its temperature drops rapidly, accelerating the rate of dew formation until the polyethylene film and the radiator underneath it reach the stagnation temperature of an ordinary exposed radiator.

Another point to consider in evaluating the expected quantitative effect of wind screening when the radiator is being used, and not under stagnant conditions, is the dependence of the wind screening effect of the temperature drop of the radiator below the ambient level. With small temperature differences the reduced convective gain may be balanced by the reduced radiant loss, because the polyethylene is only partially transparent to long-wave radiation.

In practice, when the nocturnal radiator is utilized to cool a heat-transfer medium (for example, air), which elevates the temperature of the radiator itself, the expected temperature difference between the radiator and the ambient air might be about 2 to 4°C. With such a temperature difference the effect of the wind protection above becomes smaller, relative to the convective heat transfer from the air flowing underneath the radiator. In this situation the net effect of the windscreen may be so small as to be insignificant or even counterproductive.

Another factor that may reduce the effectiveness of polyethylene windscreens is dust accumulation on the film. Polyethylene tends to exhibit "static" attraction of dust, and even without it dust that ordinarily exists in the air may settle on the film and reduce its transmissivity. Furthermore, dust particles may serve as nucleation centers for condensation and thus enhance formation of dew over the film.

It should also be stressed that the provision of a polyethylene windscreen (on specialized radiators) in a full scale building greatly increases the cost of the cooling system owing to support requirement and the need to keep the film under constant tension. In reality there is a quantum jump in the whole system cost when an exposed radiator is transformed into a screened radiator. This factor may determine the economic viability of the system.

When the storage mass itself (that is, a concrete roof or a roof water pond) is being cooled directly at night the effect of wind screening is actually negative. In this case the surface temperature during most of the night is usually above that of the ambient air. Application of a polyethylene film in this case reduces the performance in two ways: Firstly, emitted radiation is reduced because of the partial transmissivity of the film. Secondly, the lower convective exchange, which in that case is a heat loss, reduces further the cooling rate of the heat-storing mass.

References

American Society of Heating, Refrigerating and Air-Conditioning Engineers (ASHRAE) *Handbook 1989.*

Berdahl, P. and Fromberg, R. 1982. "The Thermal Radiance of Clear Skies." *Solar Energy.* Vol. 29, pp. 229–314.

Berdahl, P. and M. Martin. 1978. "The Resource for Radiative Cooling." *Proc.* 2nd Nat. Passive Solar Conf. Philadelphia, vol. 2, pp 684–686.

——— 1979. "Spectral Measurements of Infra-red Sky Radiance." *Proc.*, 3rd Nat. Passive Solar Conf. San Jose, California, pp 443–447.

——— 1984. "Emissivity of Clear Sky." *Solar Energy*, V–332, pp. 663–664.

Berger, X., and M. Schneider. 1981. "New Techniques of Natural Cooling." Int. Passive and Hybrid Cooling Conf. Miami Beach, Florida. AS/ISES. pp 261–265.

Blanpied, M. 1979. *A Model of Radiative Cooling Using I.R. Transparent Windscreens.* Trinity University, Department of Engineering Science.

Catalonotti, S., V. Cuomo, G. Piro, D. Ruggi, V. Silvestrini and G. Troise. 1975. "The Radiative Cooling of Selective Surfaces." *Solar Energy*, V 17, pp. 83–89.

Clark, E. 1981. "Passive/Hybrid Comfort Cooling by Thermal Radiation." *Proc. Int. Passive Cooling Conf.* Miami Beach, Florida. pp. 682–714.

Clark, E. and C. Allen. 1978. "The Estimation of Atmospheric Radiation for Clear and Cloudy Skies," *Proceedings of the Second National Passive Solar Conference*, vol. 2, pp. 675–678.

——— 1979. "Supplemental Cooling Requirements of Passively Cooled Building in Selected American Cities." *Proc. 3rd Nat. Passive Solar Conf.*, San Jose, Calif., pp. 344.

Clark, E., and M. Blanpied. 1979, "The Effect of I. R. Transparent Windscreens on Net Nocturnal Cooling from Horizontal Surfaces," *Proc. 4th Nat. Passive Solar Conf.*, Kansas City, MO. p. 509.

Clark, E., and P. Berdahl. 1980. "Radiative Cooling: Resource and Applications," *Proc. 5th Nat. Passive Solar Conf.*, Amherst, Mass.: *Passive Cooling Handbook*, U.S. Department of Energy Publication 375, pp. 167–201.

Clinton, J. Pala Passive Solar Project Final Report. *Solar Energy Analysis Laboratory. San Diego Gas & Electric.* San Diego, California. No date.

Cook, J., Editor. 1989. *Passive Cooling*, MIT Press.

Duffie, J., and W. Beckmann. 1974. *Solar Energy Thermal Processes.* New York: J. Wiley and Sons.

Energy Technology Engineering Center. 1984. *Roof Pond Systems*. Canoga Park, California, Special Distribution.

Erell, E., and Y. Etzion. 1991. "A Radiative Cooling System Using Water as a Heat Exchange Medium." Submitted to *Energy and Buildings*.

Etzion, Y., and A. Erell. 1991. "Radiative Cooling of Buildings." Research Report. Institute for Desert Research, Ben Gurion University, Sede Boqer, Israel.

Etzion, Y., and S. Dover. 1989. "Radiative Cooling of Buildings." Research Report. Institute for Desert Research, Ben Gurion University, Sede Boqer, Israel.

Fairy, P., and R. Vieira. 1984. "Proposal for Hybrid Building Cooling and Dehumidification Through Desiccant, Enhanced Nocturnal Radiation." Submitted by the Florida Solar Energy Center to the US DOE.

Givoni, B. 1981. "Experimental Studies on Radiant and Evaporative Cooling of Roofs. *Proc. Int. Passive and Hybrid Cooling Conf.* Miami Beach. AS/ISES, pp. 279–283.

——— 1982. "Cooling by Long-Wave Radiation." *Passive Solar Journal*, vol. 1, no. 3, pp. 131–150.

Givoni, B., and M. E. Hoffman. 1970. "Preliminary Study of Cooling of Houses in Desert Regions by Utilizing Outgoing Radiation." Research Report. Building Research Station, Technion, Israel Institute of Technology, Haifa.

Hansen, D. G., and J. I. Yellott. 1978. "A Study of Natural Cooling Processes in a Hot Arid Region." *Proc. 2nd National Passive Solar Conf.* pp. 653–657.

Hay, H. R. 1978. "A Passive Heating and Cooling System from Concept to Commercialization." *Proc. Annual Meeting of Sec. Int. Solar Energy Soc.*, pp 262–272.

Kruskopf, M., P. Berdahl, M. Martin, F. Sakkal, and M. Sobolewski. 1980. *Radiative Cooling Test Facility and Performance Evaluation of 4-Mil Aluminized Polyvinyl Flouride and White Paint Surfaces*. Lawrence Berkeley Laboratory.

Lendro, B., and P. G. McCormick. 1980. "Effect of Surface Characteristics at Atmospheric Conditions on Radiative Heat Loss to a Clear Sky." *Int. J. Heat Mass Transfer*, Vol. 23, pp. 613–620.

Marlatt, W., K. Murray, and S. Squier. 1984. "Roof Pond Systems." Report for USDOE by Energy Technology Engineering Center. Rockwell International, NTIS # DE84016401.

Martin, M. 1989. "Radiative Cooling." In Cook (1989).

McCormick, P. G., and B. Landro. 1980. "Radiative Cooling for Air-Conditioning Applications." Research Report. The University of Western Australia, Perth.

Miller, W. C., and J. O. Bradley. 1979. "Radiative Cooling with Selective Surfaces in a Desert Climate." *Proc. 4th Nat. Passive Solar Conf.*, Kansas City, V.4, p.480.

Mitchell, D. 1976. *Selective Radiation Cooling: Another Look*. Victoria, Australia: University of Melbourne.

Mostrel, M., and B. Givoni. 1982. "Windscreens in Radiant Cooling." *Passive Solar J.*, V1, #4, pp. 229–238.

Sakkal, F. et al. 1979. "Experimental Test Facility for Selective Radiative Cooling Surfaces." *Proc. 4th Nat. Passive Solar Conf.*, Kansas City, V-4, p. 483–487.

* *

5

EVAPORATIVE COOLING SYSTEMS

INTRODUCTION

The amount of heat absorbed in the process of water evaporation (its latent heat) is very high in comparison with the other modes of heat transfer common in buildings. Every gram of water that is evaporated without external heat input extracts from the ambient air, or from a material over which the evaporation takes place, nearly 0.6 Kcalory (about 0.66 whr or 0.6 Btu/lb.F). Throughout the ages, this mode of cooling has been used to cool buildings in hot arid regions. Various simple techniques have been used to realize this cooling effect—people have tried such methods as watering the floors or putting wetted matrices made of twigs in windows facing the wind.

There are two basically different approaches to cooling buildings by water evaporation. The first is to cool outdoor air directly through evaporation, and then introduce that air into the building. The air is humidified while its temperature is lowered and the indoor moisture content is elevated above the outdoor level. This is direct evaporative cooling of the residents and the interior materials in the cooled space. The air flow for direct evaporative cooling can be induced either by mechanical devices—fans—or passively by natural processes—utilizing the wind, temperature difference, or water spray in passive evaporative cooling towers. Various direct evaporating cooling systems are discussed in Section 1 of this chapter.

The second approach is to cool a given element of the building, such as the roof or a wall, by evaporation. The cooled element, in turn, serves as a heat sink and absorbs, through its interior surface, the heat that penetrates into the building through its envelope or that is generated indoors. This is indirect evaporative cooling. With such systems the indoor radiant and air temperatures are lowered without elevating the indoor moisture content of the air. Indirect evaporative cooling systems are discussed in Section 2 of this chapter.

The Physical Process of Evaporative Cooling

When water evaporates within a stream of ambient air without a supply of external heat the temperature of the air is lowered and its moisture content is elevated, while its "Wet Bulb Temperature" (WBT) remains constant. Such a change in air temperature, which occurs without addition or extraction of heat from the system, is called "adiabatic," meaning that the total energy content of the air (sensible plus latent) is kept constant. The changes in the temperature, moisture content (vapor pressure), and relative humidity of the air, which occur during the process of direct evaporative cooling, can be seen in Figure 5.1. Air with a given dry bulb temperature (DBT) and moisture content is shown in point A: DBT = 35°C (95°F), WBT = 20°C (68°F), relative humidity = 22%, moisture content = 8 gr/kg, vapor pressure = 10 mmHg. When the air passes at low speed through a wet pad of fibers that has a large surface area, its temperature is lowered and its vapor content increases along the constant WBT of 20°C (68°F).

In practice, the humidification process is not complete and the DBT of the air emerging from the evaporative cooling device is higher than the WBT by about 20 to 40% of the difference between the initial DBT and the WBT. In this example it emerges with DBT of 24.5°C (76°F), relative humidity of 65% and moisture content of 12.7 gr/kg of dry air (point B), as is seen in Figure 5.1.

The difference between the ambient DBT and the WBT is defined as the WBT depression (dWBT). The ratio between the temperature drop by the air exiting from the evaporative device (dTexit) and the WBT depression (dTexit/dWBT) is sometimes defined as the humidifying efficiency. In all systems of evaporative cooling the WBT is the most important climatic factor to consider.

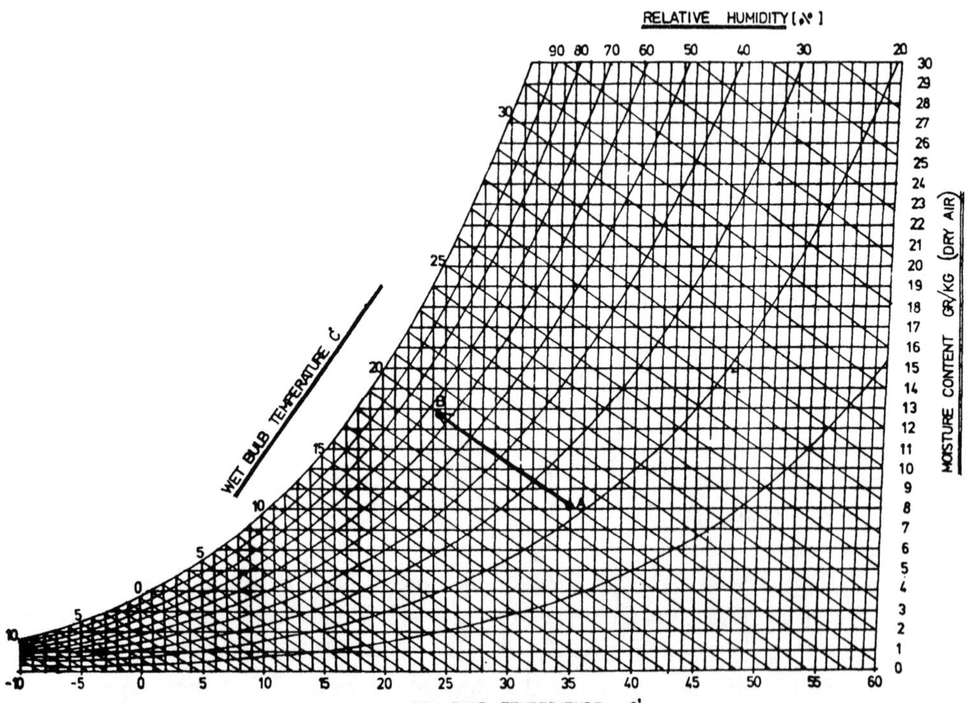

FIGURE 5.1 The humidifying process of air cooled by evaporation.

Diurnal Patterns of the Wet-Bulb Temperature

As any direct evaporative cooling takes place along a constant WBT, and cooling is needed mostly during the hot hours of the day, the diurnal pattern of the ambient WBT in a given location has an important impact on the applicability of cooling systems that rely on evaporation. The WBT of the outdoor air is not constant, but rather changes with the diurnal pattern of the ambient air Dry Bulb Temperature. The minimum and maximum WBT usually coincide in time with the corresponding values of the DBT, but the relationship between the DBT and the WBT maxima and minima depends on the climate type.

When the moisture content of the air (or its vapor pressure) is constant, the diurnal pattern of the WBT follows the DBT pattern. The diurnal range of the WBT (from minimum in the morning to maximum in early afternoon) is about one third of the DBT diurnal range. The difference between the DBT and the WBT is smallest at the outdoor minimum and largest at the outdoors maximum. This is typical in arid regions, distant from a sea. However, in most humid regions, where the soil is usually wet and there are many shallow lakes and/or trees with large canopies, evaporation increases with air temperature and solar radiation, so that the moisture content is highest in the early afternoon and lowest in early morning. These patterns of the monthly averages of the maxima and minima of the moisture content in a hot humid city (Colima, Mexico) are shown for each month in Figure 5.2, together with the ASHRAE comfort zone.

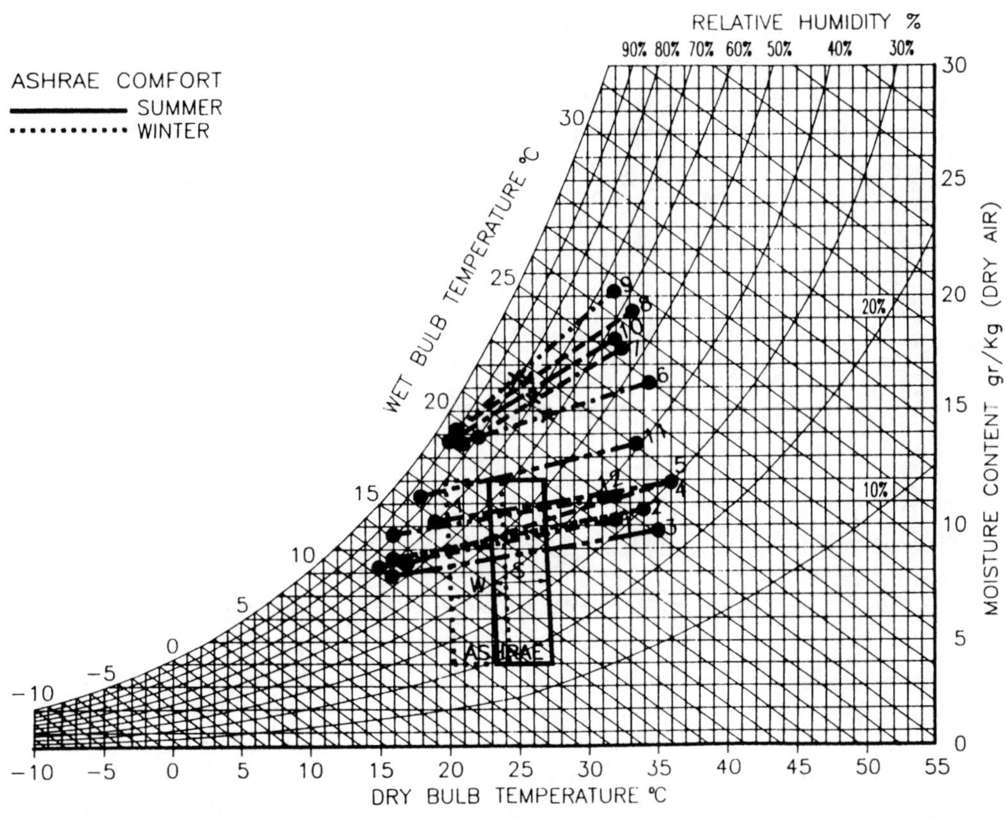

FIGURE 5.2 Maxima and minima of the vapor content for summer months in a hot, humid city—Colima, Mexico.

The diurnal pattern of the WBT, and especially the difference between the DBT and the WBT, determine the performance of direct evaporative cooling systems. The WBT is the lowest temperature theoretically attainable with air cooled by evaporation, as well as by indirect evaporating systems. In practice the temperature of evaporatively cooled air is somewhat higher than the ambient WBT, by about 20 to 30% of the DBT–WBT difference.

Climatic Applicability of Direct Evaporative Cooling

Typical values of the WBT depression during the peak temperature in arid regions is about 10 to 15°C (18 to 27°F), and in desert regions it is about 15 to 20°C (27 to 36°F). It means that the air emerging from the evaporative cooling device, either mechanical or passive, is at a temperature about 2 to 3°C (3.5 to 5.5°F) above the ambient WBT in the regions where evaporative cooling is applicable.

The indoor air of a building cooled by an evaporative cooling system is further heated, by about 1 to 3°C (2 to 5.5°F) above the output air from the evaporative cooling system, in spite of the high flow rate typical of evaporative cooling systems, due to the heat gain by the building. Typically, the indoor air temperature is higher by about 3 to 5°C (5.3 to 9°F) above the ambient WBT. These facts set the limits for the climatic applicability of direct evaporative cooling. The combination of the elevated indoor humidity on the one hand, and the high air velocity on the other hand, sets the upper limit of the indoor comfort zone at about 25 to 26°C (77 to 79°F). Consequently, direct evaporative cooling can be applied only in regions and seasons when the WBT of the ambient air does not rise above about 22°C (72°F).

It should be noted that some desert regions experience hot-humid periods. For example, in Arizona the months of July and August are usually hot and humid. A direct evaporative cooling system, in spite of being adequate during most of the summer, would not be effective during this period. An indirect evaporative cooling system, discussed in the next section, would be more effective in such climates.

SECTION 1: DIRECT EVAPORATIVE COOLING SYSTEMS

This section deals with three types of direct evaporative cooling systems—mechanical evaporative cooling, homemade direct evaporative devices, and two kinds of passive direct evaporative cooling towers.

Although mechanical evaporative systems are not passive, nevertheless they are relatively low-energy cooling systems, and therefore they are also covered briefly in this book. More detailed discussion and monitored performance data on the various types of mechanical evaporative coolers is presented in Watt et al (1986), in a chapter of Yellott (1989), and in the papers of Wu et al (1989, 1990) and of Peck et al (1979).

In many arid and desert regions of developing countries homemade evaporative cooling devices have been developed. These often use the high speed winds that often blow during the daytime, mainly in the afternoon, from a constant general direction. In such cases it is possible to cool small buildings by simple passive evaporative cooling devices. Some of the options for utilizing the natural winds rather than fans are described below. Some of these ideas have not yet been tested in order to evaluate their real performance characteristics.

Evaporative passive towers humidify and cool ambient air without using fans. The air flow can be generated by natural processes, which include temperature difference (thermosyphonic flow), wind, and falling water droplets.

Two recently developed types of direct evaporative cooling towers will be discussed in this section. The first one was developed and tested by Cunningham and Thompson [1986]. The second cooling tower was developed by the author in the course of consulting work for the 1992 EXPO in Seville, Spain. It was tested by one of my students (Al Hemiddi 1991, 1992) at the University of California in Los Angeles (UCLA). A mathematical model, calculating the performance of these cooling towers under different climatic conditions and design details, has been developed (Givoni 1992-a, 1992-b).

Mechanical Evaporative Cooling Systems

Mechanical evaporative coolers were developed in the United States in 1906 and were very popular in the hot arid southwest in the two decades before World War II. In many hot dry developing countries these devices are still used extensively.

There are three main types of mechanical evaporative cooling: a single-stage direct evaporative cooler, a two-stage indirect evaporative cooler, and multiple-effect direct coolers.

Single-Stage Direct Evaporative Cooling

Single-stage direct mechanical evaporative coolers are compact. They are less expensive and consume less electric power than compressor air conditioning. A fan (usually a centrifugal fan) draws large quantities of ambient air through wetted pads made of fibers. The wet fibers provide a very large surface area for the vaporization process and therefore the overall size of the device is rather small. Small evaporative coolers for residential and small office applications consist of a metal or plastic box containing the fan and three or four fiber pads on the sides, through which water is circulated. Excess dripping water is collected and recirculated by a pump. Mineral deposits resulting from the evaporation may clog the pads. Therefore, depending on the water quality and the evaporation rate, the pads must be cleaned and replaced at intervals.

The cooled and humid air emerging from the device flows through the building and then is exhausted. (This is known as once-through flow). The indoor air can not be recirculated through the evaporative cooler because of its high humidity.

Single stage mechanical evaporative cooling reduces the DBT by about 60 to 80% of the difference between the DBT and the WBT. It involves very high rates of outdoor air flow through the building, about 15 to 30 air changes per hour. This high air change results in high indoor air speed, which extends the upper temperature and humidity limits of comfort.

These high rates of air flow are required because the temperature of the air, at the exit from the evaporative cooler, is usually only slightly below the required indoor air temperature. Therefore a high air flow rate is required to carry away the heat that has penetrated into the building and the heat that has been generated inside it. The resulting indoor vapor pressure thus is close to the elevated vapor pressure of the evaporatively cooled air emerging from the cooling device. The high rate of ventilation also means that the air is only slightly heated on its way to the outlets. Consequently, air at temperatures significantly below the outdoor air is exhausted from the building. This cooled air is an untapped potential for structural cooling.

Direct evaporative cooling is used extensively in arid regions where the elevated humidity of the air does not affect the comfort of the occupants nor the activities and operations taking place in the building. Such systems have to be operated mainly during the day, when the cooling is most needed, and subsequently have to cool large quantities of hot air in the range of 30 to 40°C (86 to 104°F). From the physiological viewpoint this approach is not always desirable for two reasons:

1. The cooled indoor air may be excessively humid, especially when "direct effect" evaporative coolers are used in regions where the humidity is not very low, or even in regions that are mostly arid, during the relatively short spell of more humid climates (for example, July and August in Phoenix, Arizona).
2. The high rate of airflow and air change, which are necessary for effective cooling, may cause very great variations in the indoor air speed, and the associated thermal sensation, within the cooled building.

However, in desert regions that are arid throughout the hot period, where humidity is often too low from the comfort viewpoint, direct single-stage evaporative cooling can be an inexpensive, physiologically desirable system. Peck et al (1979, 1981) at the University of Arizona in Tucson have developed a design scheme that extracts the cooling effect from the air blown through the building before it is exhausted. According to this scheme the air leaves through an air space between two sheets of double glazed windows, thus cooling the inner sheet. This pattern of the air flow effectively eliminates the solar energy absorbed in the glazing and thus prevents heating of the interior by convection and longwave radiation from the inner glazing.

Two-Stage Indirect Mechanical Evaporative Cooling

Mechanical indirect evaporative cooling is possible when the cooling process is divided into two stages. In the first stage, "return" or outdoor air is cooled by a conventional evaporative cooler, blown through an air-to air heat exchanger, and exhausted. A counterflow of outdoor air then passes through the heat exchanger and is cooled convectively. In this way, the air is cooled without its vapor content being elevated. Two-stage machines are larger and consume more power than single-stage direct evaporative coolers.

Multiple-Effect Direct Evaporative Cooling

When air is cooled by a heat exchanger, which in turn is precooled by evaporatively cooled air or water, without being humidified, its WBT is lowered by about one-third of the drop in its DBT. The pre-cooled air can subsequently be further cooled down by a direct evaporative device to a temperature close to its new (reduced) WBT (a sequence of evaporative-convective-evaporative cooling processes). A multiple-effect evaporative system can produce air at lower temperatures and WBT than is achievable by simple direct evaporative cooling. However, the several stages of such systems require much larger machines and higher fan power to overcome the air flow resistances of the three stages.

Experimental Studies with Mechanical Evaporative Coolers

Wu (1989, 1990) has measured the performance of single- and multiple-stage direct evaporative coolers, cooling full-scale houses in Scottsdale and Phoenix, Arizona. In the 1989 study, the average efficiency (relative air temperature drop) of the single stage unit was 0.74 of the WBT depression. The average EER (ratio of cooling energy to electricity usage) was 35.7. Concerning the EER values it should be realized that the cooling provided by the system is a function of the ambient WBT depression, while the power consumption for a given machine is constant. Therefore the actual EER is approximately proportional to the ambient instantaneous WBT depression.

In the multiple-stage system the outdoor air was cooled in the first stage, without humidification, by an evaporatively cooled heat exchanger. The cooling efficiency at this stage was 0.54 of the WBT depression. In the second stage the air was further cooled by evaporation, reaching a final cooling efficiency of 0.88 of the WBT depression. The EER relation to the WBT depression was about the same as for the single-stage machine.

In the 1990 study Wu compared the performance of a compressor air-conditioner (a heat pump) with those of single-stage and two-stage evaporative coolers. Figure 5.3 (Wu

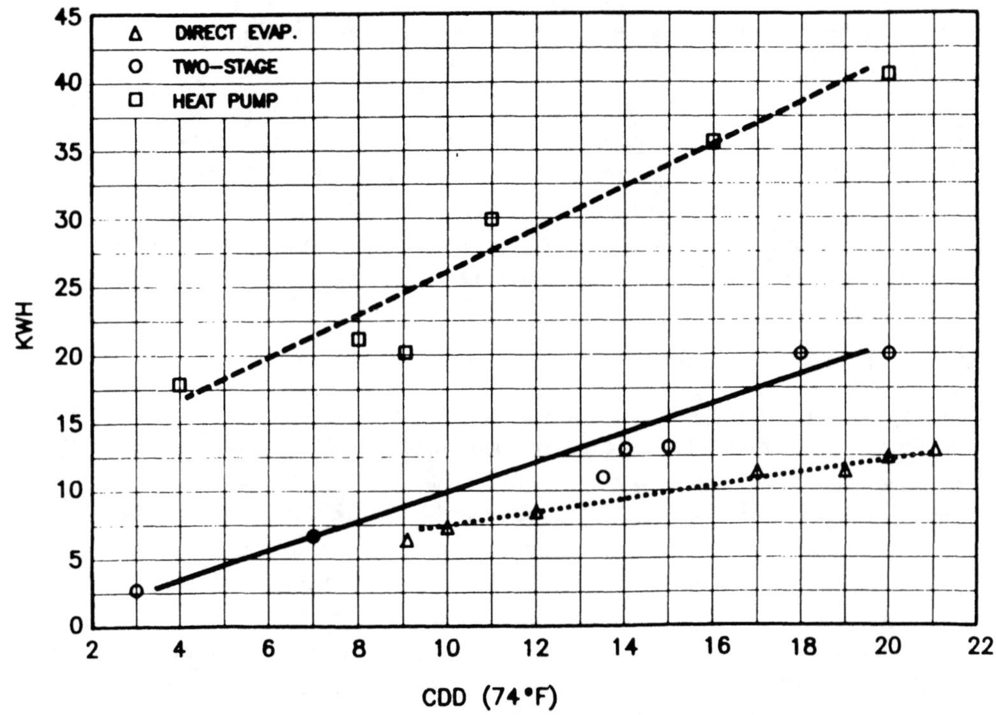

FIGURE 5.3 Energy consumptions of a heat pump, a single-stage evaporative cooler and a two-stage evaporative cooler. (Wu 1990).

1990) shows the energy consumptions of the three systems as a function of the Cooling Degree Days (CDD). The single stage cooler has used about ¼ and the two stage system about ⅓ of the A.C. system.

HOMEMADE PASSIVE DIRECT EVAPORATIVE COOLING DEVICES

Evaporative Pads in Window

When wetted pads made of fibers, or some other commercially available pads used in mechanical evaporative coolers, are installed in windows facing strong winds they can, with suitable design details, provide natural direct evaporative cooling in the same way as the mechanical coolers. In order to be effective, these pads should provide a large surface area for evaporation, without creating too much resistance for the wind to penetrate through them. It is also necessary to provide a way to wet the pads at the top and collect and recirculate the runoff water from the bottom of the pads. Such devices are also used in commercial mechanical coolers.

In many desert regions the afternoon winds are laden with dust. A secondary benefit from these passive evaporative cooling window devices is that the wetted fibers are very effective in filtering the dust out of the air that penetrates through the window. In such dusty places the water flow down the pads should be rather high, in order to flush out the trapped dust. Filtering the recirculated water is then necessary, unless the water is diverted from the bottom of the pads and is used for irrigating a garden (this is a "once through" water flow over the pads). The degree of saturation of the air passing through these passive pads is usually lower than is common in the mechanical devices. Temperature reductions of 40 to 50% of the DBT–WBT differences can be expected.

These pads block the view through the window, of course, and they reduce the air flow in seasons and times when ventilation without evaporative cooling is desirable. One way to solve this problem is to design a **fixed** frame in front of the openable glazing. This frame would contain removable lightweight pads, and the sprinkling, drainage and water recirculation systems would be an integral part of the device. When cooling would not be needed the pads could be removed and the openings would then function as conventional windows.

Use of Porches for Direct Passive Evaporative Cooling

Any system of passive direct evaporative cooling provides significant resistance to air flow and can reduce the wind speed through the opening down to about 10 to 20% of the speed through an unobstructed opening. One way to increase the air flow is to have evaporating pads with an area much greater than the area of the openings through which the air enters the building.

When a porch is available on the windward side of the building it is possible to put the evaporative pads in front of it. Windows can be integrated within the pads to allow a view from the porch.

It is important to "force" the air passing through the wet pads to enter the building through the openings—doors and windows—connecting the porch to the interior, and not to be discharged outward. The cooled air should not be diverted to the sides of the porch and/or above the roof of the building. Therefore the sides and roof of the porch should be

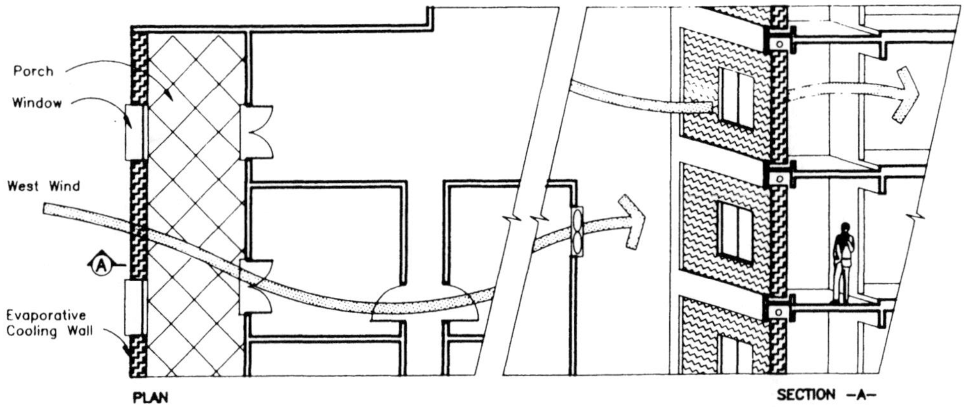

FIGURE 5.4 Use of porches for direct passive evaporative cooling.

closed, so that after passing the pads the air has no other way but to flow into the building openings. As is the case with window pads, the water wetting the pads can be collected at the bottom and recirculated. (See Figure 5.4.)

THE COOLING TOWER OF CUNNINGHAM AND THOMPSON

The passive evaporative cooling system that has been developed and tested by Cunningham and Thompson (1986) in Tucson, Arizona, consists of a downdraft passive evaporative cooling tower, 1.8 × 1.8 m. in cross section and 7.6 m high, attached to a well insulated frame construction building, with a floor area of about 100 m^2. The cooling tower has at its top four vertical wetted cellulose pads (CELdek) impregnated with antirot salts and rigidifying saturants, 10 cm thick, facing the four cardinal directions. A plywood "X" baffle at the top of the tower catches the wind and directs it downward. Water is distributed at the top of the pads and is collected at the bottom by a sump and recirculated by a pump.

The complete system in Tucson also included a solar chimney at the opposite side of the building, which was used to enhance the flow of cooled air through the cooling tower and the building. Outdoor air flows down the cooling tower through the building, up to the attic, and then out through the solar chimney. During the daytime the air is heated in the attic, which thus serves as part of the solar chimney. A cross-section of the building with the towers is shown in Figure 5.5.

A description of this system and a table showing the experimental data collected during a two-day-long test (August 22–23, 1985) was published (Cunningham and Thompson 1986). The measurements included the ambient conditions (DBT, WBT, and wind speed), tower exit air speed and temperature, indoor air temperature, and solar chimney data. The performance of the system was very impressive, as can be seen in Figure 5.6. At 4 p.m. on the second day, when the outdoor temperature reached a maximum of 40.6°C and the WBT 21.6°C, the tower exit air temperature was 23.9°C and the indoor temperature only 24.6°C. The corresponding air speed at the tower exit at that time was 0.75 m/sec.

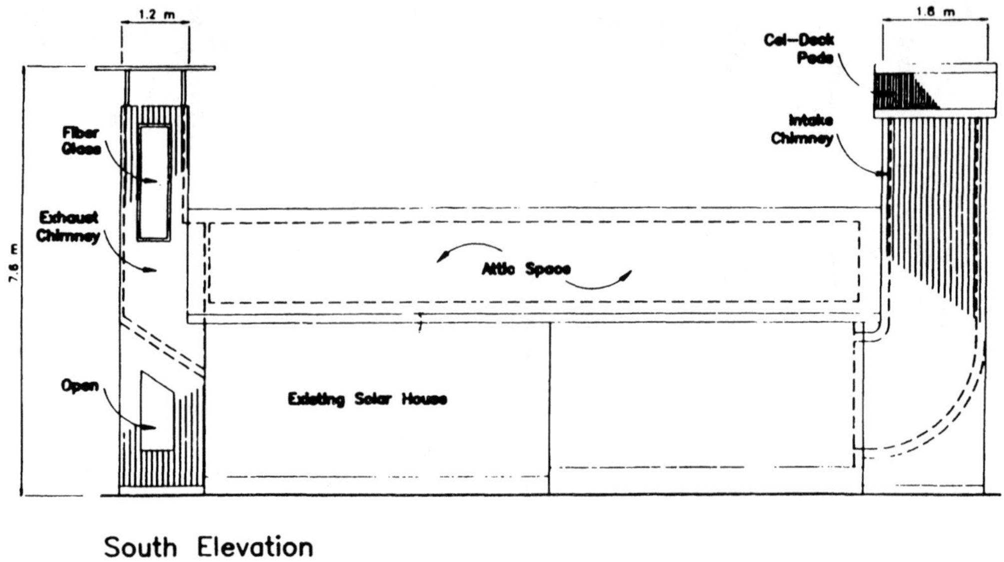

FIGURE 5.5 Cunningham and Thompson's cooling tower.

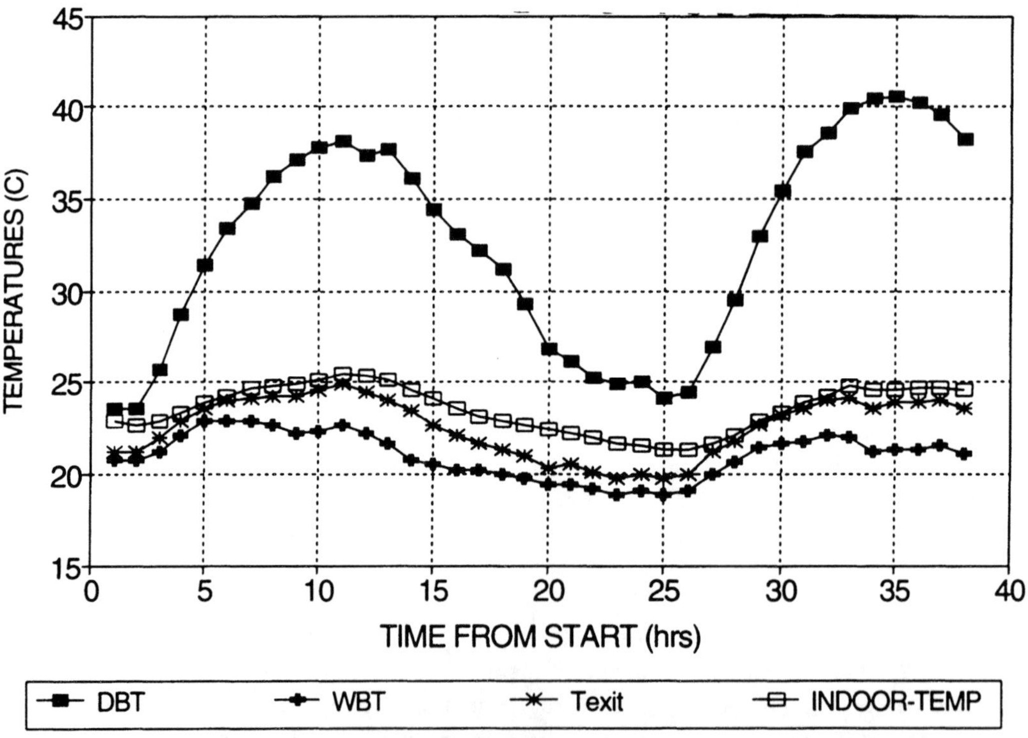

FIGURE 5.6 Performance of Cunningham and Thompson's tower.

Mathematical Modelling of the Arizona Tower

A complex theoretical model of this system was presented in a research report by Cunningham et al (1987). According to this model the air flow through the cooling tower increases with increasing wind speed and the solar chimney enhances the flow rate. This report also contained patterns of the measured ambient DBT, WBT, and indoor air temperature during three months (June to August), for a total of 16 days (without performance details of the tower itself). However, no comparison was done by the authors between their predictions and the experimental data reported in the first and second publications.

An analysis of Cunningham and Thompson's test results, as performed by Givoni (1991), has demonstrated that in reality the solar chimney did not affect the performance of the cooling tower and that the air flow rate was practically independent of the ambient wind speed. The analysis concluded that the exit air temperature from the tower and the flow rate (for a tower with given details) are controlled almost exclusively by the temperature difference between the (cooler) air inside and the (warmer) air outside the tower. This temperature difference is proportional to the WBT depression (DBT-WBT). The air flow rate through the pads of the tower thus seems to be proportional to the WBT depression and to the design details of the tower.

On the basis of these conclusions a different, simpler, mathematical model of this system has been developed (Givoni 1992). The model, and additional comparisons of its calculations with experimental data of indoor temperatures of a light-weight building cooled by such a tower, are presented in the appendix to this chapter.

Figure 5.7 shows detailed diurnal patterns of the tower's exit and the building's indoor

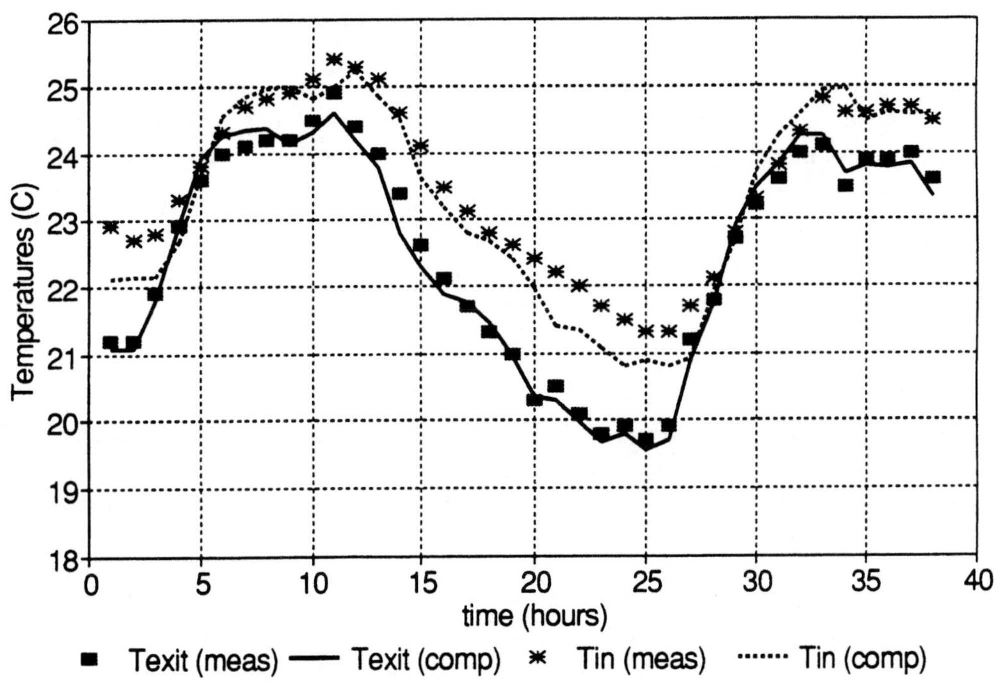

FIGURE 5.7 Measured and calculated (by Givoni's Model) exit and indoor temperatures in the building cooled by the Cunningham and Thompson tower.

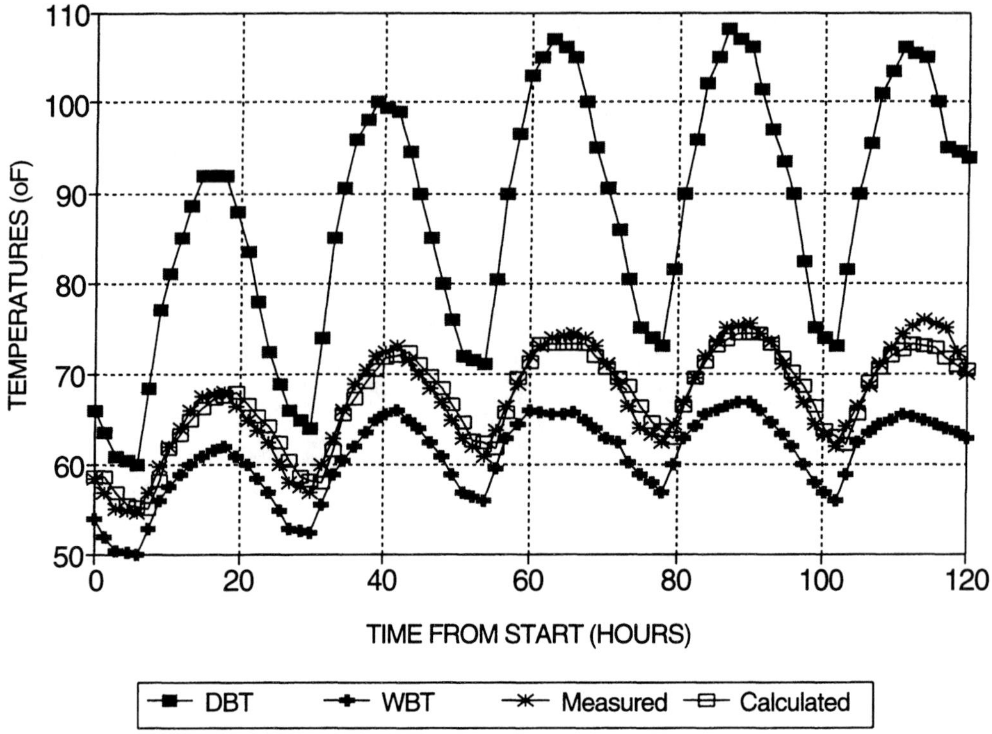

FIGURE 5.8 Comparison between computed and measured indoor air temperatures in the building cooled by the Cunningham and Thompson tower, June 1986.

air temperatures, measured during two days by Cunningham and Thompson (1986), together with temperatures calculated by the mathematical model of this system developed by Givoni. Figure 5.8 shows indoor air temperatures measured during a sequence of five days in June 1987 (Cunningham et al) together with the indoor temperatures calculated by the model.

Effect of the Tower's Design Details on its Performance

The processed air reaches its temperature while crossing the pads. Therefore the height of the tower (provided that it is reasonably insulated) and the area of the wetted pads are not expected to have any appreciable effect on the temperature of the air in the tower in a given combination of ambient DBT and WBT. However, these two system design factors affect the air flow rate, and hence the total cooling effect generated by the system.

The cooling tower tested in Tucson has a height of 7.6 meters, cross area of about 3.2 m^2 and pad area of about 6.4 m^2. With DBT of 40 and WBT of 20°C the exit air speed was about 0.7 m/s; that is, there was a flow rate of about 2.4 m^3/s. Table 5-1 shows predicted air flow rates generated by Tucson-type towers of different heights and pad areas, when the ambient DBT is 40 and the WBT is 20°C. It is assumed that the tower is square in shape, that the height of each pad is one half of its length, and that the shaft's cross area is one half

TABLE 5-1. EFFECT OF TOWER'S HEIGHT AND PADS' AREA ON AIR FLOW

Tower Height	*Pads Area*	*Air Flow Rate*
(m)	(m^2)	(m^3/s)
2.0	1.0	0.190
4.0	1.0	0.268
6.0	1.0	0.329
8.0	1.0	0.379
2.0	2.0	0.379
4.0	2.0	0.537
6.0	2.0	0.657
8.0	2.0	0.759
2.0	4.0	0.759
4.0	4.0	1.073
6.0	4.0	1.315
8.0	4.0	1.518
2.0	6.0	1.138
4.0	6.0	1.610
6.0	6.0	1.972
8.0	6.0	2.277

of the pads' area, as was the case in the Tucson tower. The height of the tower is measured to the center of the pads.

Assuming a small apartment in a multi-story building, with a volume of 250 m^3, and a minimum of 12 air changes per hour for a direct evaporative cooling system, the required minimum flow rate would be 3000 m^3/hr, or 0.833 m^3/s. A tower with a height of 2 meters and pad area of 4–5 square meters (cross area of 2 m^2), which can be accommodated in a balcony, can thus provide significant cooling.

INERTIAL CONVECTIVE/EVAPORATIVE "SHOWER" COOLING TOWER

This cooling system was originally developed by Givoni while he served as a consultant on cooling outdoor rest areas for the '92 EXPO in Seville, Spain. However, this system can also be applied, and has been tested, as a cooling system for buildings.

When fine drops of water, which have a very large surface area, are sprayed vertically downward, like a shower, from the top of an open shaft, the falling water entraps a large volume of air. The water is collected in a small pond at the bottom of the shaft and pumped back to the shower head. The momentum of the falling water is thus transmitted to an air stream, creating an inertial air flow down the shaft. The evaporation from the fine drops cools the water, as well as the air in the shaft, to a level close to the ambient Wet Bulb

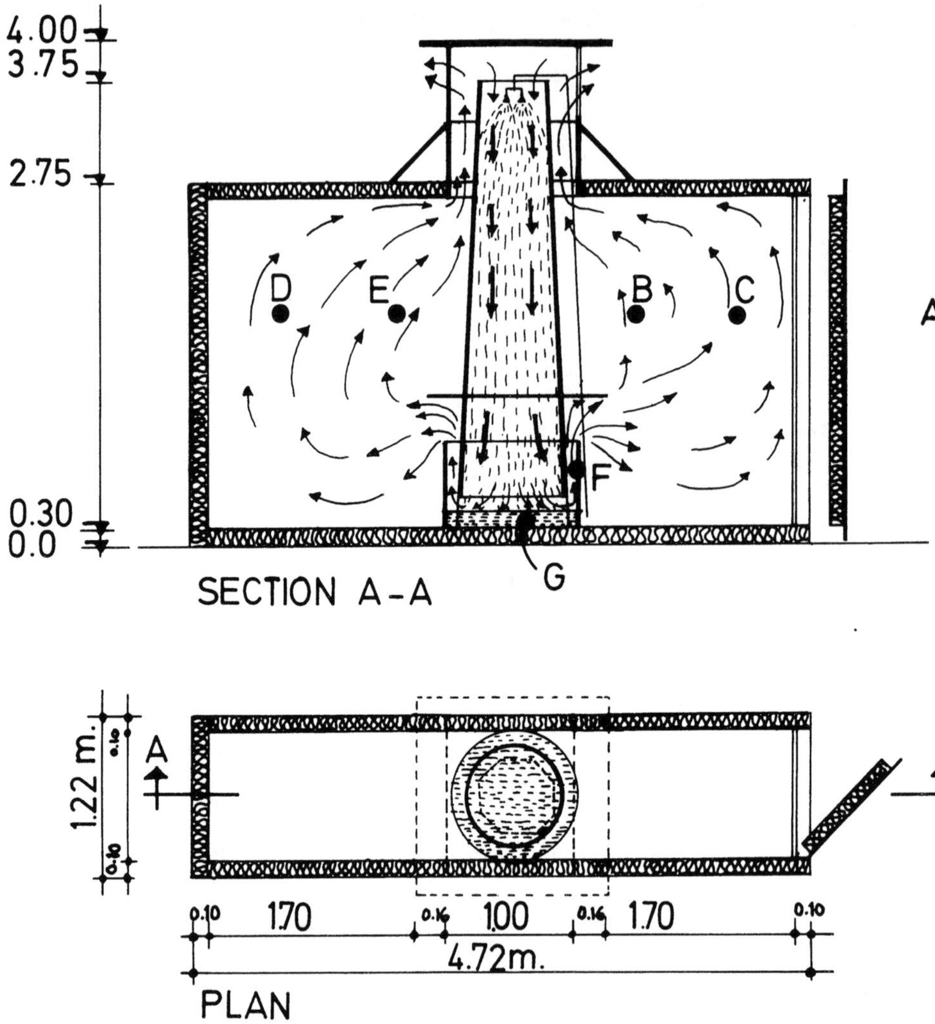

FIGURE 5.9 The Inertial convective/evaporative "shower" cooling tower.

Temperature (WBT). A wind catcher can be placed above the shower head to supplement the inertial air flow by wind effect.

Because any evaporation from the droplets takes place in the free air stream any type of water, even brackish or sea water, if available, could be used for this system. In this respect this system can be applied in places where other evaporative cooling systems, which need high quality water, are not applicable. These issues were tested, and they are discussed in the analysis of the experimental data.

To be used as a cooling device for a building, the shaft could be installed adjacent to an opening in the external walls, or it could even be placed within the building, with its top projecting above the roof and open to the outdoor air. A pond at the bottom of the shaft would collect the sprayed water for recirculation by a small pump.

The performance of this system was tested in a test cell of 4.75×1.2 meters, 2.5 m high, built of plywood/polystyrene panels. When 60% of the roof was open, it represented a patio; when it was covered (except for the shaft opening), it represented a room. The tests were performed, under my direction, by a UCLA student, Mr. Nasser Al Hemiddi.

The experimental shaft has an average cross area of about 0.38 m^2. The height of the shower head within the shaft was changeable and in the study it was at three levels: 2.1 (main tests) and 1.05 m (in the patio configuration), as well as 3.0 m (in the room configuration). The pond diameter was 90 cm, leaving an outlet space for the cooled air exiting from the shaft with an area of about 0.26 m^2. A pump circulated the water between the pond and the shower head. A removable crossed plywood wind catcher, 90 cm high, was installed above the shaft during most of the experiments.

The performance of the cooling tower and the indoor temperatures in the room configuration is discussed in detail in Appendix 2 in this chapter. The modelling and measurements of the space temperature in the patio configuration is discussed in Chapter 7.

A section of the test cell and the cooling tower is shown in Figure 5.9.

The Cooling Performance of the Shower Inertial Tower

The cooling effect of a passive direct evaporative cooling tower is the product of the temperature drop times the air flow rate (times the volumetric specific heat of air, which in the relevant temperatures is about 0.33 whr/m^3/C). The factors that may affect the performance of this type of tower are the climate (DBT, WBT and wind speed), the height of the shower head, the water flow rate, and the water type (fresh, brackish, or sea water).

The performance of the system is discussed with respect to two types of water, fresh and sea water; two heights of shower, 2.13, and 1.07 meters; and two rates of water flow, 14.4 and 9.6 liters/minute.

The Exit Air Temperature

The air temperature drop, as a function of the WBT depression with both fresh and sea water, is shown in Figure 5.10. The figure shows all the data with both fresh and sea water, with a shower head of 2.13 m height and a water flow rate of 14.4 1/min for a total of 17 experimental days. The data with fresh and sea water are marked differently.

It can be seen that with both types of water the relative temperature drop was about the same, as is expressed also by the almost identical regression lines. The slope of the regression lines (with zero origin—as with zero WBT depression no cooling is expected) is about 0.72, meaning that the air temperature drop is about 72% of the WBT depression (72% saturation efficiency).

The fact that the shower system provides the same cooling performance with sea water as with fresh water has profound implication for its applicability in vast arid regions and desert coastal regions, where fresh water is in short supply while brackish and sea water often are plentiful.

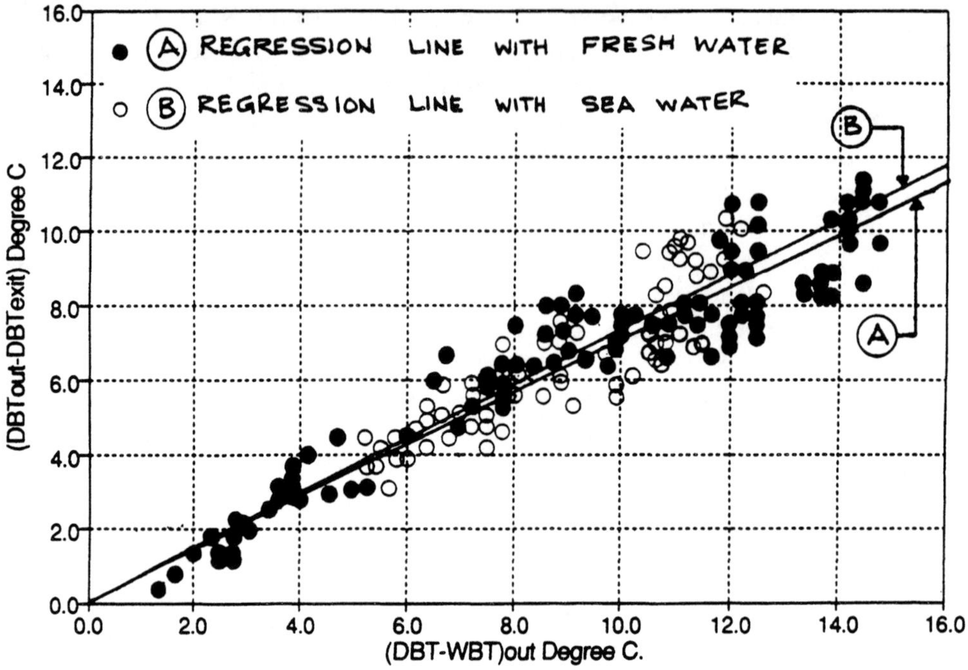

FIGURE 5.10 Correlation between ambient WBT depression and exit air temperature drop of the "shower" cooling tower for fresh water and sea water.

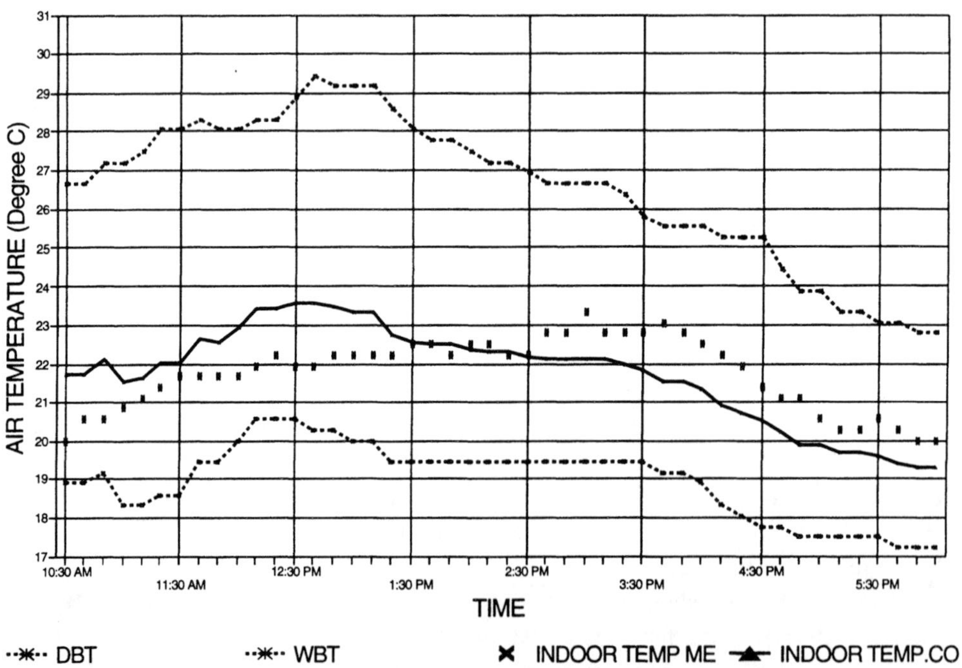

FIGURE 5.11 Performance of the "Shower" cooling tower, October 8, 1992.

Performance of a Building Cooled by the Tower

The development of the formulae for calculating the expected indoor temperature of a building cooled by the shower tower are described in Appendix 2 to this chapter.

Figure 5.11 shows the measured and computed indoor temperatures of the test cell when the roof was covered (representing a room) on October 8, 1992. The indoor temperatures during three more days (October 2, 7, and 11, 1992) are shown in Figures 5.30–5.33. The shaft in these experiments was without a windcatcher and its height was 3 meters. By and large, a reasonable agreement exists between the computed and the measured indoor temperatures.

Section 2: Indirect Evaporative Cooling by Roof Ponds

Introduction to Roof Ponds

Evaporative cooling of a building can also be passive and indirect (that is, it can occur without elevating the indoor humidity) by cooling a given element of the building, such as the roof or a wall, evaporatively. The cooled element, in turn, serves as a heat sink and absorbs, through its interior surface, heat penetrating into the building through its envelope or heat generated indoors. With such systems the indoor radiant and air temperatures are lowered without elevating the indoor moisture content of the air. Passive evaporative cooling can be provided, for instance, by a *shaded* water pond over an *uninsulated roof*. The roof's surface temperature closely follows the ambient wet bulb temperature, while the ceiling, thermally coupled to the pond, acts as a radiant/convective cooling panel for the space under it. Consequently, the indoor air and radiant temperatures can be lowered without elevating the indoor humidity level.

By shading a roof water pond both the heating effect of solar radiation as well as the cooling effect of nocturnal radiation, are eliminated. The only cooling process is by evaporation. The pond gains heat mainly from the ambient air by convection. It gains some heat, as well, by conduction from the indoor space and by long-wave radiation from the shade elements. The balance between the heat gain and loss results in an average water temperature a little higher than the average ambient wet bulb temperature. Heat flow from the interior to the pond, which usually is small in comparison with the convective gain from the ambient air flowing over the pond, tends to elevate the water temperature. But this elevation raises the vapor pressure of the water and increases the evaporative cooling rate, reducing the temperature elevation. Consequently, the average water temperature does not deviate greatly from the average WBT.

When the water is in thermal contact with the roof, and the roof itself is made of materials with high thermal mass and conductivity, such as concrete, the combined mass of the water and the roof structure serves as an integrated "cold" storage. The ceiling temperature follows the water temperature closely, although with some elevation. Consequently, the soffit of the system (the ceiling) serves as a passive cooling panel for the space below. This space is cooled both by radiation and by natural convection, which is maximized when the cooled air under the ceiling sinks down. Thus, the total heat exchange between the ceiling and the indoor space is also maximized, enabling satisfactory cooling even with a small temperature difference of about 2 to 3°C (3.5 to 5.5°F) between the ceiling and the space below it.

To bring the ceiling temperature as close as possible to that of the pond, the water must be in direct thermal contact with the roof structure below—no insulation may be installed between the water and the ceiling. This last detail has obvious repercussions for application of roof water ponds in regions with cold winters, when the roof should be insulated. It also has implications for the options of ceiling treatment in spaces cooled by roof ponds. Any applied ceiling finish, such as acoustic treatments, should be detailed so as to allow maximum exposure of the structural slab to radiation. Convective downward air flow should be restricted as little as possible.

The water temperature is much lower in a shaded pond than it is in an exposed pond. Water consumption by evaporation in a shaded pond is much lower than in an exposed pond since the effect of the absorbed solar radiation on the evaporation is eliminated. The additional cooling potential is achieved together with conservation of water. In dry regions the cooling potential of roof ponds is higher than for most other passive cooling systems. Therefore, these systems deserve detailed analysis and discussion.

Types of Evaporative Roof Ponds

In order to maximize the cooling of the space by the pond, the water should be shaded during the daytime to eliminate solar heat gain. The thermal conductance between the pond and the space below should be as high as practical. In summer this design offers the highest cooling efficiency. However, in many regions that require summer cooling the winters are cool or even cold. A pond without insulation, even if it is drained in winter, will result in a high rate of heat loss through the roof and therefore would be unacceptable. Three types of insulation/shading that can provide winter insulation for roof ponds will be discussed below:

- **A** Fixed insulation, with an air gap, above the pond.
- **B** Insulation floating over the water.
- **C** Insulation embedded within the water in a pebble layer.

The different forms of the insulation/shading element result in different types (systems) of roof ponds. In all of these systems the roof must structurally support the water load—about 200 to 400 kg/m^2 (40–80 lb/sf), depending on the system—and provide perfect waterproofing. However, it should be noted that once the waterproofing membrane is installed, the water and the insulation protect its material from the sun and the effects of sharply changing weather conditions. Therefore, the factors that usually destroy the material are largely eliminated, ensuring a long lifetime for the membrane.

In all the different types of evaporating roof ponds the problem of algae growth and other forms of biological activity that can produce smells and health hazards should be considered, although there is no direct contact between the water and the cooled space. When necessary, experts should be consulted and methods to control biological activity in the pond water should be applied.

VENTILATED ROOF PONDS WITH FIXED INSULATION

This type of roof pond includes, in addition to the structural roof supporting the pond, a secondary lightweight insulated roof over the pond, shading the water, with large openings between the water and the shade, to permit permanent air flow over the water and enhance the evaporation. The openings should be kept open all through the summer and be closed during the winter. The cooling process in this type of roof pond is only by evaporation. In winter the pond should be drained and the openings closed. The roof then becomes an ordinary insulated roof. A scheme of a ventilated roof-pond is shown in Figure 5.12.

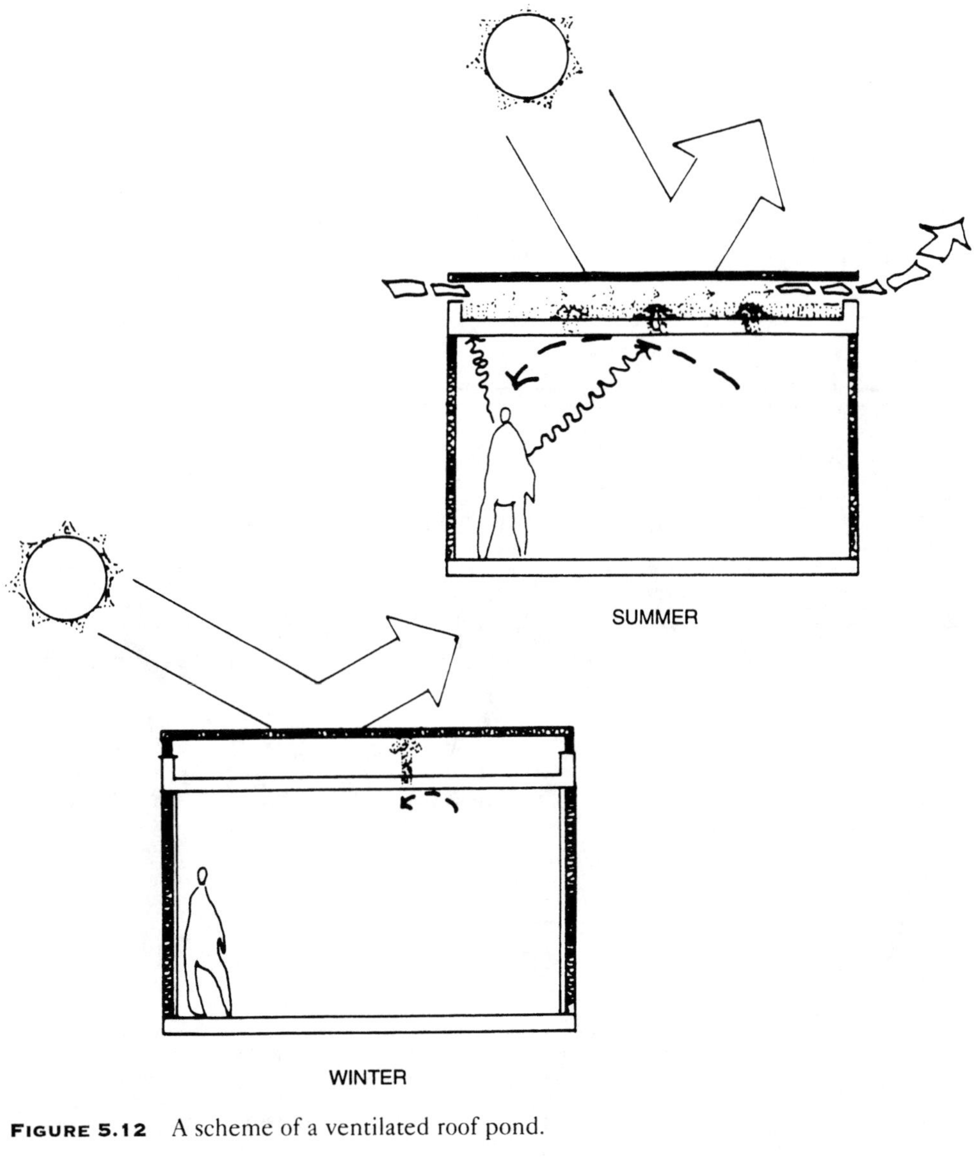

FIGURE 5.12 A scheme of a ventilated roof pond.

The pond's temperature follows, with some elevation, the average ambient WBT. In dry regions, with maximum WBT not exceeding about 24°C, this cooling system can provide very effective cooling even when the daytime temperature is very high, above 40°C (104°F), as is illustrated in the next section. The ceiling temperature of a concrete roof under the pond follows that of the pond, again with some elevation, and thus serves as a radiant/convective cooling element for the space below. The indoor temperature is determined by the balance between the heating and cooling processes. On one hand you have heat gain through the envelope and internal heat generation, and on the other hand conductive heat loss from the space, through the roof, into the pond. A mathematical model of this system is presented in Appendix 3.

Experimental Testing of Ventilated Roof Ponds

Extensive experimental studies of the thermal performance of thermal models cooled by this type of roof pond, built with conventional building materials, has been carried out by Givoni at the Institute for Desert Research of Ben Gurion University in Israel (Givoni 1980). Two types of thermal models were used: The first was a large, room-size, low-mass model with insulated stud walls and a wooden floor. There were two models representing the second type—smaller high-mass "mini-rooms" with concrete walls, roofs, and floors.

The large low-mass model had internal dimensions of 304×390 cm, and a height of 230

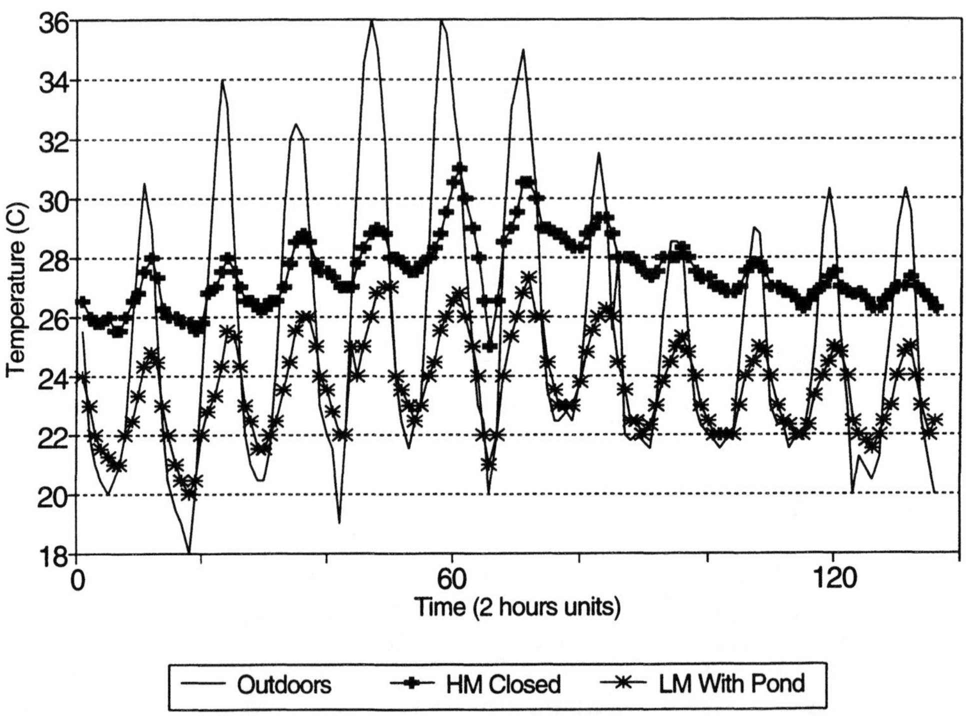

FIGURE 5.13 Diurnal temperature patterns of a low-mass room with ventilated roof pond.

cm. The walls were insulated panels, like stud-wall construction: 5 cm of expanded polystyrene sandwiched between layers of plywood. The roof was of 20 cm concrete, supported by 4 concrete columns within the indoor space. The floor was 2 cm plywood over a gravel-filled space (for heat storage in solar heating experiments).

The two small high-mass test cells were built of concrete walls, floors, and roofs, 20 cm thick. The walls were insulated externally by 5 cm expanded polystyrene and plastered. Their size was 206×166×202 cm (internal dimensions). One cell had a ventilated roof pond and the other had a roof pond with floating insulation (to be discussed in the following section). A third identical cell, without a pond and with highly insulated roof, remained closed all the time and served as a control to evaluate the temperature reduction derived from the ponds. More details on these two rooms is given in Appendix 3, when their observed performance is compared with calculations by the mathematical model.

Diurnal patterns of the water, the ceiling, and the indoor air temperatures of the room-size low-mass model during four days, together with the DBT and the WBT, are shown in Figure 5.13. Figure 5.14 shows the corresponding temperatures of the smaller high-mass model with the ventilated pond.

More data on the performance of the high-mass test cells is presented in the next section, when the average temperatures obtained with a ventilated roof pond are compared with those obtained in the model cooled by a pond with floating insulation during a sequence of 18 days.

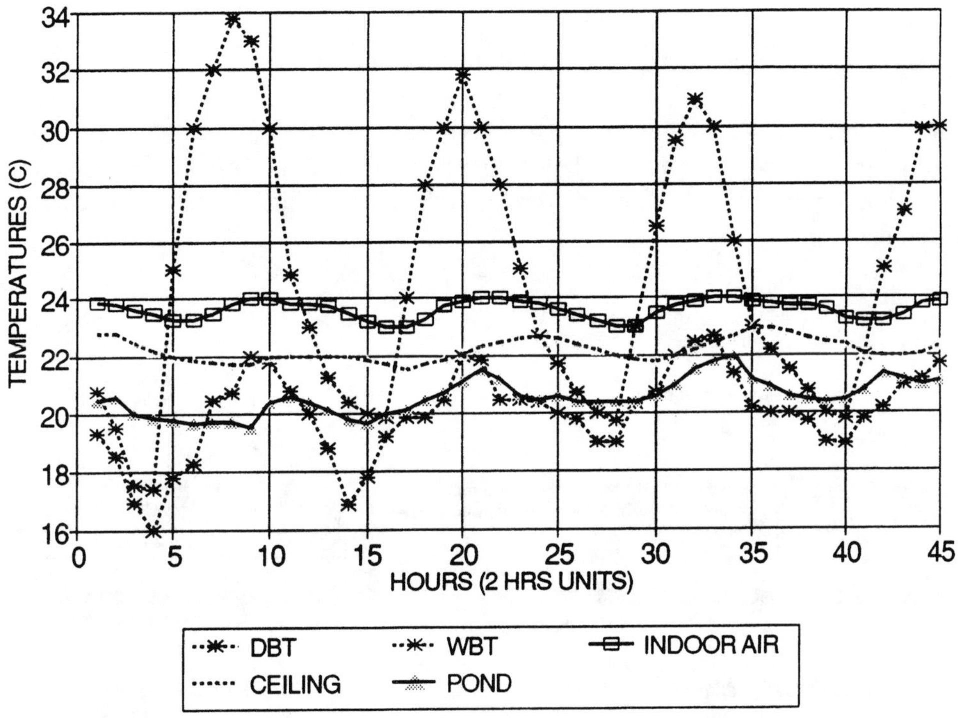

FIGURE 5.14 Diurnal pond, indoor air, and ceiling temperature patterns of high-mass test cell with ventilated roof pond.

ROOF PONDS WITH FLOATING INSULATION

The idea of floating the insulation over the water and circulating the water above the insulation during the night was conceived by Dick Bourne of Davis, California (personal communication, 1980). His objective was to overcome some of the problems encountered with moving insulation panels as in the Skytherm system of H. Hay. (See the chapter on Radiant Cooling.) Bourne called his system the "CoolRoof." A scheme of the CoolRoof system is shown in Figure 5.15.

The performance of this system was tested extensively with thermal models by Givoni in Israel (1987, 1991), and with his participation in Mexico (Sandoval et al. 1991). In 1991 Bourne (1992) has demonstrated the performance of this system on a large-scale industrial building (Bourne and Hoeschele 1992), and Givoni was asked to review and evaluate the Monitoring Report of the demonstration project.

Bourne's idea was to overcome the difficulties posed by the need to move the large panels insulating the pond during the daytime by installing permanent floating insulation plates over the water.

In this system the water is circulated at night over the floating insulation and thus is cooled by a combination of convection, evaporation, and longwave radiation. During the day the water is insulated from the sun and the warmer ambient air by the tightly assembled floating insulation. The insulation panels should be of some material that is impermeable to water, such as extruded polystyrene.

Performance of Roof Ponds with Floating Insulation

The performance of this type of roof pond was measured under a variety of climatic summer conditions: in the Israeli desert (Sede Boqer Campus of the University of the Negev), in the hot humid climate of Colima, Mexico, and in the semidry climate of Sacramento, California.

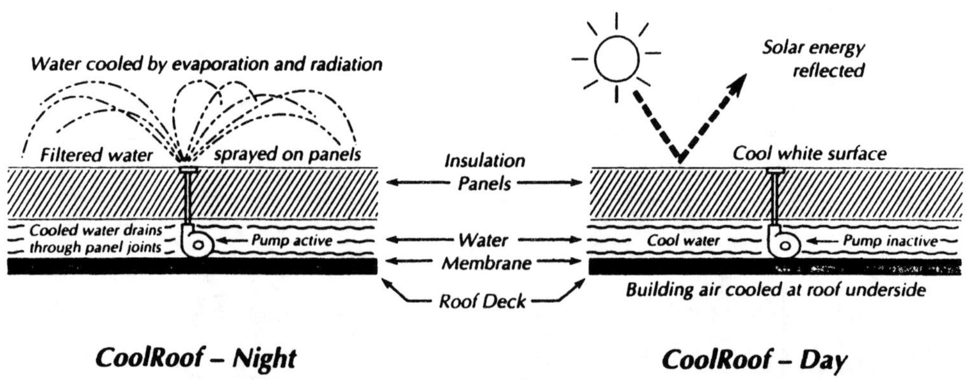

FIGURE 5.15 A scheme of the CoolRoof system (roof pond with floating insulation).

Experiments in Israel

Figure 5.16 shows diurnal patterns of the indoor air temperatures of the two high-mass test-cells described in the previous section, one with the ventilated pond and the other with a pond with floating insulation, during the four days shown in Figures 5.12 and 5.13, together with the ambient DBT and the WBT. It can be seen that, in spite of the different type of roof pond, the temperature patterns in the two cells are almost identical.

It can be seen that in both models the pond temperatures were very similar and followed the average WBT, with an elevation of about 1°C and diurnal swings of about 1°C. The average ceiling temperatures in both models were also similar, about 1.5 to 2°C above the pond temperatures. Also the *average* indoor temperatures of the two models were similar, about 2°C above the ceiling temperatures, but with different diurnal swings. In the high-mass small model, with completely impermeable walls, the diurnal swing of the indoor temperature was about 1°C. The swing in the low-mass model, with more permeable walls, was about 3°C. The outdoor temperature swing during these four days has ranged from 16 to 10°C.

Figure 5.17 shows a sequence for 18 days of the daily *average* temperatures of the ponds, ceilings, and indoor air of the two cells with the different roof ponds, together with the average indoor temperature of the control high-mass cell. (The control cell was identical to the two cooled cells, but without cooling.) It can be seen that the average

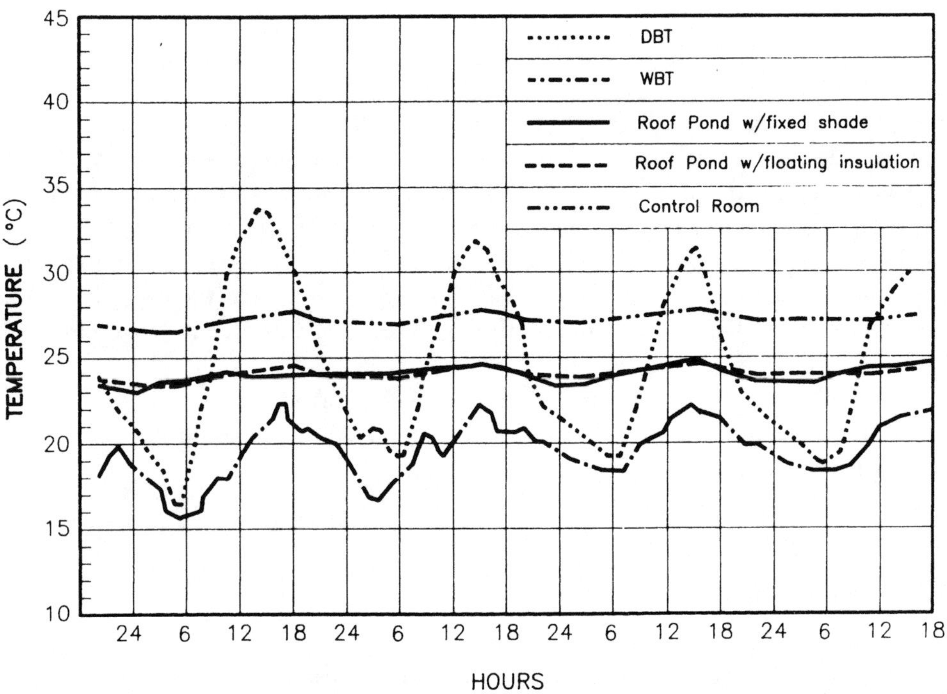

FIGURE 5.16 Comparison of the diurnal patterns of the indoor air temperatures of the two high-mass test cells, one with the ventilated pond and the other with a pond with floating insulation.

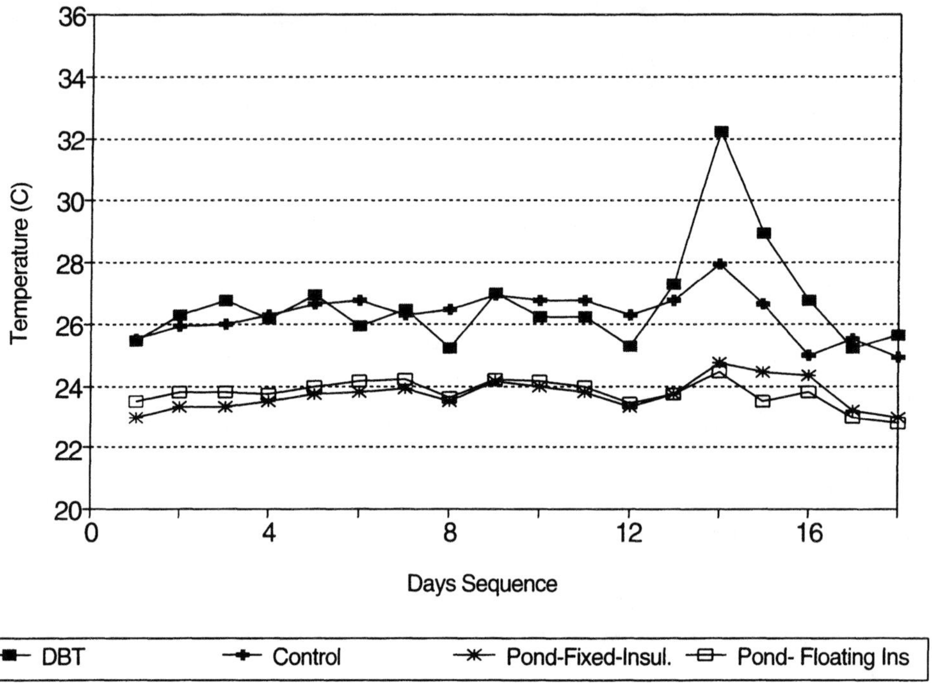

FIGURE 5.17 Sequence of 18 days of the averages of the indoor air, ceiling, and pond temperatures of two cells with different roof ponds.

temperatures in the two cells were very close. The average temperature of the control cell followed the pattern of the average DBT, except for a three-day heat wave (August 1 to 3) in which the average indoor temperature of the high-mass cell was significantly lower than the outdoor average. The ponds and the indoor temperatures were almost unaffected by the heat wave and consequently the cooling effect of the ponds was then maximized. The difference in the indoor air temperature between the control and the cooled cells was about 2.5°C throughout most of the period and reached about three degrees during the heat wave.

Figures 5.16 and 5.17 demonstrate that the cooling effect of the pond with floating insulation is practically identical to that of the ventilated roof pond, in spite of the very different heat loss processes governing the two systems.

Experiments in Colima, Mexico

Colima is a city in a hot humid region near the Pacific Ocean in Mexico. Almost all of the buildings in the city have flat concrete roofs. While visiting the University of Colima I was asked to advise the School of Architecture about research on passive cooling systems that may be effective in this climate and I suggested an experiment that would use roof ponds

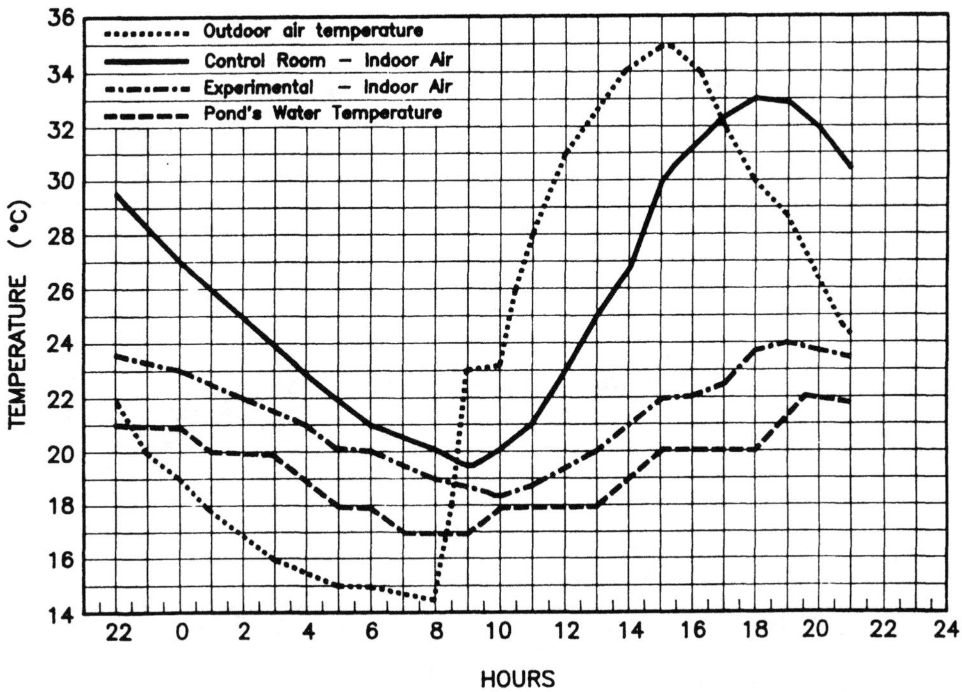

FIGURE 5.18 Temperature patterns of a control and a test cell with roof pond with floating insulation in Colima, Mexico.

with floating insulation for cooling. Consequently the school has built two thermal models, 1.60 × 1.60 × 0.95 (height) m, built of uninsulated adobe blocks and with concrete flat roofs, 8 cm thick. One model served as a control and the other one had a roof pond with floating insulation. The temperatures of the water and the indoor in the two models were measured under several "treatments" of the insulation panels (Sandoval et al. 1991).

Figure 5.18 shows diurnal temperature patterns of the control and the cooled model. With outdoor maximum of about 35°C the indoor temperature of the control model reached about 33°C and the temperature of the model with the roof pond was about 22°C, almost parallel and about 2°C above the water temperature. The water temperature ranged between 17 and 21°C.

After the impressive performance of the models the principal investigator, Mr. Leandro Sandoval, decided to apply this system in his own house. He applied a pond with floating insulation over the roof of one room and took indoor air temperature measurements in this room and in a control room. The cooled room remained closed while the control room was ventilated after midnight.

Figure 5.19 (private communication) shows diurnal temperature patterns of the indoor air in the two rooms, the ambient air, and the water. With outdoor temperature range of 18.5 to 32°C the indoor range in the cooled room was from 21.8 to 24.2 while the range in the control room was from 21.8 (with night ventilation) to 28°C. The water temperature range was from 17 to 21.5°C.

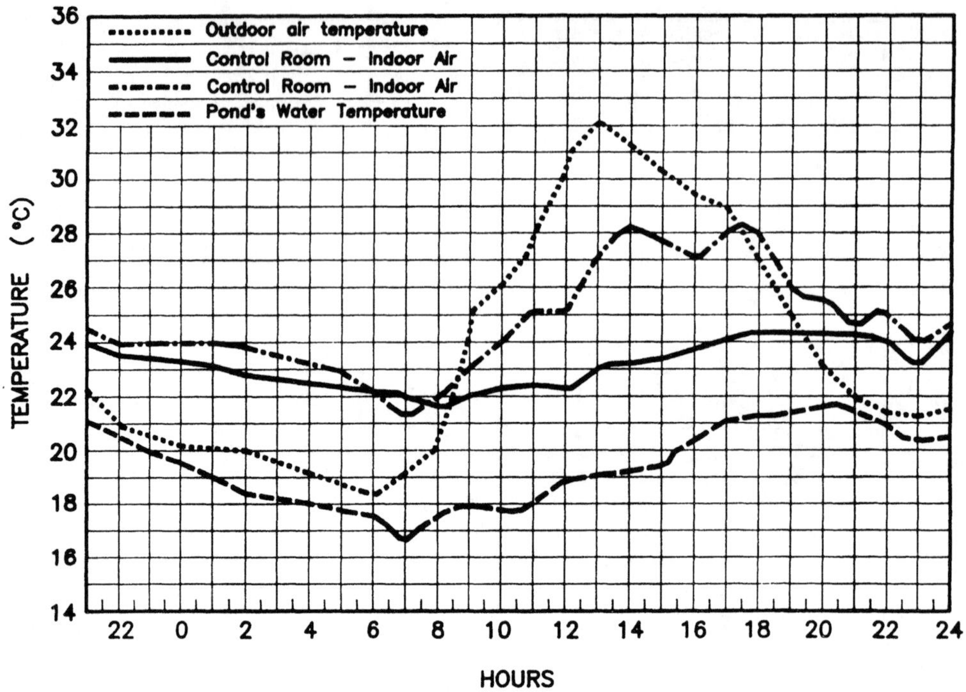

FIGURE 5.19 Temperature patterns of indoor air in the two rooms, one serving as a control and one with a roof pond with floating insulation, in Colima, Mexico.

Testing in Sacramento, California

Richard Bourne installed his system, a demonstration project, on a one-story book storage warehouse at the California Office of State Printing in Sacramento. The pond covered a section of 590 m^2 of the 3600 m^2 roof of the building. As the roof was insulated, the thermal coupling between the pond and the interior space was provided by a fan-coil system that circulated the water from the pond through a water-to-air heat exchanger. In this respect the system was a hybrid one: the cooling of the water was achieved by passive processes while the transfer of the cooling from the pond to the building was by a mechanical subsystem. Givoni was asked to evaluate Bourne's monitoring report and the following evaluation of the system's performance is based on both the Monitoring Report and my subsequent evaluation.

Measurements were taken from September 6 through November 14, 1991, under two schedules:

- **A** The fan-coil was running to achieve the thermostat setting 24 hours of the day, maximizing the fan-coil operation time and cooling delivery (no fan-coil lockout mode).
- **B** The fan-coil was locked out after 4 p.m., when occupancy of the building ended.

Figure 5.20 (reproduced by permission from Bourne and Hoeschele, 1992) shows diurnal patterns of the outdoor, indoor, and water temperatures during two sequences of

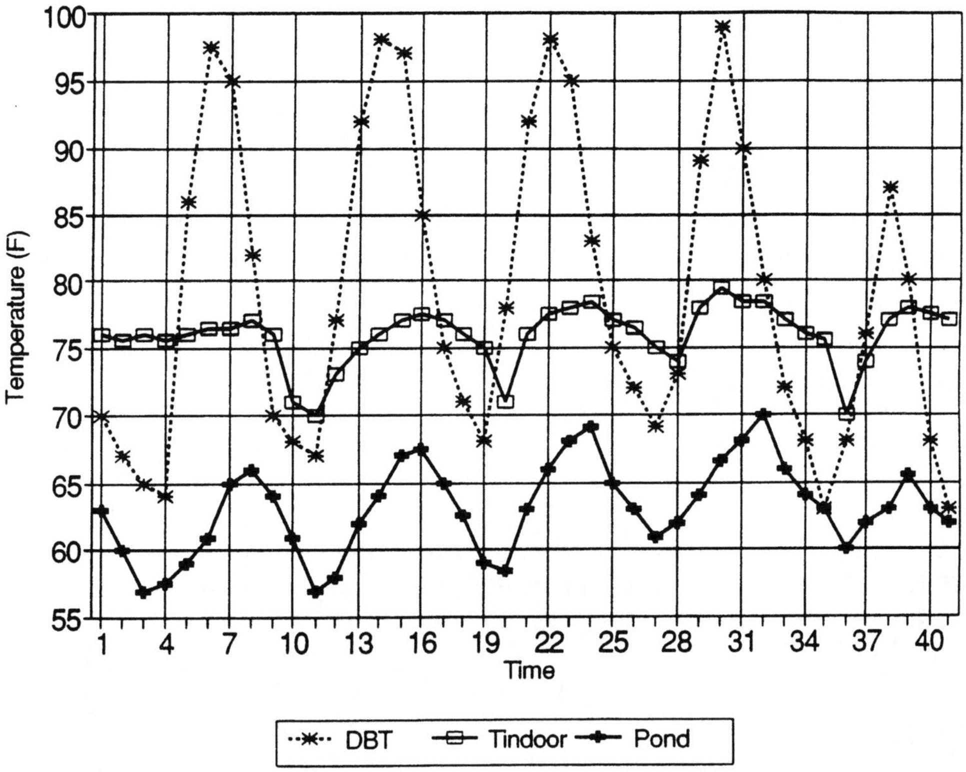

FIGURE 5.20 Temperature patterns of a warehouse with the CoolRoof system (roof pond with floating insulation) in Sacramento, California.

five and 7 days, without and with fan-coil lockout (Bourne and Hoeschele 1992). The performance in the mode without fan-coil lockout is the closest to what would be obtained with a passive coupling between the pond and the interior. With outdoor maxima of about 98 to 99°F (about 36.5 to 37°C) the indoor temperatures during the daytime were around 77 to 78°F (25 to 25.5°C). The average water temperature was about 63°F (17.5°C). During the daytime the water was heated by about 5°C, but its maximum remained at about the same level as the outdoor air minimum.

Effect of the Pond's Daytime Heat Gain on its Minimum Temperature

The temperature rise of the water during the daytime indicates the total heat gain of the pond, both from the outdoors through the floating insulation and from the building by conductance and through the fan coil. The subsequent pond water minimum temperature deviation from the WBT minimum was only weakly related to this temperature rise. The pond minimum temperature had a tendency to rise with respect to the WBT minimum as the daytime heat gain of the water increased, but the quantitative effect of this factor was rather small. The range of the daytime temperature rise, during all of the experimental period, was between 2.5 and 13°F, while the range of the changes in the relationship between the pond's minima and the WBT minima was about 2 degrees, from –1 to +1°F.

Thus, a fivefold increase in the total daytime heat gain by the pond has not produced a marked effect on the final temperature of the pond at the end of the night. A similar relationship between the outdoor temperature and the water temperature rise was also observed by Givoni in the experiments in Israel.

These observations have important implications in evaluating the potential of a roof pond with floating insulation to store cold night energy. They mean that the initial temperature of the pond, at the beginning of the evening, has only a minor effect on its final temperature, which is determined almost exclusively by the climatic conditions prevailing during the nights.

Insulation Embedded Within the Pond

This system has been developed by Givoni (1984) in order to allow some form of utilization of the roof area. It was designed specifically for buildings with reinforced concrete flat roofs in desert regions with mild winters (minimum temperatures usually above freezing).

Description of the System

The pond consists of the following layers (see Figure 5.21):

1. Reinforced concrete flat roof, capable of supporting a static load of up to 400 kg/m^2, with slight inclined planes (about 1%) to enable drainage of rain water in winter.
2. A "perfect" waterproofing membrane, preferably with low thermal resistance.
3. A lower layer of pebbles, about 2 cm thick at the highest point of the roof. This layer is leveled at its top.
4. Impermeable insulation plates, about 1×1 meters, 5 to 10 cm thick (depending on the severity of the summer and/or the winter), laid with gaps of about one cm between the plates. The recommended material is extruded expanded polystyrene.
5. An upper layer of pebbles, about 12 cm thick.
6. Openable drains at the lowest points of the pond, tightly closed in summer and open in winter.
7. A float that governs the level of the water, about 3 cm above the insulation at its highest point.

During the summer, the pond should be filled with water to a level of about 3 cm above the insulation. The insulation plates thus divide the water into two layers—below and above—with gaps permitting thermosyphonic circulation.

This system has been applied and tested in a full-size semidetached house in the city of Beit-Shean in Israel's Jordan valley. The building had two identical units, each of about 50 m^2, with uninsulated walls built of conventional hollow concrete blocks, 20 cm thick, with concrete beams and columns. One unit got the roof pond while the other unit (the

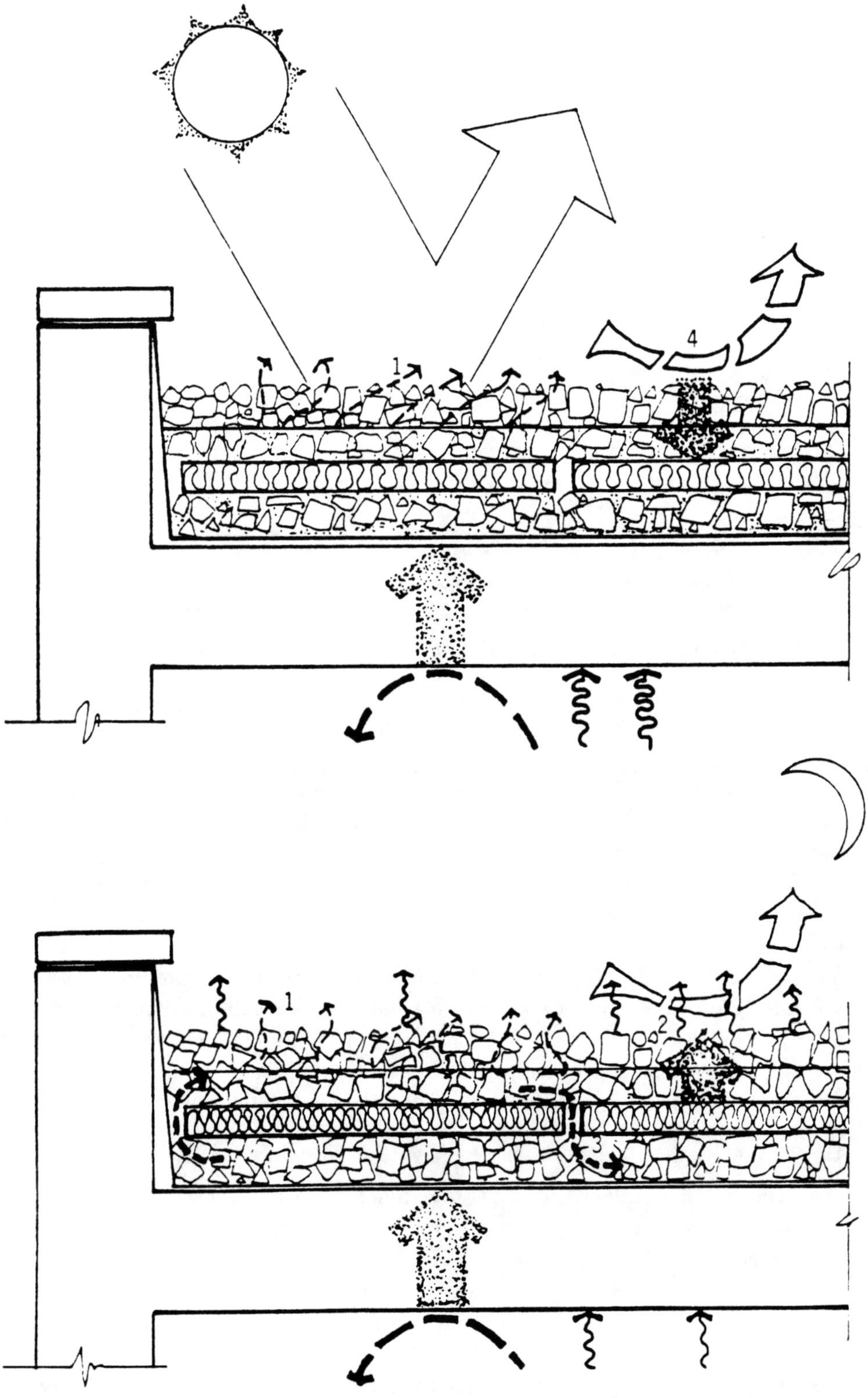

FIGURE 5.21 A roof pond with insulation embedded within a pebble layer.

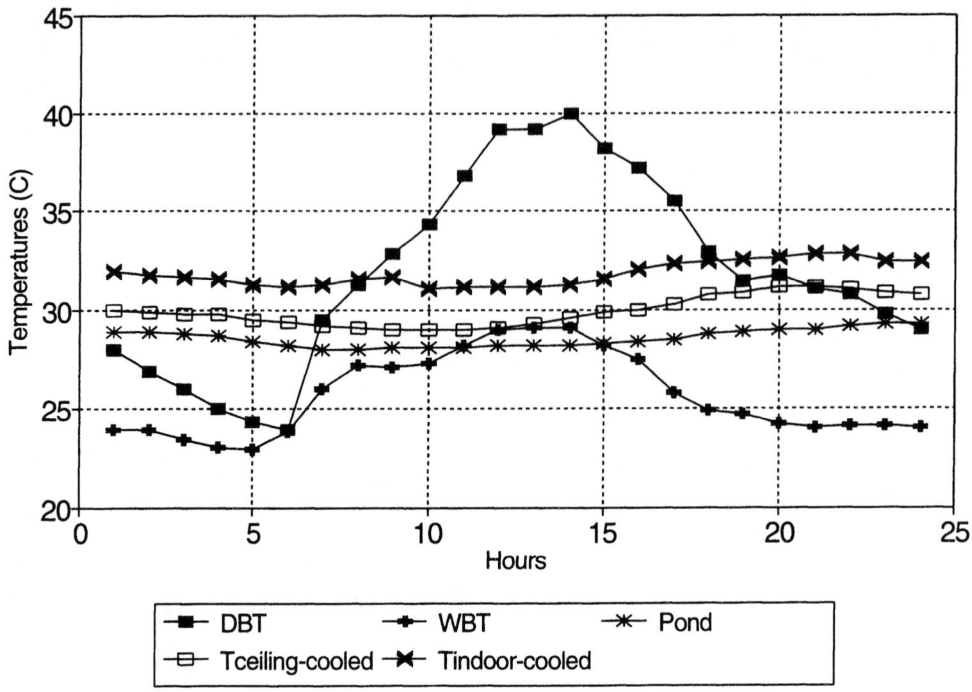

FIGURE 5.22 Diurnal temperature in a control and a house cooled by a roof pond with embedded insulation.

control) was insulated with a continuous layer of 3 cm polystyrene plus waterproofing above the insulation. The volumetric heat loss/gain coefficient ("G") of each unit was about 2 (watts/m^3/C), a rather high value for this hot region.

The waterproofing of the pond was a layer of EPDM, which is an excellent waterproofing material but has a rather high thermal resistance. The insulating panels were 3 cm thick.

Figure 5.22 shows diurnal patterns of temperatures in the two units during a "typical" day. Figure 5.23 shows a sequence of the indoor and outdoor maximum temperatures from August 12 through September 18, 1983. The indoor maximum temperature in the cooled unit was lower by about 2°C as compared with the control unit.

In evaluating the observed performance of the system the following factors should be taken into account:

- **A** The building, built according to the common standards for low cost housing at the time, had a rather high heat gain through the walls and the windows, a fact which reduces the relative effect of the pond.
- **B** Because of budgetary limitations the insulation layer embedded within the pebbles was only 3 cm thick, and of material that absorbs water (expanded rather than extruded polystyrene).
- **C** Because of widespread irrigation in the region around the site the humidity was higher than what is "ideal" for evaporative cooling.

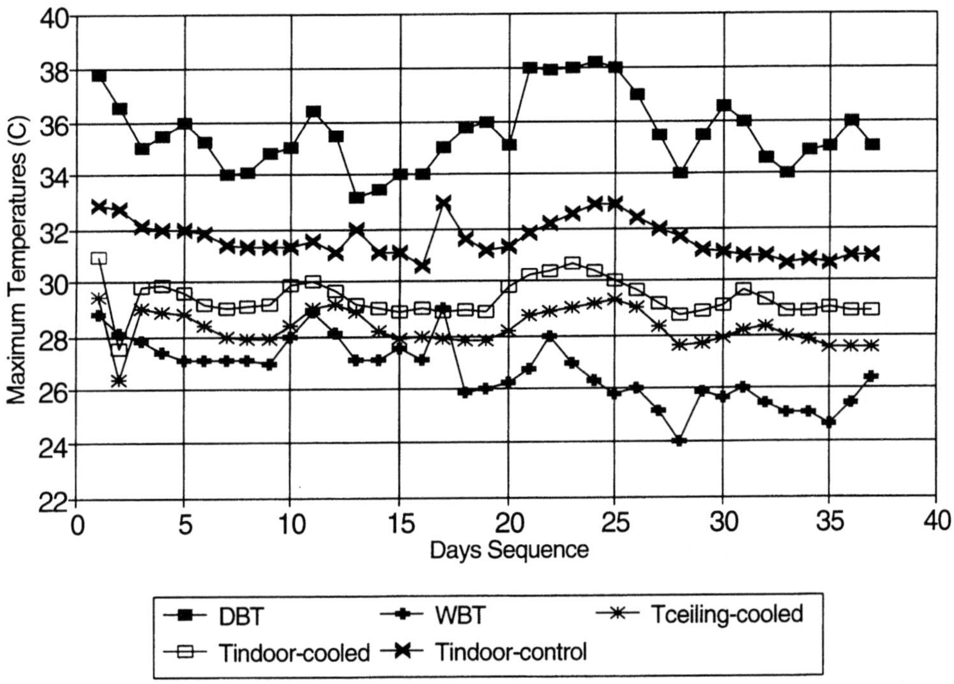

FIGURE 5.23 Maximum temperatures in the two housing units, one serving as a control and the other cooled by a roof pond with embedded insulation, during a sequence of 17 days.

EFFECT OF BUILDING DETAILS AND CLIMATE ON THE THERMAL PERFORMANCE OF BUILDINGS COOLED BY ROOF PONDS

A mathematical model predicting the indoor air, ceiling, and water temperatures of buildings cooled by ventilated roof ponds has been developed by Givoni (1982), and it is described in Appendix 3. It enables researchers to estimate the effects of various building and pond design details, and of the climatic conditions, on the performance of a building cooled by such ponds. However, in view of the fact that the observed performance of the test cell with the floating insulation was almost identical to that of the test cell with the ventilated pond, it is reasonable to assume that this model also allows researchers to estimate the effects of climate and building design on the performance of buildings cooled by roof ponds with floating insulation.

The main parameters affecting the indoor temperatures of buildings cooled by roof ponds are the thermal quality of the building (building's heat gain/loss coefficient – BLC); solar heat gain and internal heat generation (IHG); the area of the pond relative to the floor area; the thermal conductance of the roof under the pond, which depends on the roof's thickness, the conductivity of its material, and the ceiling's surface coefficient. Of course, the performance also depends on the outdoor air DB and WB temperatures.

The wind speed, above a minimum level of about 1 m/s, has a relatively minor effect on the performance of a roof pond because its effect on the evaporative heat loss is proportional to its effect on the convective heat gain (which is the main heat gain factor for the pond).

The humidity is expressed in the model in terms of the vapor pressure because evaporation is proportional to the vapor pressure difference and also because, in a given location and day, it is more stable than the WBT.

Samples of the expected effects of some design variables of the building and the ventilated roof ponds, and of the climatic conditions, on the performance of a building (the indoor air and ceiling average temperatures), in comparison with the performance of a "base case" building without a roof pond, were calculated by the model and are shown below in a tabular form.

The Climate and the "Base case" Building

Climate:

Average DBT of 30°C (Max. 40 and Min. 20°C)
Vapor pressure of 10 mmHg (typical in hot arid regions)
Average wind speed = 2 m/s

Building (floor area 150 m^2) and Pond:

Building's Loss Coef. (BLC) = 6000 (wh/day.C)
Solar and internal heat generation (IHG) = 10000 (wh/day)
Roof's thickness (under pond) = 0.2 m
Roof's conductivity (concrete) = 1.5 (w/m.C)
Ceiling's surface coefficient, $h_{ceiling}$ = 10 (w/m^2.C)
Basic Pond area = 100 m^2 (2/3 of the floor area)

Effects of Building's BLC and IHG

Table 5-2 shows the expected average indoor air and ceiling temperatures under three levels of BLC: 3000 wh/day.C ("super insulated"), 6000 (base case) and 9000 (poorly insulated), and three levels of total daily internal heat generation (including solar penetration): zero, 10 (base case) and 20 (Kwh/day). All other factors remained under the base case conditions.

TABLE 5-2. EFFECTS OF BUILDING LOSS COEFFICIENT AND INTERNAL HEAT GENERATION

$Ta=30°C$, $dTsa=2°C$, $VP=10$ mmHg, $V=2m/s$, $A_{pond}=100\ m^2$, $d_{roof}=0.2m$

BLC (wh/d.C)	*IHG (wh/d)*	*Tceil*	*Tin*
3000	0	21.6	22.7
6000	0	22.7	24.6
9000	0	23.4	25.8
3000	10,000	22.1	23.6
6000	10,000	23.1	25.2
9000	10,000	23.8	26.3
3000	20,000	22.6	24.4
6000	20,000	23.6	25.9
9000	20,000	24.1	26.9

Pond's Area, BLC and Vapor Pressure

Table 5-3 shows the effects of changing the area of the pond (from ⅓ to ⅚ of the floor area—50 to 125 m^2), under two levels of the BLC (3000 and 6000 w/day.C) and two vapor pressures (10 and 15 mmHg). Temperature, IHG, and the other factors remain under the base case conditions.

Table 5-3. Effects of Pond's Area on Indoor Temperature Under Two BLC and Two Vapor Pressure Levels

BLC		*3000*		*6000*	
VP	(mmHg)	10	15	10	15
Pond	50	25.9	27.5	27.6	28.7
Area	75	24.4	26.3	26.2	27.6
m^2	100	23.6	25.5	25.2	26.8
	125	22.9	25.0	24.5	26.2

Outdoor Averages of DBT and Vapor Pressure

Table 5-4 shows the indoor temperatures of the "base case" building under three levels of the average ambient DBT (25, 30 and 35°C) and three levels of the vapor pressure (5, 10 and 15 mmHg).

Table 5-4. Effect of DBT and Vapor Pressure on Indoor Temperature (Base Case Building)

VP	*DBT=25°C*	*DBT=30°C*	*DBT=35°C*
5	18.9	21.6	24.2
10	20.9	23.6	26.2
15	22.9	25.5	28.2

Ground-Based Ponds as Cooling Sources for Buildings

Roof ponds require a flat load-bearing roof to support the water weight, a feature not found in many building types. Often an open area is available next to or even under a building, where a pond can be installed. When the pond is not located over a roof it can be much deeper, providing a much higher heat capacity per unit area. Such ponds also may be much cheaper than roof ponds and their temperature can be stabilized near the weekly rather than the daily average WBT. As a result, the difference between the water temperature and

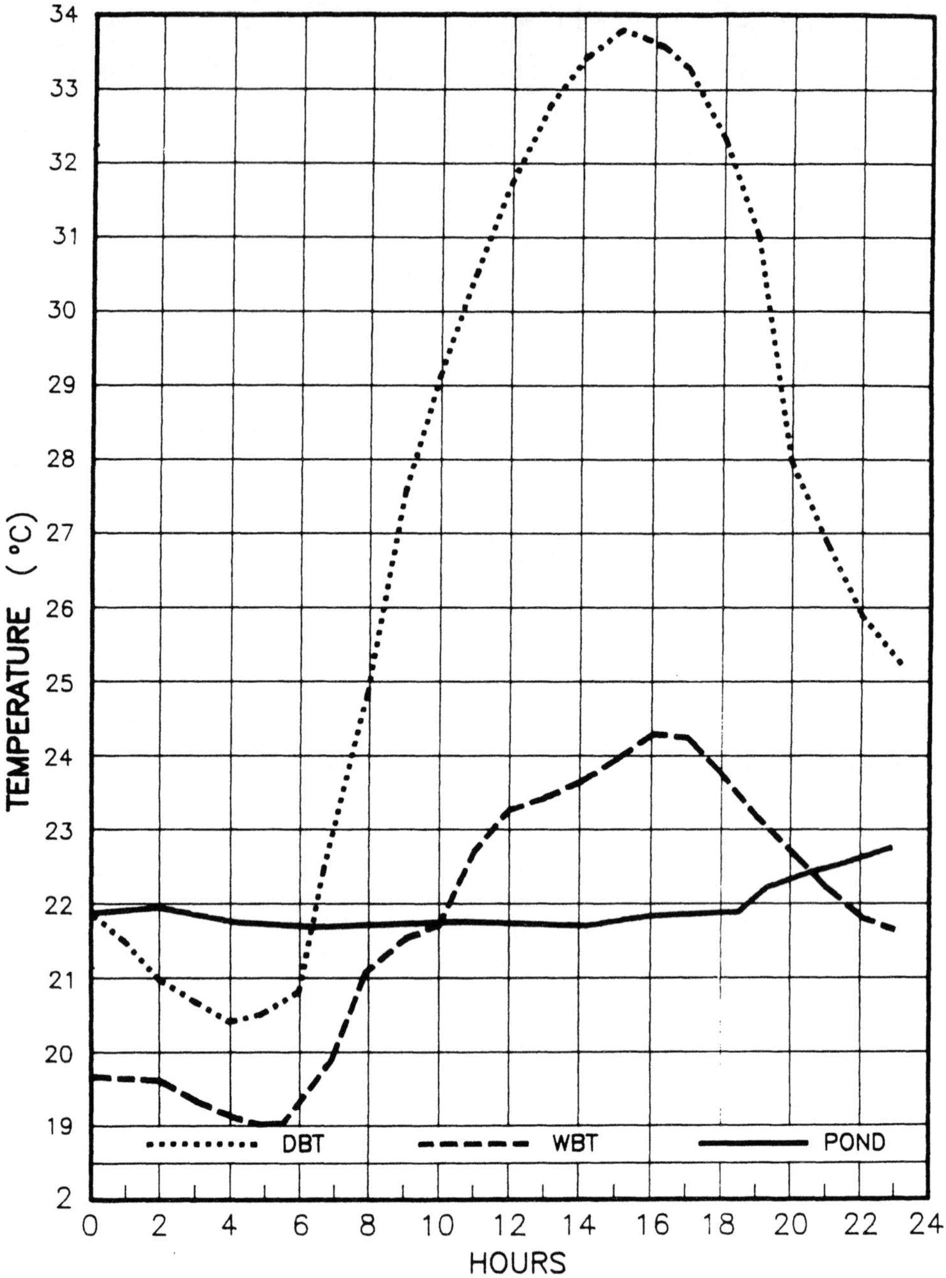

FIGURE 5.24 Temperature patterns of a deep pond with floating insulation, cooled by water circulation during the night hours over the insulation in Tallahassee, Florida.

the daytime ambient air temperature increases as the days become hotter and, in turn, the cooling potential increases with the needs.

The potential for using a deep pond with floating insulation as a cooling source has been tested, with the participation of the author, at the School of Architecture of A&M University in Tallahassee, Florida. The pond was about 3×3 meters, with a depth of 1.0 m. Water was circulated over the floating insulation during the night hours.

The diurnal swing of the water temperature was extremely small and the average followed the daily average of the WBT. Consequently it was below the daytime WBT and well below the daytime air temperature. Figure 5.24 shows diurnal temperature patterns of the DBT, the WBT, and the pond water on June 16, 1983. The water temperature is about 1°C above the average WBT, and about 11° below the maximum air temperature. Thus it seems that even in humid regions cooled water ponds can serve as a cooling source for buildings.

Thermal Coupling Between the Building and a Ground-Based Pond

The major design issue with ponds that are not located directly over the roof is that a choice of the method of "heat transfer" from the room to the water must be made. Two main modes of heat transfer are possible—air or water.

Heat Transfer by Airflow

The cool water of the pond may be utilized to cool air (either outdoor air or the indoor air in a closed circuit). The cooled air can then be blown into the building that is to be cooled.

The main advantage of cooling the indoor space by air flow is that no heat exchanger is required within the cooled space. All the heat exchange occurs at the pond. This provides more freedom for the design details of the interior space. Cooling the outdoor (ventilation) air has a special advantages in public buildings that require a high rate of ventilation (for example, schools).

Heat Transfer by Water Flow

The pond's cool water can provide direct cooling by water to air heat exchangers, such as a fan-coil system, as was done in Bourne's demonstration project in Sacramento. (See the section on floating insulation above.)

Cooled Concrete Ceiling and Walls with Embedded Water Pipes

In regions where the average WBT in midsummer is about 15 to 20°C it is possible to utilize concrete ceilings and walls, in conjunction with a ground-based pond, as cooling panels. A network of cool water pipes embedded within the concrete with distances of about 15–20 cm between the pipes can cool the concrete to a temperature about 3 to 4°C above the water temperature. Concrete roofs and external cooled walls should be well insulated from the outdoors, so that most of the heat flow to the cooled elements will be from the interior space. Of course, internal cooled walls should not be insulated. The heat transfer between the cooled surfaces and the interior space can be increased by providing ceiling fans, which increase the convective surface coefficient.

Summary of the Design Issues Posed by Cooling Ponds

Five major issues are involved in the design of ponds that are intended to provide cooling for buildings.

1. **The location of the pond.**
   Cooling ponds can be located over the roof of the building, under the building, or near the building.
2. **The mode of heat transfer between the pond and the building.**
   It can be either passive (by conduction through the ceiling) or active (by air or water flow).
3. **Mode of coolth distribution within the cooled space.**
   It can be either passive (by radiation/natural convection to the ceiling) or active (by forced cool air flow, or by specialized cooling panels).
4. **The shading details of the pond.**
   It can be a movable shade, a shading/insulating secondary roof, or floating insulation. In the case of an outdoor pond (coupled with the building) the shade can be provided by a pergola over the pond or by shade trees.
5. **Control of undesirable biological activity in the pond water.**
   In some cases, the chosen design feature is dictated by the building structure. For instance, when the roof cannot support the load of the pond a roof pond is inappropriate, and other locations should be considered. The choice to use one design option usually limits the choices available with respect to the other features of the cooling system. Thus, several combinations of solutions for the various design issues are possible with different implications for the performance characteristics of the system.

Location of Cooling Pond

Originally cooling ponds were placed only over the roof of a building. In this way one could cool the space below by direct contact, as the bottom of the pond would be in thermal contact with the building's roof, or even be the ceiling itself. As a result, the thermal coupling between the pond and the space to be cooled is maximized. The ceiling acts as a radiant panel over the space below. The heat exchange between the cool ceiling and the space below is both by longwave radiation and by natural convection. The convective mode of heat transfer is maximized when the cold area is above the space to be cooled. The resulting heat transfer coefficient, h, is about 10 watts/m^{2}C and can even be increased, by ceiling fans directing the air flow upwards, to about 15 watts/m^{2}C.

Locating the pond over the roof involves two main issues:

1. Waterproofing:
   When the pond is located over the roof it requires absolutely perfect waterproofing. Technically this problem is resolvable, but the cost usually is high.

2 Thermal Insulation:
In order to obtain passive cooling of the space by the pond, the thermal conductance between the pond and the space below should be maximized. In summer this design offers the highest cooling efficiency.

However, in most regions where summer cooling is required, the winters are cool or even cold. A pond without insulation, even if it is drained in winter, would be intolerable then. There are four design options for providing winter insulation for roof ponds.

1. Operable insulation above the pond.
2. Fixed insulation above the pond.
3. Insulation embedded within the pond.
4. Insulation floating over the water.

The problems involved with the application of each one of these design options for the winter insulation are discussed below.

Operable Insulation above the Pond

Operable insulation above the pond can be provided in two forms:

1. Horizontally movable insulation panels, such as in Harold Hay's Skytherm system. (See the chapter on radiant cooling.)
2. Hinged insulating reflective plates, put in vertical position during the night and closed during daytime. (See the chapter on radiant cooling.)

By providing the pond with operable insulation during the summer, it can be exposed to the sky during the night hours and insulated from the sun and the hotter ambient air during the daytime. Cooling is then provided simultaneously by both evaporation and longwave radiant heat loss. This combination maximizes the cooling rate of the pond.

During the daytime, the heat gain from the environment is minimized by the closed insulation. Thus, the heat absorption from the building is maximized.

From the point of view of a physicist this is the most efficient design. However, at present, all the systems with movable insulation are expensive and require extensive maintenance. As a result of the high cost and maintenance problems of operable insulation this option, in the opinion of the author, is still not ready for practical application for evaporative roof ponds.

Fixed Insulation above the Pond

This design involved building a secondary lightweight insulated roof with large insulated openings at the sides over the pond. These openings should be kept open all through the summer to allow air flow over the water. The cooling process is only by evaporation.

In winter the pond should be drained and the openings closed. The roof then will become an ordinary insulated roof. (See Figure 4.2b.) This design may be less efficient than the operable insulation in its cooling performance since it cannot utilize the radiant cooling process. The pond temperature follows the average ambient WBT. Elevations above this temperature depend mainly on the wind speed. In dry windy regions this design may yield very effective cooling even when the daytime temperature is very high, above 40°C (104°F). In regions without high wind speed the elevation of the pond temperature above the average WBT may be about 3 to 4°C (5.5 to 7°F). This design is less expensive and requires less maintenance than the operable insulation; however, the cost is still quite high.

Building Design Implications of Roof Ponds

From their impact on the design of the building, the various types of roof ponds have some common features and some differences. The main common feature, of course, is the need to have a flat roof capable of supporting a load of about 300 kg/m^2.

One of the major differences between roof ponds with movable insulation on the one hand, and roof ponds with fixed, floating, or embedded insulation on the other hand, is their impact on the architectural design of buildings. From the practical point of view, operable insulation can be applied only to buildings with strict rectangular shapes, so that the insulation panels can be moved along the roof.

Buildings with immovable roof ponds can be of almost any shape. Shading, in the form of a secondary lightweight roof, can have various forms. Spraying the water over floating insulation can be accomplished with a variety of roof shapes, and in the case of ponds with embedded insulation nothing moves at all.

SECTION 3: APPENDICES

APPENDIX 1: *Model of the Cunningham and Thompson Tower*

The Temperature of the Air Exiting from the Tower

The temperature drop of evaporatively cooled air is proportional to the WBT depression. The proportionality constant depends on the type of the pads and can be estimated experimentally. Based on the two days' experimental data from Cunningham and Thompson, the temperature of the air exiting from the tower (Texit) can be expressed by the formula:

$$\mathbf{Texit = DBT - 0.87 * (DBT - WBT);\ ^\circ C} \qquad (1)$$

Figure 5.7 shows the diurnal patterns of measured and computed exit air temperatures. It can be seen that even with this very simple formula, taking into account only the air and the wet bulb temperatures and not the wind speed, the agreement between the measured and computed temperatures is quite good. This formula would be particularly useful in predicting the temperature of the exit air in cases when wind speed data, at the local site, is not available—which may be the case in most locations.

Analysis of the experimental data has shown that the wind speed has only a minor effect and actually reduces the cooling effect of the tower, but the effect is rather small and can be disregarded. Detailed discussion of this point is presented in Givoni (1992a).

The Flow Rate and the Exit Air Speed

The flow rate (Flow) and the exit air speed (Speed) were calculated as functions of the area of the cooling pads, the tower's height, and the WBT depression:

Flow = 0.033*Aevap * sqrt(H* (DBT–WBT)); (2)

Speed = Flow / Atower; (m/s) (3)

where: **Flow = The tower's air flow rate; (m^3/s)**
Speed = Tower's exit air speed; (m/s)
Aevap = Area of the wetted pads; (m^2),
Atower = Cross area of the tower; (m^2)
H = Height of the tower; (m).

Figure 5.25 shows the computed and measured exit air speeds.

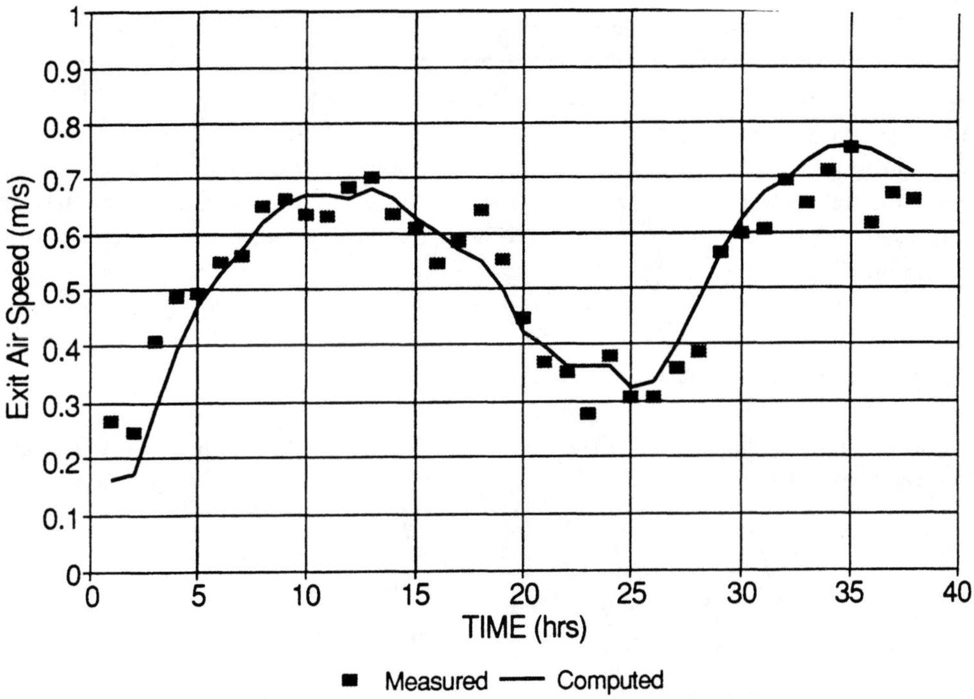

FIGURE 5.25 Computed (by Givoni's model) and measured exit air speeds of the Cunningham and Thompson cooling tower.

The "constant" (0.033) in formula 2 represents the total pressure drop through the whole system—the cooling tower and the building. For any other configuration the constant should be either calculated or determined experimentally.

Once the air flow rate through the tower is calculated, the air change rate per hour, of a building with a given volume (Vol), can be calculated by the formula:

AirChange = 3600*Flow/Vol + Infiltration; (4)

Effect of the Flow Rate on the Temperature Drop

In a passive cooling tower the flow rate changes greatly during the day. Theoretically, a higher flow rate through the wet pads may reduce the air temperature drop relative to the wet bulb depression. However, the fact that the relative temperature drop has remained about constant throughout the two days, in spite of the changes in the measured exit air speed (and the flow rate), indicates that with this type of tower the relative temperature drop is practically independent of the flow rate.

The Indoor Temperature of a Building Cooled by the Tower

Because of the high flow rate typical of direct evaporative cooling systems the indoor air temperature in a building cooled by such a cool tower should follow, with some amplitude decrement and time lag, the temperature of the air exiting from the tower.

The calculations of the expected indoor temperature are performed in two stages by the present model. In the first stage indoor temperature for a hypothetical building without any thermal mass (a "no-mass" building) is calculated. In this case the thermal response of the building to both heat flow across the envelope and the air flow from the cool tower is instantaneous. Calculations of the indoor temperature in this stage are on the basis of a steady-state heat balance. In the second stage "corrections" are applied, taking into account the time lag and the diurnal swing decrement resulting from the thermal mass of the building.

Indoor Temperature in a "No-Mass" Building

For a building without any thermal mass the instantaneous heat gain and loss can be calculated as follows:

Let: Tin = Average indoor air temperature, °C
Tout = Temperature of air flowing out of the building, °C
Ta = Outdoor air temperature, °C
Q = Internal and solar heat gain, watts
0.33 = Volumetric heat capacity of the air, Wh/C.m^3
UA = Hourly heat gain coefficient of the building, W/C.
Tout = Texit+2*(Tin–Texit), °C.

Then:

Gain=UA*(Ta–Tin)+Q
Loss=Flow*0.33*(2*(Tin–Texit))=0.67*Flow*(Tin–Texit)

Hence:

UA*(Ta–Tin)+Q = 0.67*Flow*Tin–0.67*Flow*Texit, and
Tin(nm)=(UA*To+0.67*Flow*Texit+Q)/(0.67*Flow+UA) (5)

(Hypothetical instantaneous indoor temperature for a building without any mass.)

Corrections for Diurnal Swing Decrement and Time Lag

In a building with a high rate of air change the indoor temperature is controlled mainly by the temperature of the air flowing through the building. However, during the daytime heat is absorbed in the internal surfaces of the building's mass and is released back to the interior air during the night, resulting in some time lag behind the pattern of the flowing air and a decrease in the diurnal swing.

Experimentally it is possible to estimate the indoor swing decrement (amplitude damping) and the time lag for specific types of buildings, with different mass characteristics and given rates of ventilation, by comparing a measured diurnal pattern with calculations based on steady state.

The Indoor Average Temperature

The average indoor temperature in an unconditioned building does not depend on its mass. However, even in the absence of any direct solar penetration, it is usually higher than the average outdoor air temperature due to solar radiation absorbed in the external surfaces of the building (the sol-air elevation, dT (s-a).

T^{sa}av(nm) = Tav(nm)+dT(s–a); (C)

The Instantaneous Indoor Temperature

The formula describing the indoor temperature of a building with real thermal mass, at a given time t, T_t, is:

T_t=T^{sa}av(nm)+D*((T($_t$nm–L)–T^{sa}av(nm)): (C) (6)

where:

D,L = Amplitude damping and Time-lag factors, depending on mass and insulation characteristic of the building.

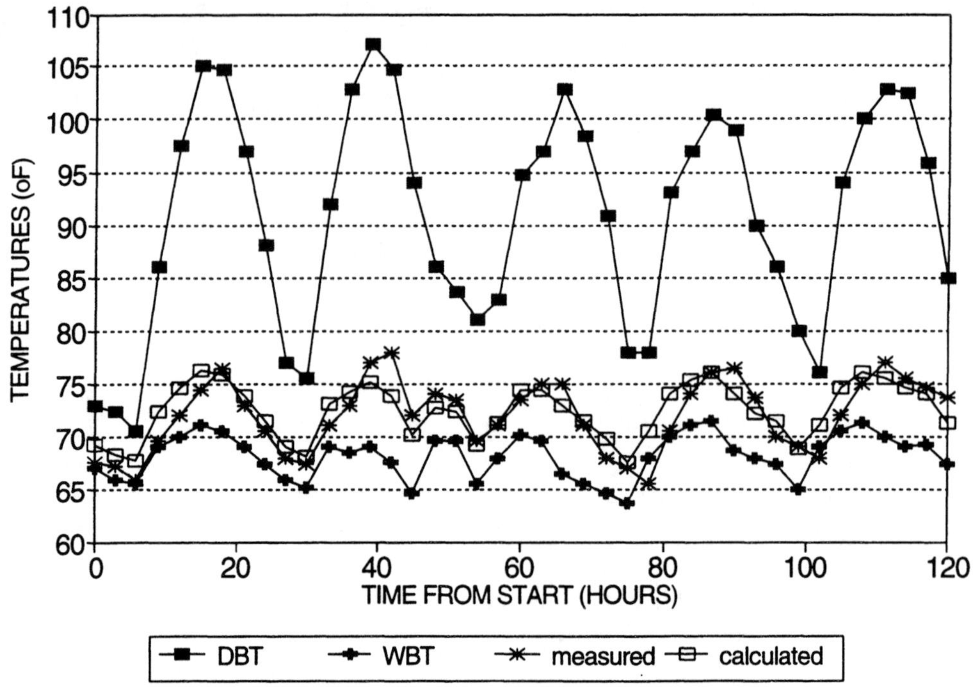

FIGURE 5.26 Comparison between computed (by Givoni's model) and measured indoor air temperatures in a building cooled by the Cunningham and Thompson cooling tower, July 1986.

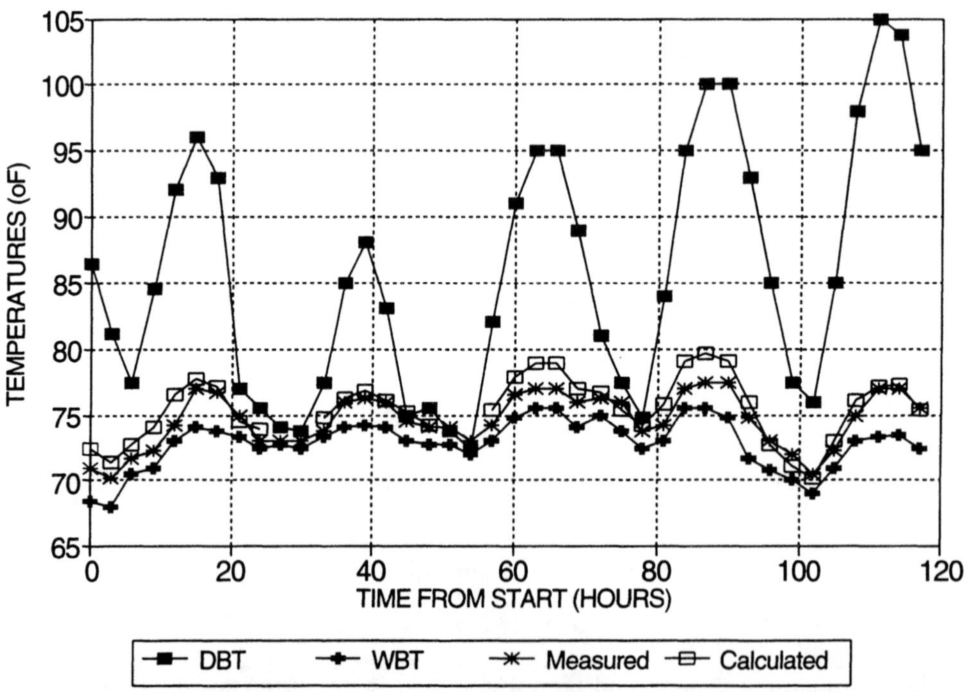

FIGURE 5.27 Comparison between computed (by Givoni's model) and measured indoor air temperatures in a building cooled by the Cunningham and Thompson cooling tower, August 1986.

Partial Validation of the Model

The model was validated by comparing calculations made from it with additional measurements taken during sequences of five to six days in June, July, and August (Cunningham et al. 1987). During these periods the climatic conditions were different from those under which the experimental model was generated. Thus, the ability of the model to take into account the effect of the ambient conditions on the indoor temperature can be checked. But the building and the tower are the same which were used for the derivation. This fact limits the scope of the validation.

A comparison between the computed and measured exit temperature and air speed in June has been shown in Figure 5.8. Additional comparisons shown in Figures 5.26 and 5.27 are for five days in July and August. It can be seen that, overall, the predicted curves are very close to the measurements.

Applicability and Limitations of the Model

The performance (exit air temperature and flow rate) of a cooling tower with pads similar to the ones applied and tested at the University of Arizona can be predicted by the model for buildings of different heat gain characteristics. This is so because the basic thermal properties of a building, the heat loss coefficient, the decrement factor, and time lag, as well as the sol-air temperature elevation, are incorporated in the model.

Considering the limited validation of the model its applicability in estimating the *indoor* temperature of a building is limited at present to relatively lightweight residential buildings, such as the stud-wall wood construction that is common in the United States.

APPENDIX 2: *Model of the Inertial "Shower" Cooling Tower*

The empirical formulae presented below describe quantitatively the effect of the climatic conditions and various design details on the performance of the "shower" cooling tower. The formulae are based on general principles of evaporative cooling and a generalization of the experimental performance data. The following formula describe the effect of the climatic conditions and system design details on the flow rate and temperature of the air exiting from the shower shaft. First the formula were derived to express the observed performance obtained with the specific detail of the prototype. Then they were generalized to allow estimation of the expected performance under any climatic conditions and system design details.

Factors Effecting the Exit Air Temperature

The temperature drop of the exit air below the ambient level, in any direct evaporative cooling system, is determined by the DBT-WBT difference (WBT depression). In the "shower" system it is also affected by the height of the shower head and the water flow rate.

Effect of WBT Depression on the Exit Temperature Drop, with "Standard" Shower Height and Water Flow Rate

Most of the tests with the "shower" cooling tower were performed with one level of shower height (2.13 m) and one level of water flow rate (14.4 1/min). As could be seen in Figure 5.10, the exit air temperature drop for these "standard" conditions was the same for sea water as that for fresh water. Therefore, the exit air temperature, given "standard" test conditions at the shower cooling tower, can be expressed, **for both fresh and sea water,** by the same experimental formula:

$$\mathbf{T_{exit} = Ta - 0.72*(Ta-WBT)} \quad (1)$$

Effect of the Shower Height and Water Flow Rate on the Exit Temperature Drop and Air Flow Rate

The height of the shower head above the water level in the pond determines the length of the contact zone between the water and the air in the shaft, as well as the final speed of the falling water drops. This, in turn, affects the cooling potential of the system—the temperature drop and the speed of the exit air.

The slopes of the regression lines, expressing the relationship between the air temperature drop and the WBT depression, have decreased as the shower height was lower: for the 2.13 m height (Figure 5.11) the slope was 0.72. For the 1.07 m. height it was 0.55. These slopes are in effect the relative temperature drop of the exit air (relative to the WBT depression).

The effects of shower height and water flow rate on the relative air temperature drop have a limit: the temperature of an evaporatively cooled air, T_{exit}, generally can not drop below the ambient wet bulb temperature[1]. With higher height and higher water flow rates the temperature of the exit air approaches asymptotically the WBT. Eventually, the upper limit of the relative temperature drop could be expressed as a function of all the system's design conditions.

With a shower height of 2.13 m at a flow rate of 14.4 1/min, the relative temperature drop was 0.72. At a flow rate of 9.61/min it was 0.55. With a shower height of 3 meters and a water flow rate of 12.2 liters/minute the relative temperature drop of the exit air was 78%.

Considering the saturation limit and the experimentally observed effects of the shower height and water flow rate, the exit air temperature (T_{exit}) can be expressed as functions of the ambient DBT and WBT, of the shower height (H) and water flow rate (WF), by the following experimental formulae:

$$\mathbf{T_{exit} = DB-(DB-WB)*(1-\exp(-.8*H))*(1-\exp(-.15*WF))} \quad (2)$$

---

[1]An exception to the WBT limit is when the pond water temperature is lower than the ambient WBT. This situation may occur when the outdoor WBT rises relatively fast. The water temperature, due to its high thermal inertia, rises more slowly. In this case the exit air temperature, which is controlled mainly by the water temperature, can also be below the ambient WBT. This phenomenon can be very helpful during the hours around the peak air and WBT temperatures. This factor is not yet included in the model.

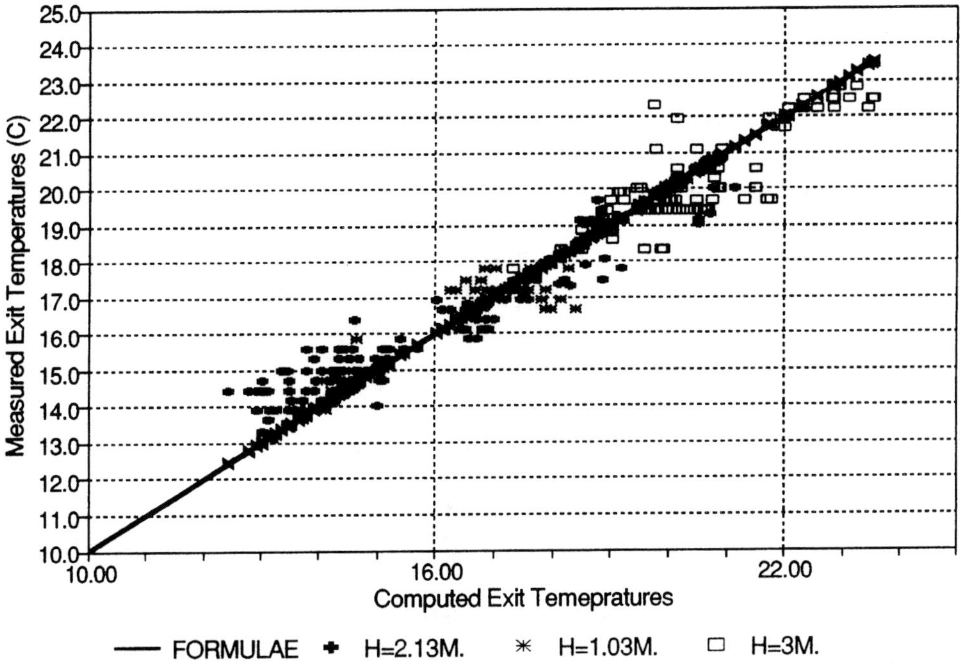

FIGURE 5.28 Comparison between computed and measured air temperature exiting from the "shower" cooling tower.

This formula is used in calculating the indoor air temperatures of a building or a patio cooled by this system. (See both the discussion below and Chapter 7, Cooling of Outdoor Spaces.) Eventually, other factors have to be taken into account in calculating the expected exit air temperature, especially the water drop size at a given flow rate.

Figure 5.28 shows the correlation between the computed and measured exit air temperatures. It includes all the data with fresh water and sea water, shower heights of 3, 2.13 and 1.07 meters, and water flow rates of 14.4 and 9.6 liters per minute.

Factors Affecting the Air Flow Rate

In this particular system the air flow is generated mainly by the falling water drops. The potential of the water drops to generate air flow depends, in turn, on the height of the shower head and the water flow rate.

The experimental air flow rate can be derived directly from the measured average exit speed (in m^3/hr):

Flow = 3600*0.26*Speed = 936*Speed

(Note that the exit opening area was 0.26 m^2.) Assuming that the air flow is directly proportional to the water flow rate (WF) and to the square root of the shower height (H), it can be calculated by the formula:

Flow = 45*WF*H^0.5; m^3/hr, or
Flow = 0.0125*WF*H^0.5; m^3/sec (3)

Effect of the Wind Speed on the Exit Air Speed

When a windcatcher is placed above the shaft it is expected that the air flow, as measured by the exit air speed, will increase as a function of the ambient wind, as is demonstrated in Figure 5.29.

With ambient winds below about 1 m/s the wind speed had no apparent effect on the air flow, and the exit speed was almost constant at about 1 m/s. This indicates that the air flow was generated mainly by the falling water drops. When the wind speed rose above 1 m/s, some effect on the exit speed is suggested: an increase of the wind speed from 1 to 2.5 m/s caused, on the average, an increase in the exit air speed from about 1 to about 1.7 m/s.

The observed wind effect, when its speed (WS) is above 1 m/s, can thus be expressed by the formula:

$$\textbf{Speed} = \mathbf{0.7 * h^{0.5} + 0.47 * (WS-1)}$$

The line representing this formula is also marked in Figure 5.29.

Taking into account the actual size of the air channel (0.26 m^2), the wind speed (WS) effect can be included in the prediction formula of the air flow rate:

Flow = 45*WF*H^0.5 + 440*(WS–1) ; m^3/hr

The wind coefficients in the above formulae should be considered only as an estimation until more experimental data is available.

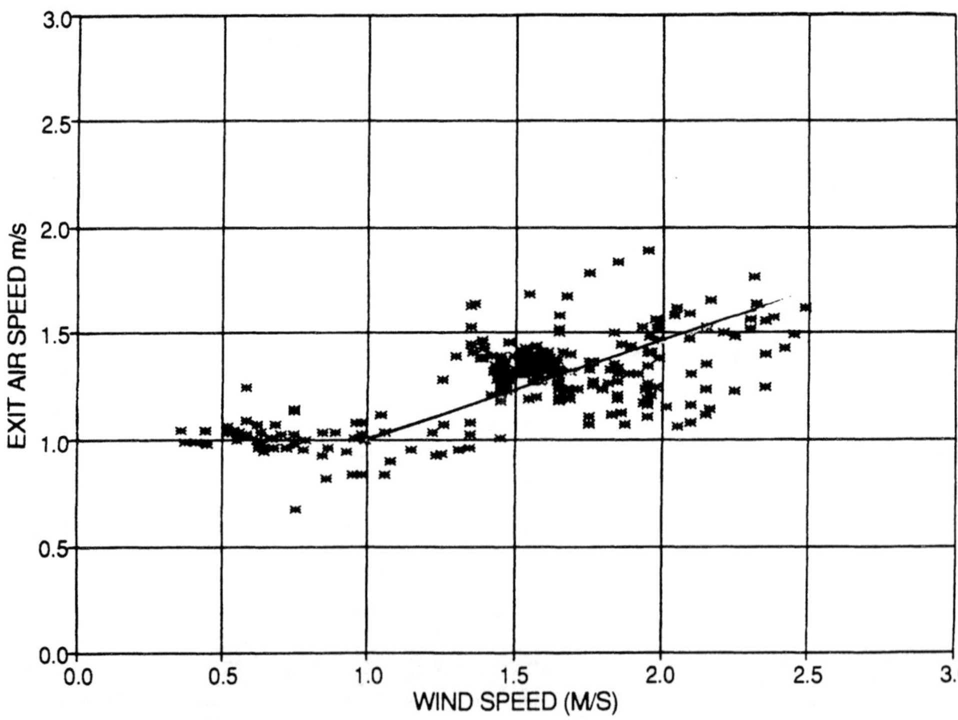

FIGURE 5.29 Effect of ambient wind speed on the exit air speed in the "Shower" cooling tower.

Once the air flow rate through the shower cooling tower is calculated, the air change rate per hour and indoor air temperatures of a building with a given volume (Vol), cooled by the shower cooling tower, can be calculated by the same relevant formulae developed for the Arizona cooling tower:

$$\textbf{AirChange/Hr = Flow/Vol + Infiltration} \tag{4}$$

The Indoor Temperature of a Building Cooled by the Tower

As in the case of the Arizona Tower, the calculations of the expected indoor temperature are performed in two stages. The reasoning behind these predicting formulae is discussed in Appendix 1.

Indoor Temperature in a "No-Mass" Building

For a building without any thermal mass the instantaneous heat gain and loss can be calculated as follows:

Let: T_{in} = **Average indoor air temperature, °C**
T_{out} = **Temperature of air flowing out of the building, °C**
T_a = **Outdoor air temperature, °C**
Q = **Internal and solar heat gain, watts**
0.333 = Volumetric heat capacity of the air, Whr/C.m^3
UA = Hourly heat gain coefficient of the building, W/C.
$T_{out} = T_{exit}+2*(T_{in}-T_{exit})$, °C.

Then:

$$Gain=UA*(T_a-T_{in})+Q$$
$$Loss=Flow*0.333*(2*(T_{in}-T_{exit}))=0.67*Flow*(T_{in}-T_{exit})$$

Hence:

$$UA*(T_a-T_{in})+Q= 0.67*Flow*T_{in}-0.67*Flow*T_{exit}, \text{ and}$$
$$T_{in}(nm)=UA*T_a+0.67*Flow*T_{exit}+Q)/(0.67*Flow+UA) \tag{5}$$

(Instantaneous indoor temperature for a building without any significant mass).

Testing the Formula for the Indoor Temperature

Figures 5.30 through 5.33 show the measured and computed indoor temperatures of the very lightweight test cell of a room described on page 150. The measurements were taken on October 2, 7, 8, and 11 of 1992. By and large there is a good correspondence between the measured and computed temperatures.

The following table summarizes the measured and computed average temperatures during the experiments for the four days shown in Figures 5.30 through 5.33. They also

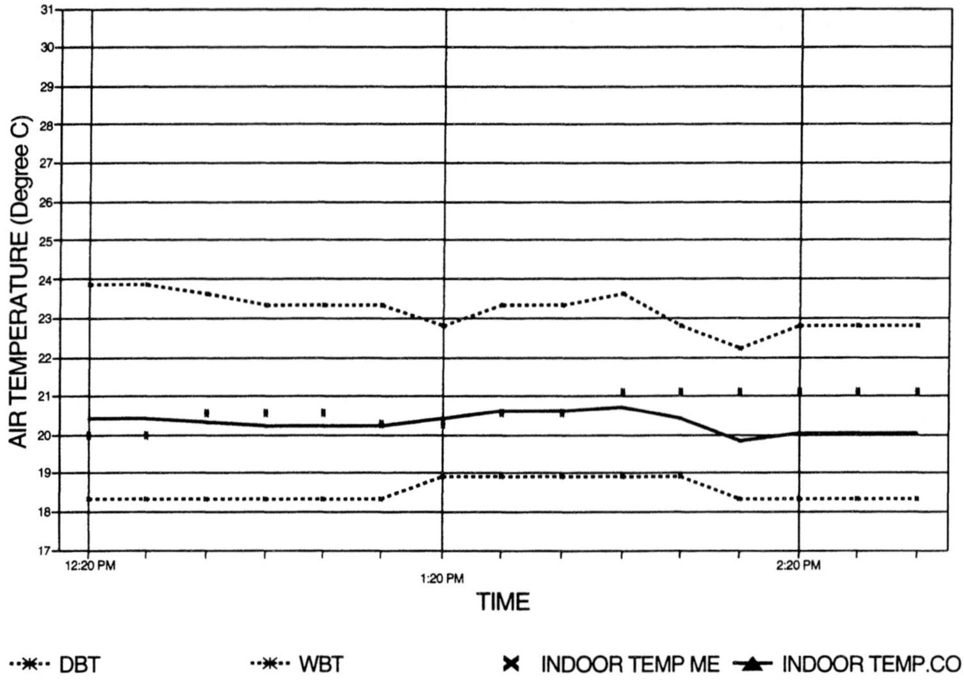

FIGURE 5.30 Performance of the "Shower" cooling tower, October 2, 1992.

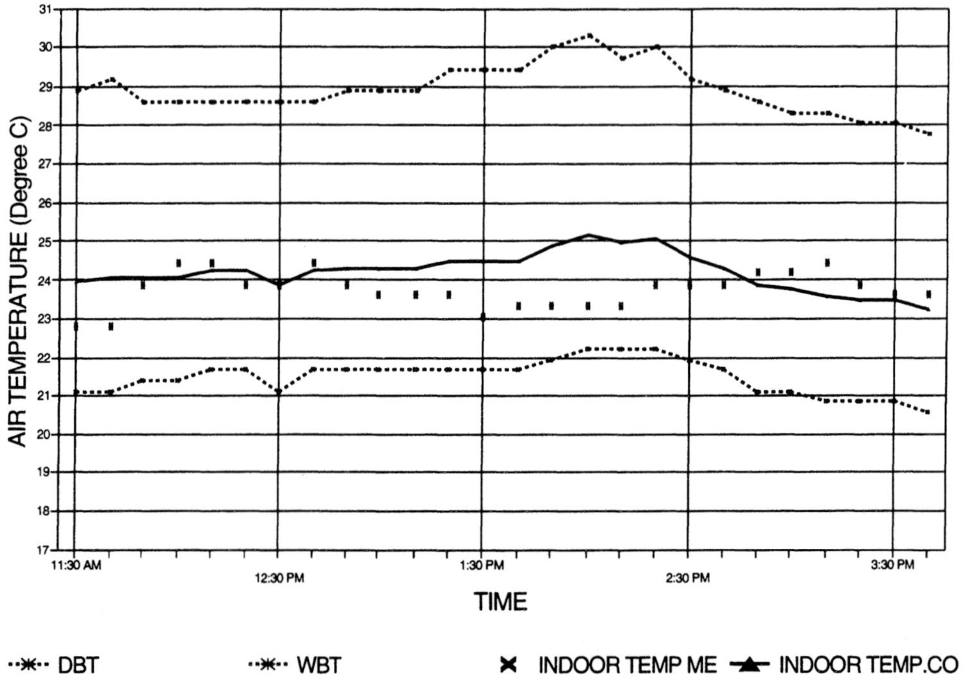

FIGURE 5.31 Performance of the "Shower" cooling tower, October 7, 1992.

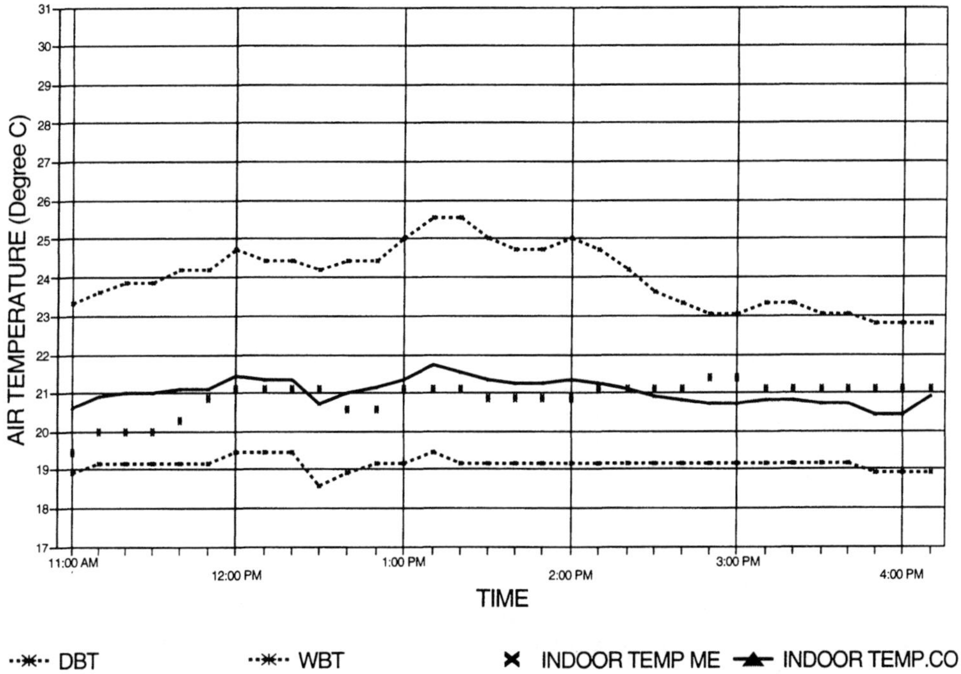

FIGURE 5.32 Performance of the "Shower" cooling tower, October 8, 1992.

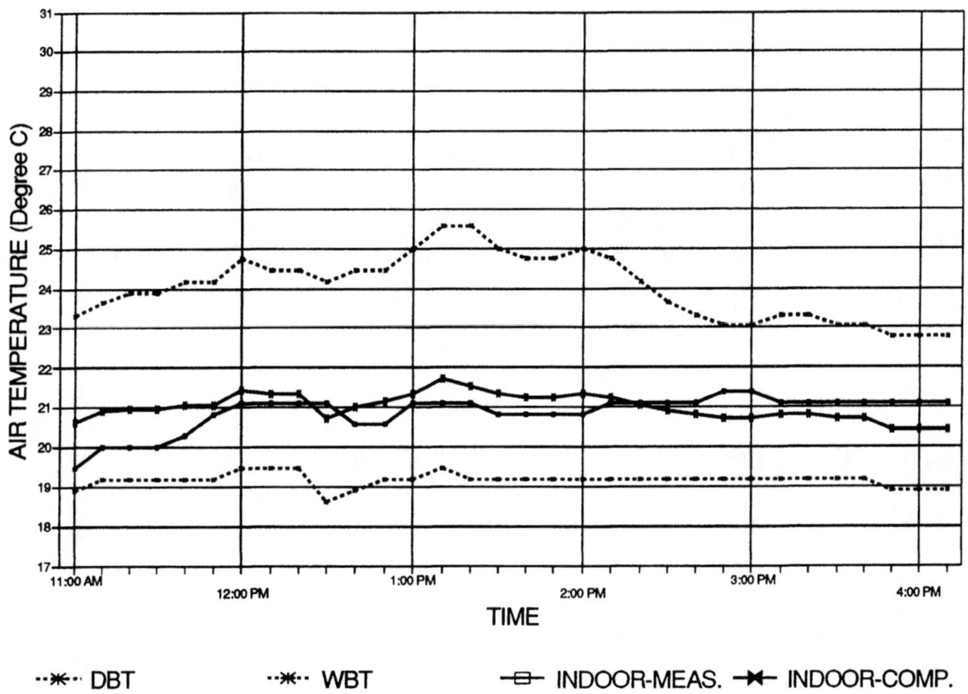

FIGURE 5.33 Performance of the "Shower" cooling tower, October 11, 1992.

show the temperature drop of the indoor temperature relative to the Wet Bulb Depression. The measured average relative indoor temperature drop (dT) was about 60% of the WB Depression (dWBT).

Measured and Computed Temperatures in a Cooled Close Test Cell

Date	DBT	WBT	Tin-Meas.	Tin-Comp.	(dT)/(dWBT)
Oct. 2	23.2	18.5	20.7	20.3	0.53
Oct. 7	28.9	21.5	23.7	24.2	0.70
Oct. 8	26.6	19.0	21.7	21.7	0.64
Oct. 11	24.0	19.1	20.9	21.0	0.63
Averages	25.7	19.5	21.7	21.8	0.62

Summary

The shower cooling tower, although developed originally to cool such outdoor spaces as patios and rest areas, can also be used to cool buildings. In such applications the shaft and the pond under it can be located either attached to the space to be cooled or within that space. In view of the very limited experimental data, the model should be considered only as a "model of a model." It is not possible at this stage to develop it further until more experimental data become available. Still, it allows an estimation of the effect of various design features and climatic conditions on the expected performance.

Corrections for Diurnal Swing Decrement and Time Lag

In a building with a high rate of air change the indoor temperature is controlled mainly by the temperature of the air flowing through the building. However, during the daytime heat is absorbed in the internal surfaces of the building's mass. That heat is released back to the interior air during the night, resulting in some time lag behind the pattern of the flowing air and a decrease in the diurnal swing. The same procedure applied in the case of the Arizona cooling tower (Appendix 1) can be used also with this system.

APPENDIX 3: *Model of Buildings with Ventilated Roof Ponds*

Mathematical modelling of the temperature of a shaded pond (without a building) was developed by Givoni (1983). This section presents an extension of the model, predicting also the performance of a building cooled by a ventilated roof pond (including the pond, ceiling, and indoor temperatures). The modelling is based on a detailed analysis of the various components of the heat balance of the building and the pond. The predictions by the model are compared with data measured in two types of test buildings, one high mass and the other low mass (Givoni 1980, 1992c).

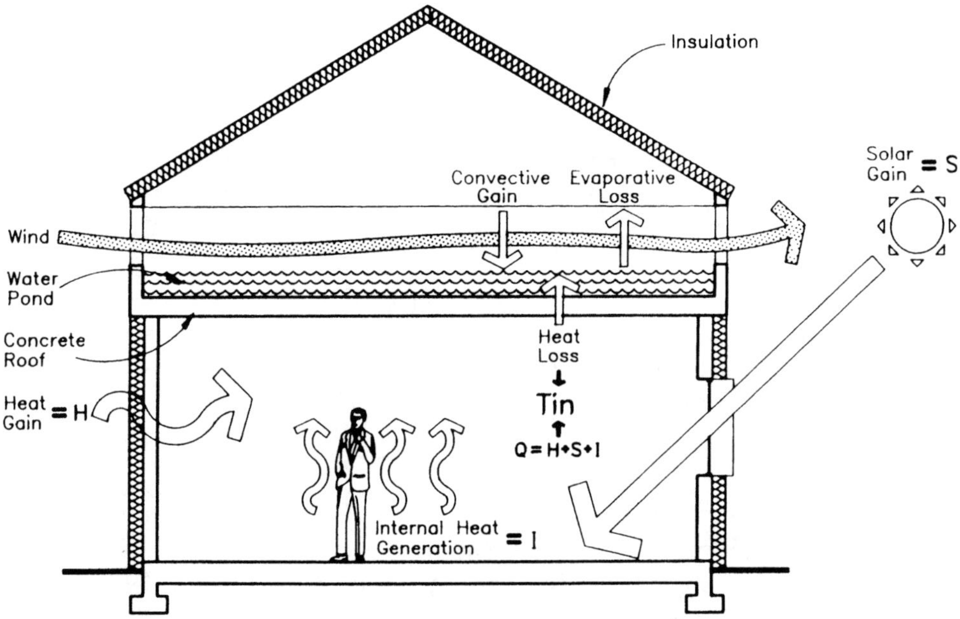

FIGURE 5.34 Heat balance of a building cooled by a roof pond.

The Heat Balance of a Building Cooled by a Roof Pond

A prediction of the indoor average temperature of an unventilated building cooled by a water pond can be derived from the daily average heat balance of the building. (See Figure 5.34.)

On a daily balance, the total heat gain through the envelope, plus internal heat generation and solar gain, is lost by conduction to the pond. The average indoor temperature of an unventilated building that is neither heated nor cooled is equal to the weighted average sol-air temperature (Tsa). The average hourly heat flow into the building through its envelope, both through the opaque elements and the shaded windows, Q_{in}, is proportional to the temperature difference (T_{sa} – T_{in}) and to the daily total Building Loss Coefficient (BLC) of the building.

$$\mathbf{Q_{in} = (BLC/24)*(T_{sa}-T_{in});} \qquad (1)$$

Q_{in} is thus the average cooling load, in the absence of solar penetration and internal heat generation. In calculating T_{sa} the roof under the pond is not taken into account, as the heat flow through it is from the indoors outward.

In addition to the conductive heat transfer, the total amount of penetrating solar radiation, and the internal heat generation, the internal heat gain (IHG) has to be dissipated through the pond, so the total cooling load, Q, is:

$$\mathbf{Q = Q_{in} + IHG/24;} \qquad (2)$$

As the shaded pond temperature is lower than the indoor average temperature (unless the building is ventilated during the late hours of the night) heat flows continuously from the interior through the roof into the pond.

The daily heat transfer to the pond (Q_{out}) is determined by the area of the ceiling under the pond (A_p), the conductance of the roof/ceiling (U_R) and the temperature difference between the average indoor and ponds' water temperature ($T_{in} - T_p$):

$$Q_{out} = A_p * U_R * (T_{in} - T_p) \tag{3}$$

(U_R is an hourly rate while BLC is a diurnal rate).

The ceiling/roof conductance, U_R, depends on the thermal resistance of the roof itself (r_R), the resistance of the waterproofing layer (rwp), and the surface coefficient of the ceiling, ($h_{ceiling}$).

$$U_R = 1/(r_R + 1/h_{ceiling} + rwp); \tag{4}$$

The resistance of the roof, r_R, is its thickness, d, divided by the conductivity of its material, k (r_R=d/k). The surface coefficient of a cooled ceiling under still indoor air conditions, $h_{ceiling}$, is about 10 (w/m^{2}C). It can be increased to about 15 by ceiling fans circulating the air upward.

The Indoor Temperatures

The average indoor temperature can be extracted from the heat gain and loss through the pond of the cooled building, Q_{in} and Q_{out}.

$$Q = Q_{in} = Q_{out};$$

Hence, the average indoor temperature is:

$$T_{in} = (24A_pU_RT_P + BLC * T_{sa} + (IHG))/(24.A_PU_R + BLC) \tag{5}$$

Thus, the pond temperature is needed to calculate the indoor temperature. This calculation is discussed later on.

Indoor Temperature Swing, Maximum and Minimum

The indoor temperature swing of a building cooled by a roof pond is smaller than that of a similar building without a roof pond. This relative stability results from the relatively stable temperature of the pond and its large impact on the indoor temperature. The swing of such buildings can be estimated in relation to the building's specific mass characteristic, and it can be expressed as a fraction, MASS, of the outdoor air temperature range:

$$dT_{in} = (MASS) * (T_{a(max)} - T_{a(min)}) \tag{6}$$

Taking into account the high mass of the roof/pond combination, the value of MASS for a well insulated high mass building (for example, one with concrete walls and floor) can be about 0.1 to 0.15. For a medium mass building (one with walls of hollow concrete blocks or that has only a high mass floor) the value of MASS would be about 0.25, and for a low mass insulated building (except for the pond) it would be about 0.3 to 0.4. In an experimental study the value of MASS of a given building or test cell can be evaluated by comparing the indoor and the outdoor temperature swings.

The indoor maximum and minimum are then:

$$T_{in(max)} = T_{in} + 0.5*(T_{a(max)} - T_{a(min)}) \quad (7)$$

$$T_{in(min)} = T_{in} - 0.5*(T_{a(max)} - T_{a(min)}) \quad (8)$$

Time Lag and Diurnal Pattern of the Indoor Temperature

Experimental studies by the author have demonstrated that the indoor temperature patterns in buildings that are cooled by roof ponds and not ventilated during the night hours are determined mostly by the pond temperature, with little difference between high mass and low mass buildings (see Figures 5.13 and 5.14). The indoor maxima occur at about 6 p.m. (18.00). As the main interest in summer is in the maximum temperature the diurnal pattern can be constructed with the indoor maximum at that hour.

The indoor temperature diurnal cycle, in reality, does not follow a simple sine or cosine function. In particular, the temperature drop after the peak is faster in the evening than in the late hours of the night. However, for evaluating the overall performance of a building, and to greatly simplify the model, a simple cosine function has been applied in the model. The temperature patterns are calculated from the minimum in the morning of one day (T_{min1}), through its maximum, to the minimum in the following day (T_{min2}). Often the two minima are not the same. This procedure enables one to predict the temperatures of a sequence of days, with different minima and maxima. The temperature at time t, T(t) is then given by:

$$T_{(t)} = T_{in} + 0.5*(T_{in(max)} - T_{in(min)})*Cos[15*(t-18)] \quad (9)$$

T_{min1} is taken into account in the hours between the minimum and the maximum of the same day. T_{min2} is used in the hours after the maximum of a given day until the minimum in the next day.

The Pond Temperature (T_p)

The average temperature of a shaded pond equals the average ambient WBT, plus a given elevation that depends on how effectively radiant heat gain from the underside of the shade is excluded. It is only slightly affected by the heat flow into the water from the building below the pond, which usually is much smaller than the convective heat gain from the hotter ambient air.

The average pond temperature can be calculated from the energy balance of the pond. It loses heat by evaporation (E) and gains heat by convection from the ambient air (Conv), by longwave radiation from the underside of the shade (Rad), and from the building below (Q). Thus:

$$(Conv) + (Rad) + Q - E = 0$$

Convective Heat Gain

The convective heat gain (Conv) is proportional to the convective coefficient (h_c) and the temperature difference between the ambient air (T_a) and the pond's water temperature (T_p).

$$\textbf{Conv} = h_c(T_a - T_p); \ (w/m^2)$$

h_c is a function of the wind speed near the water surface. Many formulae were suggested for h_c. Some of them are complex functions of the Reynolds Number (Re) and the Prandtl Number (Pr), especially at low speeds. However, the exact wind speed near the water is in practice not known anyway, and the outdoor wind speed is seldom below 0.5 m/sec.

Consequently the following expression is suggested for estimating the convective heat gain coefficient:

$$h_c = 1 + 6\,V^{0.75}$$

when V is given in (m/sec).

The convective heat gain is then:

$$\textbf{Conv.} = (1+6V^{0.75})*(T_a - T_p); \ (w/m^2.C) \tag{10}$$

Evaporative Heat Loss

The evaporative heat loss (E) is proportional to the evaporative coefficient h_{evap} and the difference ($P_p - P_a$) between the vapor pressures of the pond water and the ambient air.

$$E = h_{evap}*(P_p - P_a) \tag{11}$$

The evaporative coefficient, h_{evap}, like the pond's convective coefficient, is a function of the wind speed near the water. It is discussed later on. The vapor pressure of the water (P_p) is the saturated vapor pressure of water at temperature (T_p). The air vapor pressure (P_a) is one of the basic meteorological characteristics of the ambient air.

Relationship between Evaporative and Convective Coefficients

Both coefficients are functions of the wind speed near the water. They are related by a constant (g), which is in fact the slope of the wet-bulb temperature lines on a psychrometric chart.Thus:

g = 2.1 C/mmHg, and therefore:

$$h_{evap} = 2.1\,h_c \tag{12}$$

If the absolute humidity is expressed in gr/kg of dry air then the slope of the WBT, g, is 2.5 C/gr and h_{evap} will become:

$$h_{evap} = 2.5\ h_c \tag{12a}$$

Relationship between Temperature and Vapor Pressure of Water

The pond's temperatures, in the climatic conditions when roof ponds may be utilized, are between 15 and 25°C. In this range the water vapor pressure at sea level can be expressed with reasonable accuracy by the formula:

$$P_w = P_0 + j(T_p - 20)$$

where P_0 is the vapor pressure of water at 20°C (17.7 mmHg) and j is the constant of linear approximation of the changes in the vapor pressure with varying temperature. In the range from 15 to 25°C the value of j is 1.2 (mmHg/C) when the absolute humidity is expressed in terms of vapor pressure (in mmHg).

Consequently the water vapor pressure (at sea level) can be expressed by:

$$P_p = 17.7 + 1.2(T_p - 20);\ (\text{mmHg}) \tag{13}$$

If the absolute humidity is expressed in terms of specific humidity (gr/Kg of dry air) then the value of j will be 0.95 (gr/Kg-C).

Estimating the Wet-Bulb Temperature from VP

From the relationship between the convective and evaporative coefficients it is possible to derive an equation for estimating the WBT, given the DBT (T_a) and the vapor pressure (P_a). The heat balance for the wet thermometer, at high wind speed which nullifies the radiation effect, yields:

$$h_{evap}(P_{WB} - P_a) = h_c(T_a - WBT)$$

Hence, inserting the expressions for h_{evap} and P_p:

$$2.1\text{x}h_c(17.7 + 1.2(T_p - 20)) - P_a) = h_c\,(T_a - WBT)$$

and, after rearranging:

$$WBT = (T_a + 2.1 * P_a + 13)/(3.5) \tag{14}$$

This formula for the WBT is valid in the WBT range of 15 to 25°C. For example: letting DBT=30°C and VP=10 mmHg yields:

$$WBT = (30 + 2.1 * 10 + 13)/3.5 = 18.3°\text{C}$$

which can be verified by comparing with the WBT value on a psychrometric chart for DBT = 30°C and P_a = 10 mmHg,

$$WBT = 18.4°\text{C}$$

Radiant Heat Gain of the Pond (Rad)

The radiant heat gain from an uninsulated shade, with underside temperature of T_s, can be estimated by the formula:

$$Rad = 5(T_s - T_p); (w/m^2.C) \quad (15)$$

When the shade is insulated its underside temperature is about the same as that of the ambient air (T_a). In this case:

$$Rad = 5 (T_a - T_p) \quad (15a)$$

Calculating the Pond Average Temperature, T_p

Incorporating the factors discussed above the pond's average temperature can be calculated as follows:

The heat flow, per unit area of the pond, from the building to the pond, Q, is:

$$Q = U_r * (T_{in} - T_p)$$

where U_r is the thermal conductance of the roof under the pond.

Balancing the total unit area heat gain by the water and the heat loss by evaporation yields:

$$h_{evap}(P_p - P_a) = h_c(T_a - T_p) + Rad + Q$$

Where:

$$Q = [(BLC)(T_{sa} - T_{in}) + IHG] / (24 * A_p); (w/m^2)$$

Incorporating the expressions for h_{evap}, P_p, and Q in the general heat balance formula of the pond yields:

$$h_{evap}(P_p - P_a) = h_c(T_a - T_p) + 5(T_a - T_p) + U_r * (T_{in} - T_p)$$

hence:

$$2.1h_c * ((17.7 + 1.2 * (T_p - 20)) - P_a)$$
$$= h_c(T_a - T_p) + 5(T_a - T_p) + U_R * (T_{in} - T_p)$$

After rearranging and dividing by h_c:

$$T_p = [(1 + 5/h_c)T_a + 2.1P_a + 13 + (U_r/h_c)T_{in}] / (3.5 + 5/h_c + U_r/h_{conv}) \quad (16)$$

Note that the water temperature in a ventilated pond is only slightly affected by the heat gain from the building. The reason is that heat input raises the temperature and water vapor pressure of the water, resulting in an increased evaporation which, in turn, reduces the temperature elevation. Therefore it is practical and convenient first to assume some elevation of the pond's average temperature above the ambient WBT (for example, about 1.5°C), to calculate an initial indoor average temperature, then to correct the pond's temperature and to repeat the procedure to calculate more exact water and indoor average temperatures. This procedure has been applied in the model's computer code.

Summary of the Operational Formulae of the Model

INPUT DATA:

Building (without taking into account the roof under the pond):

- Heat gain/loss daily coefficient, BLC; (wh/day.C) (disregarding the roof/pond area)
- Sol-air temperature elevation, dT_{sa}, (C)
- Solar and/or internal heat generation, IHG; (wh/day)
  Building "mass Constant," MASS.
  for very high mass: MASS=0.1
  for lightweight building (stud walls) : MASS=0.3

Roof/Pond:

- Ceiling/Pond area, A_p; (m^2)
- Ceiling surface coefficient, $h_{ceiling}$; ($w/m^2.C$)
  Default values: $h_{ceiling}$ = 10 for common conditions:
  $h_{ceiling}$ =15 with "reversed" ceiling fans
- Roof thickness, d; (m)
- Roof's material conductivity, k; (w/m.C)
  Default value for concrete: 1.5
- Resistance of waterproofing layer, rwp; (m^2C/w)
  Default value 0.1.

Climate:

- Daily temperature maximum, $T_{a(max)}$; (°C)
- Sequential temperature minima, $T_{a(min1)}$, $T_{a(min2)}$; (°C)
- Average air temperature, $T_a=(T_{a(min1)}+2*T_{a(max)}+T_{a(min2)})/4$
- Average vapor pressure, P_a; (mmHg)
- Wind speed, V; (m/s)

Operational Formulae:

$T_{sa} = T_a+dT_{sa}$

$dT = (T_{a(max)}-T_{a(min)})+2*dT_{sa}$

$WBT = (T_a+2.1*P_a+13)/(3.5)$; (as a check for P_a)

$h_c = 1+6*V^{0.75}$

$U_r = 1/((d/k)+1/h_{ceiling}+rwp)$

$T_p = [(1+5/h_c)T_a+2.1P_a+13+(U_r/h_c)T_{in}]/(3.5+5/h_c+U_r/h_{conv})$

$T_{in} = (24A_pU_rT_p+BLC*T_{sa}+IHG)/(24.A_pU_r+BLC)$

$Q = [BLC(T_{sa}-T_{in})+IHG]/(24*A_p)$

$T_w = [(1+5/h_c)T_a+2.1*P_a+13+(Q/h_c)]/(3.5+5/h_c)$

$T_c = T_{in}-Q/h_{ceiling}$

$T_{in\text{-}max} = T_{in}+0.5*MASS*(T_{a(max)}-T_{a(min)})$

$T_{in\text{-}min} = T_{in}-0.5*MASS*(T_{a(max)}-T_{a(min)})$

(Note: BLC = 24 × UA)

Hourly calculations:

$T_{(t)}=T_{in}+0.5*(T_{in(max)}-T_{in(min)})*Cos[15*(t-18)]$

Procedure of Calculations

The indoor temperature of a building cooled by a roof pond is greatly affected by the water temperature. However, the water temperature of a pond coupled with a building is also affected to some extent by the heat flow from the building, which depends on its temperature. The procedure adopted in programming the model to overcome this coupling is to perform the calculations in a sequence of iterations. As the pond's average temperature is, relatively, only slightly affected by the indoor temperature, it is first estimated, without heat flow from the building, by adding 1.5°C to the average WBT. With this approximation, initial indoor temperature and the heat flow into the building are calculated. Then the water temperature is calculated again, taking into account the heat flow across the ceiling. This sequence of calculation is repeated till the changes of the indoor temperature between consecutive calculations is less than 0.2°C. It was found that three rounds are enough in most cases.

Comparison of Measured Indoor Temperatures in Two Test Cells with Calculations by the Model

As mentioned above, in the section on the thermal performance of ventilated roof ponds, measurements of indoor temperatures were carried out in two thermal models (test cells) by the author at the Institute for Desert Research of Ben Gurion University in Israel. The first "cell" was large, room-size, and low mass, with insulated stud-wall type walls and wooden floor. It had internal dimensions of 304×390 cm, and height of 230 cm. The walls were insulated panels. They were built like stud walls, with 5 cm of expanded polystyrene sandwiched between layers of plywood. The roof was of 20 cm concrete, supported by four concrete columns within the indoor space, thus contributing to the thermal mass of the room. The floor was 2 cm plywood over a gravel-filled space (which was used for heat storage in solar heating experiments). An insulated door of 0.8×2.05 m was in the northern wall. The calculated Building Loss Coefficient (BLC) of this room (disregarding the roof/pond) was 650 (w/C.Day). The ponds area was 12.1 sq.m. The measured indoor/outdoor swing ratio was 0.3.

The walls of the high mass test cell were built of filled concrete walls, floors, and roofs, 20 cm thick. The walls were insulated externally by 5 cm expanded polystyrene and plastered. The internal dimensions were 206×166×202 cm. An insulated door of 0.8×2.05 m was in the northern wall. The calculated Building Loss Coefficient (BLC) of this room was (180 w/C.Day). The sol-air temperature elevation (the difference between the averages of the indoor and outdoor temperature) was measured independently in identical test cells, and was about 1°C. The measured indoor/outdoor swing ratio was 0.1.

Figure 5.35 shows the measured and calculated indoor temperatures in the low mass room and the high mass test cell during three days—the same days that were shown in Figures 5.13 and 5.14). It can be seen that the model, which was derived on the basis of a heat balance analysis and not on curve fitting, has good agreement with the measured data. It should be noted, however, that the average outdoor temperatures were similar during these three days. (See the discussion below on the model's limitations.)

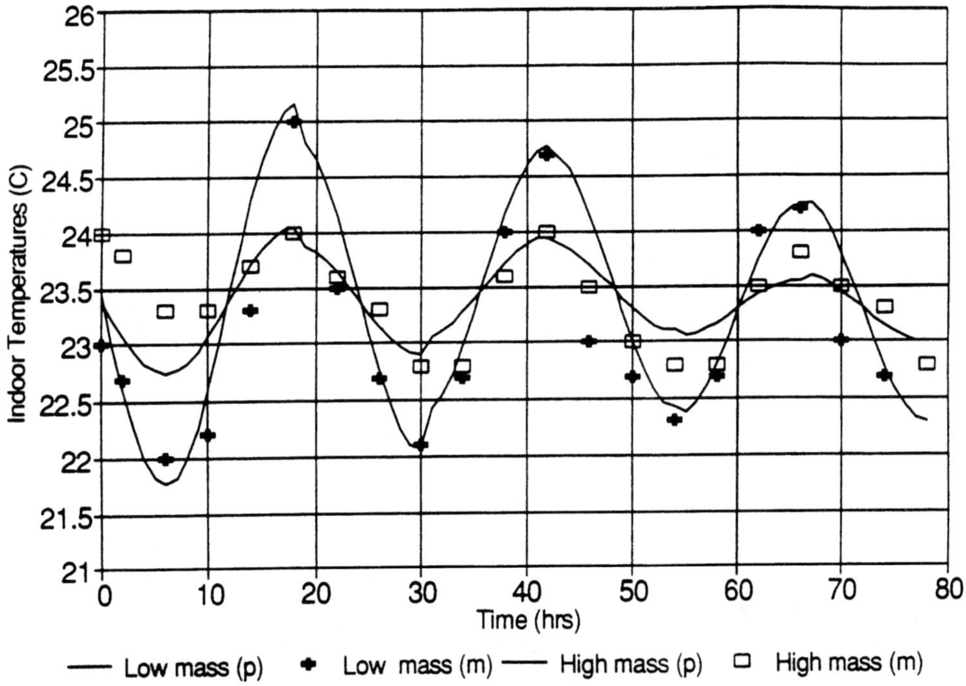

FIGURE 5.35 Comparison between computed (by Givoni's model) and measured indoor temperatures in low-mass and high-mass test models cooled by ventilated roof ponds.

Applicability and Limitations of the Mathematical Model

In its present form the model is based on the daily heat balance of a building cooled by a ventilated roof pond. It does not include the "memory" of the building's mass when the outdoor average temperature changes abruptly, as it does during a heat wave, when indoor temperatures are significantly lower than the outdoor daytime temperatures. (See Figure 5.17.) The model, in its present stage, can not take into account the inertia of a building exposed to the sharp day-to-day climatic changes common during heat waves.

Under relatively steady climatic conditions, or for "design" climates, the model is useful in evaluating the expected performance of a building cooled by a ventilated roof pond under given DBT and WBT conditions based on its BLC, mass, daily internal heat generation, and the sol-air temperature elevation (which depends on the color of the envelope).

REFERENCES

Al Hemiddi, N. 1991. "Preliminary Investigation of the Effect of a Passive Direct Evaporative Cooling System on a Courtyard at UCLA." Seville, Spain: Proc. PLEA'91.

——— 1992. "A Cooling Tower for a Passive Evaporative System in Hot Dry Climate." Cocoa Beach, Florida: ISES Congress.

Alkin, G, 1985. "Cooling Effects of Roof Ponds." Research Report, Graduate School of Architecture and Urban Planning, UCLA.

Bourne, R. C., and M. A. Hoeschele. 1992. "Cool Storage Roof, Demonstration Project Monitoring Report." Davis, CA: Davis Energy Group Inc.

Givoni, B. 1980. "Experimental Studies on Passive Cooling at the Institute for Desert Research in Israel." *Proc. Solar Energy Symposium.* Nice, France. pp. 296–300.

——— 1983. "Modelling and Validation of Shaded Roof Ponds." *4th Int. Symp. on the Use of Computers for Environmental Engineering Related to Buildings.* Tokyo, Japan. pp. 340–345.

——— 1992a. "A Semi-Empirical Model of a Building with a Passive Evaporative Cooling Tower." Accepted, *Solar Energy Journal.*

——— 1992b. "A Shower." Inertial Evaporative Cooling Tower." Submitted to *Solar Energy Journal.*

——— 1992c. "Modelling the Thermal Performance of a Building Cooled by a Roof Pond." Submitted to *Solar Energy Journal.*

Givoni, B., and Y. Morgenstern. 1984. "Performance of the Experimental Building Cooled by a Roof-pond with Pebbles Shading." Research Report to the Israeli Ministry of Energy and Infrastructure (Hebrew).

——— 1991. "Modelling A Passive Evaporative Cooling Tower." *Proceedings, Solar World Congress.* Denver, pp. 3067–3071.

Hay, H. R., and J. I. Yellott. 1969. "Natural Air Conditioning with Roof Ponds and Movable Insulation." *Trans. ASHRAE,* 75 pp. 165–177.

Peck, F. J., H. J. Kessler, and T. L. Thompson. 1979. "Monitoring, Evaluating, & Optimizing Two Stage Evaporative Cooling Techniques." Environmental Research Lab University Of Arizona, Tucson.

Sandoval, L., J. Pineda, R. Castaneda, and L. Sanchez. 1991. "Cooling Ceiling Pond in Hot-Humid Climate." Proc. ISES Solar World Congress, Denver, Colorado.

Watt, G., et al. 1986. *Evaporative Air-Conditioning Handbook.* 2nd Ed. N.Y.: Chapman and Hall.

Wu, H. 1989. "Performance Monitoring of A Two Stage Evaporative Cooler." ASHRAE Transaction, V.95, Pt.1, Chicago.

——— 1990. "Moisture Uptake and Energy Efficiencies Associated with the Use of Evaporative Coolers." Environmental Testing Lab. Arizona State University, Tempe, Arizona.

6

THE EARTH AS A COOLING SOURCE FOR BUILDINGS

The earth mass under, around, and sometimes also above a building can serve in most climatic regions as a natural cooling source for the building, either in a passive or an active way. In summer the soil temperature, at a depth of a few meters, is always below the average ambient temperature, and is especially below the daytime air temperature. It thus has the potential to serve as a heat sink with either passive or active mechanisms of heat transfer from the building. Furthermore, it is possible by very simple means to lower the earth temperature in summer well below the "natural" temperature characteristic of a given location.

Two basic approaches can be applied in principle for cooling the soil:

- **A** Surface cooling, where the surface temperature of the soil is lowered and consequently the temperature of the earth mass below the surface is also lowered, providing a large cooled mass. This cooling process takes place in the summer and the generated cold energy is concurrently utilized. The cooling method is discussed in the first part of this chapter.
- **B** Direct cooling of a large mass of the subsurface earth by increasing its cooling rate in winter relative to the heating rate in summer. In fact, this approach stores winter cold energy for summer use. This subject is discussed later in this chapter.

Shading the earth surface cools it, preventing its becoming heated by the sun while allowing evaporation. My own experiments in Israel have demonstrated that it is possible to lower the earth's average surface temperature in summer by about 8 to 10°C (14.5 to 18°F) below the temperature of exposed nearby soil. In arid regions the difference between the outdoor maximum air temperature and the cooled earth temperature, at a depth of one meter, can be up to about 15°C (27°F) in midsummer. In the hot humid climate of North Florida, soil temperature at a depth of one meter has been cooled below the ambient air maximum temperature by about 12°C.

When the soil temperature is cool enough, it is possible to utilize it as a cooling source for buildings in several ways. In the case of earth-sheltered buildings, in which the walls are bermed and the roof is covered by earth, cooling of the earth mass adjacent to the building provides a direct passive cooling of the building by conduction. In this case most of the immediate "environment" of the building is the cooled soil and not the ambient, much hotter air. Instead of a daytime heat gain the building experiences a constant heat loss to the cooler soil.

Another approach to utilize the cooled soil, an active one, is to install air pipes in the soil, and to circulate either the air from the building or the ventilation (outdoor) air through them. The cooler earth mass serves as a heat sink, cooling the air that is then introduced into the building.

This chapter provides an overview of the various aspects of soil utilization as a cooling source.

The first section discusses the temperature patterns of the "natural" soil at different depth levels in qualitative terms.

The second section discusses various surface treatments by which the soil's temperature can be lowered below the level characterizing the "natural" soil. The discussion includes a summary of various experimental studies.

The third section discusses various ways to use the cooled soil as a cooling source for buildings.

The fourth section summarizes experimental studies in which earth tubes were used to cool ventilation air.

The fifth section discusses the architectural and functional aspects of earth-sheltered buildings and their climatic applicability.

The sixth section describes ideas for cooling of the undersurface soil layers in the winter and utilizing this seasonal storage in summer.

Mathematical models of the soil temperature are presented and their limitations and applicability discussed in the appendix to this chapter.

THE UNMODIFIED SOIL TEMPERATURE

The "natural" unmodified temperature of the ground is governed by two boundary conditions: the cyclic annual pattern of the surface temperature and the constant temperature at a depth of several meters. Over large homogenous areas, the "depth" temperature equals the long-term annual average of the surface temperature. However, in any given small-scale area, the surface temperature may be affected by such local factors as color (reflectivity, albedo), ground cover, and water content. In addition, the annual pattern of precipitation (rain and snow), as well as large-scale regular irrigation, may introduce a bias in favor of the surface temperatures that prevail either in the summer or in the winter in determining the constant depth temperature.

Annual Pattern of Surface Temperature

The ground surface temperature exhibits strong diurnal and annual wave patterns, which are determined by the solar energy absorbed at the surface and the ambient air temperature. The diurnal range of the surface temperature is usually largest in the summer, especially in arid regions, when the solar radiation impinging on the surface is highest, and it is smallest in the winter, especially in cold regions with snow cover on the ground. The actual diurnal range depends mainly on the ground vegetation cover (bare soil, grass, trees), on soil moisture, and on the color (albedo) of the surface. It is also affected by the soil's thermo-physical properties, particularly its conductivity and resulting diffusivity (the ratio of conductivity to specific heat capacity). With a higher diffusivity heat is exchanged more readily between the surface and the layers below so that the surface temperature fluctuations are damped. Wet soil has higher conductivity (and diffusivity) than dry soil. Therefore, the wetter the soil, the smaller the diurnal range of the surface temperature, and the faster the propagation of heat through the ground during the day.

The diurnal pattern of the soil temperature is of interest mainly for agriculture, as it can affect plant growth and survival. However, the diurnal swing decays rapidly with depth. Even at a depth of about one meter the daily temperature is almost constant, usually with a swing of less than one degree Centigrade. When the soil is considered as a cooling source and/or as a cold storage mass, only the pattern of the monthly swing about the annual average is of interest.

The annual average of the unmodified soil temperature in a given area, below a depth of about one meter, is almost constant and independent of depth. At the surface, the average temperature of the earth is determined by the local climate pattern, which may vary from year to year, and by the balance between several heat transfer modes. Each one of these modes can be modified by some "treatments." Therefore, by appropriate treatments the soil temperature can be either elevated or lowered above or below the natural temperature of the soil in a specific site. The change in the surface temperature also affects the temperature at deeper layers. When dealing with the use of the soil as a cooling source the interest is, of course, in lowering its temperature. Therefore, the factors that determine the soil's temperature, along with hints about treatments that can modify them to enhance soil cooling, are discussed below.

Radiation Balance at the Soil Surface

The amount of solar radiation impinging on the ground is maximum during summer and minimum in winter. Absorption of the solar radiation depends on the ground cover (vegetation) and color. Shading the soil minimizes, or even completely eliminates, the heating effect of the sun. The long-wave radiant loss depends on soil surface temperature. Therefore it is also maximum in summer and minimum in winter.

The net radiant balance between solar gain and long-wave loss is usually positive (a gain) in summer and negative (a loss) in winter. This causes heat to flow down from the surface into the soil in the summer and upward to the surface during the winter. The net radiant balance also determines the relationship between the averages of the earth surface and the ambient air temperatures. In hot regions, especially the arid areas that are most often found at latitudes below 35 degrees, the annual radiant balance is positive. The summer net radiant gain is greater than the winter net loss. Therefore shading of the soil has much greater effect on the soil temperature in summer than in winter.

Permanent shading of the soil reduces or even eliminates both the solar and the long-wave radiant exchanges at the soil's surface. The effect is to lower the soil temperature. By shading the soil in summer while partially exposing it to the sky in winter—for example, with deciduous trees—it is possible to lower the temperature to a greater extent.

Convective Exchange with the Ambient Air

Large air masses may obtain their temperatures at places far away from a given site. Therefore the air over the site may have a different temperature from the soil's surface. Usually, however, an exposed earth surface loses heat by convection in summer when the average air temperature is lower than its own, and gains heat by convection in winter when the average temperature is higher than the earth's surface temperature. The convective exchange is, of course, proportional to the convective coefficient, which in turn depends on the wind speed next to the surface. This is a factor that can be modified by treatments of the soil surface.

In the case of shaded soil, when the radiant exchange is eliminated, and especially if moisture is available and evaporates from the surface, the soil surface temperature is likely to be below the ambient air level even in midsummer. The average convective heat flow in summer is therefore from the air to the soil. In this case it is desirable to minimize the convection coefficient by appropriate treatments of the soil surface. Such treatments may include covering the soil with a vapor-permeable layer that introduces thermal resistance between the soil's surface and the hotter ambient air. For example, when the soil is covered by some type of permeable mulch, such as wood chips or pebbles, the convective exchange between the soil surface and the ambient air is greatly reduced. Whenever the soil surface temperature is lower than the air level a reduction of the convective exchange is desirable if the soil is to provide cooling because it lowers further the soil's temperature.

Evaporative Heat Loss from the Soil

Heat loss by water evaporation from the surface of the soil is determined by the availability of moisture in the soil, by the evaporation coefficient, and by the difference in the vapor pressure between the moist soil surface and the air above the soil. In most cases the evaporation coefficient is proportional to the convective coefficient—both are increasing with the air speed in a similar way. Lower wind speed reduces the convective exchange as well as the evaporative heat loss. However, with some soil treatments it is possible to reduce the convective coefficient much more than the evaporative one.

Whenever moisture in the soil is plentiful, as is the case in hot humid regions where the summer is the rainy season, the evaporation rate is roughly proportional to the amount of solar radiation reaching the earth from the surface. Shading the soil thus greatly reduces the evaporation rate.

In modelling the annual soil temperature two approximation assumptions are often made: (1) that the monthly ground surface temperature is the same as the ambient air average, and (2) that the annual pattern can be represented by a sine wave about the *annual average* of the ambient air temperature. In fact, in summer the ground surface daily average temperature in many hot regions is higher than the average of the ambient air, while in winter it is somewhat below it, as is discussed below.

The fundamental sine wave approximation may not be very accurate for describing the surface monthly average temperature but it becomes quite useful at depth levels of one meter and below. Therefore, for applications concerning utilization of the earth as a source for cooling it seems to be sufficiently accurate.

Relationship between Soil Surface and Ambient Air Annual Ranges

Often information is available about the annual patterns of the ambient air monthly average temperatures, but such information about the soil surface temperature is lacking. As the data about the soil's annual average surface temperature and the annual range between the average of the hottest month and the coldest month is needed for predicting soil temperature patterns at different depths, the known data on the ambient air temperature is often substituted without modifications for the unknown data on the soil surface. In such cases it would be useful to know the actual relationship between the ambient air and the soil's surface annual temperature patterns, so that the soil temperature could be estimated more accurately on the basis of the air temperature data. This relationship is not unique but depends on the climate and on the natural soil cover.

In arid and desert regions the natural soil is usually bare except in landscaped areas. In such regions the average of the soil surface temperature in summer is about 5°C (9°F) above the average air temperature. Figure 6.2 on page 201 shows measured average temperatures of bare soil in the arid region of Israel's Negev from July 27 through August 31, 1979, together with the maxima, minima, and averages of the ambient air temperature. The soil temperatures were measured at depth levels of 10, 30, and 60 cm.

The average ambient air temperatures during this period were Maximum, 30.3°C; Minimum, 21.6°C; and Mean, 25.7°C. The *average* bare soil temperatures at the different depth levels were at 10 cm, 30.9°C; at 30 cm, 29.1°C; and at 60 cm, 28.1°C. By extrapolation the *average* surface temperature can be estimated to be about 32°C, which is above the ambient maximum and about 6.5°C above the ambient air average mean temperature.

This elevation results from the fact that during soil temperature peaks there is almost no evaporation loss from the dry soil. The absorbed solar radiation in summer is more than the combined long wave radiant loss and the heat flow downwards into the ground, so there is a net convective loss to the ambient air.

In winter in desert regions with bare soil the average soil surface temperature is closer to the average air temperature, although it still may be slightly higher because the long wave radiant loss almost equals the combined solar gain and the upwards heat flow from the deep soil to the surface. As a result, the annual range of the soil surface temperature in deserts is larger, by about 6°C, than the air temperature range.

In humid regions, where the soil is usually moist and shaded by vegetation, the average soil temperature is about the same as the average air temperature during the other months of the year. In such regions the annual pattern of the average air temperature can be used as a close approximation of the earth's average surface temperature.

During the winter in cold high latitude regions, the average ambient air is significantly colder than the ground surface, which is covered by an insulating snow layer. On the other hand, in the summer the average surface temperature closely follows the ambient air. Thus, in measurements taken in St. Paul, Minnesota, cited by Bligh (1975), the average air and surface temperatures in January were –11°C and –3°C (12.2 and 26.6°F) respectively, while in July the corresponding temperatures were 22°C and 23°C (71.6 and 73.4°F).

Subsurface Earth Temperature

In almost all cases, at a depth of about 30 cm the diurnal range of the earth temperature wave is damped to less than about 2°C (3.5°F). Heat or cold waves that last a few days are damped to a small range at a depth of about 60 cm. At depths of more than about 60 cm only the annual temperature pattern is of significance, and the average monthly temperature alone could be considered in estimating the underground temperature. A discussion of the mathematical models describing the annual pattern of the surface and subsurface earth temperatures is presented in the appendix to this chapter.

Effect of Precipitation and Irrigation Patterns on Subsurface Temperature

Heat flow downward within the earth mass may take place not only by conduction, but also by what may be termed convective heat transfer—rain and irrigation water permeating down through the ground. The effect of rains and irrigation on the subsurface temperature, whether to elevate or to depress it, depends on the season in which the watering occurs. Winter rains and irrigation lower the temperature of the deeper layer while summer daytime rains and irrigation may raise it. Furthermore, the earth's thermal conductivity increases greatly when it is wet, proportionately increasing the heat transfer through it. The enhanced conductive and convective modes of heat transfer increase the relative weight of the temperature prevailing during the rainy or irrigation seasons in determining the stabilized level of the subsurface temperature.

Thus, some bias exists in favor of the rainy and irrigation seasons in a given region in affecting deviations of the subsurface temperature from the exact annual average of the earth surface temperature. Daytime irrigation in summer instantly lowers the surface temperature while elevating the temperature of the lower layers to which the water permeates. On the other hand, irrigation during the night and early morning hours lowers the temperature of the soil down below that of the surface. Thus, programmed timing of the irrigation can be used either to lower or to elevate the subsurface temperatures above and around an earth-integrated building, according to the comfort requirements. It should also be taken into account that the temperature at a depth of several meters is determined by the long-term (several-years-long) pattern of the surface temperature, while surface pattern and average can vary from one year to another.

COOLING THE SOIL IN SUMMER BY SURFACE TREATMENTS

In regions where the natural summertime soil temperature at a depth of 1 to 3 meters is above about 22°C, the earth is not effective as a source of cooling unless it is cooled by some treatment. An exception is when the soil is utilized for a partial cooling of the ventilation air—a precooling—when the ambient temperature is very high (for example, about 35 to 40°C). In this case it can still be utilized with temperatures up to about 26°C.

My experiments (which are described later) have demonstrated that it is possible, using very simple methods, to lower the soil temperature below the level that characterizes the region.

In hot regions with dry summers, shading the soil alone will not lower its temperature enough to allow it to be used as a cooling source. A combination of shading and watering, which can provide evaporation from the soil's surface, is needed in such cases. When water evaporates from the surface of a shaded soil no evaporative cooling is wasted to overcome the heating effect of solar radiation and more energy is extracted from the soil, lowering its temperature. In this way the soil surface temperature can be lowered below the ambient air average temperature by a combination of the following treatments:

- **A** Shading the soil.
- **B** Introducing or increasing heat loss by water evaporation *from the surface.*
- **C** Reducing the convective heat gain from the hotter ambient air.

The following three soil cooling methods have been tried successfully:

1. Shading by vegetation in a way that does not reduce the evaporation from the soil surface—that is, enough water must be used to allow unrestricted evaporation both from the soil surface and from the leaves of the plants, and soil surface conditions must permit evaporation from the surface.
2. Covering the soil with a wood or gravel mulching layer at least 10 cm thick and, in regions with dry summers, irrigating it. Solar radiation is intercepted and absorbed within about 3 to 5 cm at the top of that layer, and the conduction of the heat down the layer of pebbles, and especially down a bark mulch, is rather low.
3. Utilizing soil under a building which is raised off the ground, as it is when there is a crawl space, and permitting evaporation from the surface of water, either from irrigation or from summer rains.

Comments on Shading by Vegetation

Shading by some vegetation—grass or shrubs—also reduces the rate of evaporative heat loss from the soil surface. The effectiveness of plants in cooling the soil may be reduced by the following factors:

- **A** Moisture is taken from the soil by the roots and most of the evaporation takes place by evapotranspiration from the leaves; thus it cools the leaves and is not effective in cooling the soil below the ambient air temperature.
- **B** Dense vegetation, such as thick lawn and dense low shrubs, maintains a layer of near-saturated air above the soil, thereby minimizing both the driving force for evaporation from the soil surface itself and its cooling effect.

Once the soil surface temperature in summer is lowered the temperatures of the layers below the surface are also lowered. Consequently it is possible to consider using the soil, even at relatively small depths, as a heat sink.

Kusuda's Experiments with Different Surface Treatments

In Washington, D.C., over the course of a year, Kusuda (1975) measured the soil temperatures under patches with five different surface conditions, down to a depth of 9 meters. The patches were squares, 15×15 meters each. The surface cover conditions were:

1 Black asphalt, to a depth of 12.5 cm.
2 Asphalt, painted white.
3 Bare soil surface
4 Grass, maintained at 10 cm height.
5 Grass, left unmowed.

Table 6-1 derived by the author from the graphs presented in Kusuda's paper, shows the maximum temperatures measured at different depth levels in the patches with the different surface treatments.

Kusuda's experiments have demonstrated that the maximum surface temperature of soil covered by long grass was lowered about 4°C (7.2°F) as compared with the bare soil. When the long grass-covered soil is compared to an impermeable black asphalt cover, the total effect of partial shading by grass and evaporation from the soil (usually wet during the summer in Washington) was to lower the average surface temperature in midsummer by about 11.5°C (21°F). At a depth of 30 cm the corresponding reductions in soil temperature were still similar to those at the surface.

TABLE 6-1. MAXIMUM AND MINIMUM SOIL TEMPERATURES (F) WITH DIFFERENT SURFACE TREATMENTS (AFTER KUSUDA).

		Depth Level (meters) of Soil				
Treatment		*0*	*0.3*	*1.2*	*3.0*	*9.0*
Black Asphalt	Max	91	87	79	68	56
	Min	29	33	37	48	52
	Mean	60	60	58	58	54
White Asphalt	Max	77	74	67	60	55
	Min	27	28	37	45	51
	Mean	52	51	52	52.5	53
Bare Soil	Max	77	74	72	63	56
	Min	28	32	34	43	51
	Mean	52.5	53	53	53	53
Short Grass	Max	75	69	64	61	56
	Min	31	35	39	47	52
	Mean	53	52	51.5	54	54
Long Grass	Max	70	65	62	61	54
	Min	28	34	38	46	51
	Mean	49	49.5	50	53.5	52.5

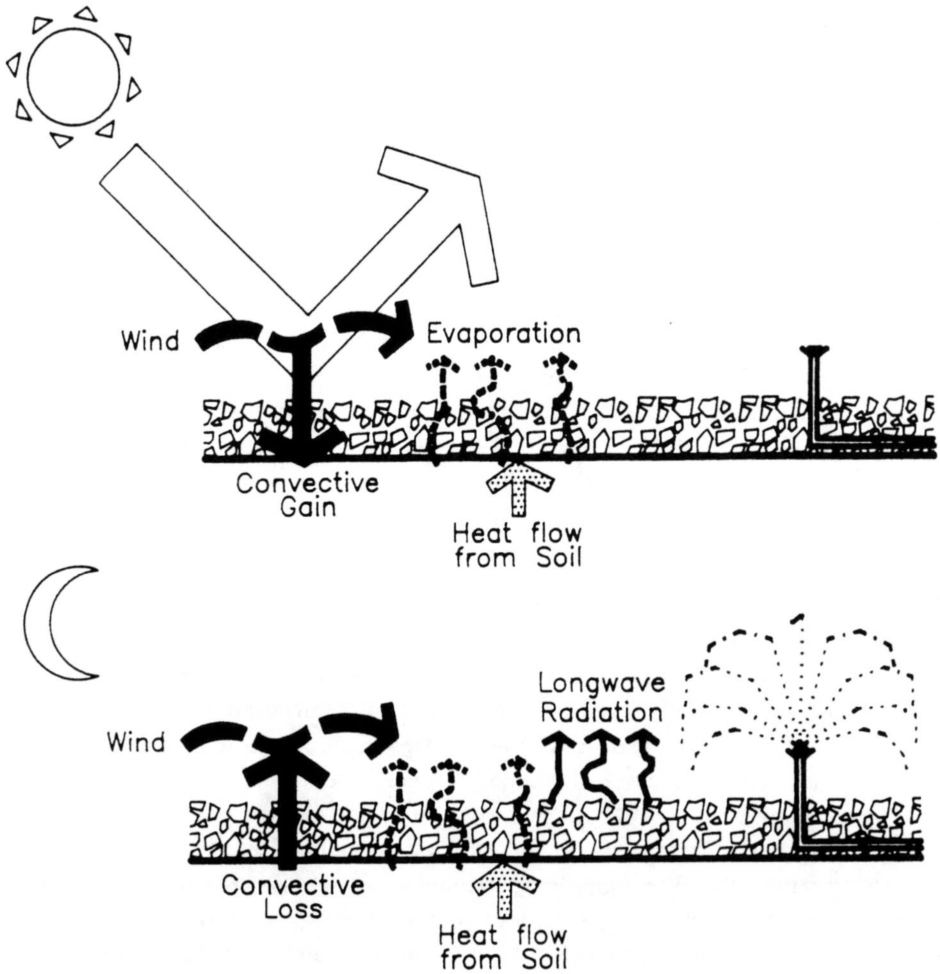

FIGURE 6.1 Thermal balance of wetted soil covered by a mulching layer (e.g. pebbles).

Soil Cooling by Mulching Cover and Wetting It

When moist soil is covered by a mulching layer such as bark chips or pebbles, the thermal balance at the soil's surface is modified in several ways. (See Figure 6.1.) They include radiant balance, convective exchange, and evaporative cooling of the soil.

Radiant Balance

Solar radiation is intercepted at the upper surface of the cover layer. As the thermal conductivity of the cover layer is lower than that of the soil itself, less solar heat reaches the soil material. However, long-wave radiant heat loss occurs also at the surface of the cover layer. The rock layer thus reduces also the effective long-wave radiant cooling of the soil. In summer, when the solar gain exceeds the long-wave radiant loss, the result is less radiant gain and consequently less heating of the soil.

On the other hand, in winter, when the long-wave radiant loss is usually greater than the lower solar gain (due to the low altitude of the sun and the shorter days) the cover layer reduces the radiant loss, and consequently results in less winter cooling. However, as cold air can sink within the rock's layer and reach the soil, the net result is only a small difference between exposed and rock-shaded soil temperatures in winter.

The modified radiant balance results in seasonally different relationships between the soil temperatures of exposed and "covered" soil in summer and in winter. In summer shaded soil is significantly cooler than exposed soil while in winter the difference is rather small, although shaded soil may still be cooler.

Convective Exchange

The rock layer, although not impermeable to wind, nevertheless reduces the rate of convective exchange between ambient air and the soil. As the surface of shaded and moist soils is at a lower temperature than the ambient air, the convective exchange in summer is positive: the soil gains heat by convection from the air. The mulching layer provides resistance to this heat gain, and thus reduces it.

Evaporative Cooling of the Soil

If and when moisture is available for evaporation, it produces a significant cooling effect. If the amount of available moisture is practically unlimited by water availability in the soil, as is the case in regions with rainy summers or when the soil is often watered, then the rate of evaporation from exposed soil is much greater than from soil covered by rocks. The higher temperature of the exposed soil elevates its vapor pressure and the higher wind speed at the surface of the soil increases the evaporative coefficient.

However, this higher evaporation rate is mostly used to overcome the solar heating of the exposed soil. On balance, the much smaller evaporation from shaded soil results in a lower temperature than the higher evaporation rate from exposed soil. In regions with a limited water supply the cooling of exposed soil by irrigation is a wasteful practice with low effectiveness. When the soil is shaded it is possible to obtain lower soil temperatures with less water supply.

Experimental Studies of Soil Cooling by Rock Cover

Several experimental and analytical studies have been conducted by the author and his colleagues and students at the Blaustein Institute for Desert Research, Ben Gurion University, Israel, on the practical possibilities of cooling the soil in summer by covering it with a layer of rock and watering it.

Austi Brown's Study

The first preliminary experimental study was conducted by Austi Brown, a graduate student at UCLA, in the summer of 1979 (Brown 1980). In this study the effect of the thickness of gravel layers and watering on soil temperatures were measured. Only the data from the bare soil and the plot covered by 12 cm gravel will be discussed here.

The test plots were each 250 × 300 cm. Measurements were taken at three depth levels at the untreated plot (10, 30 and 60 cm) and at levels (adding the interface between the soil and the rocks) in the treated soil. The main results from this study are shown in Figures 6.2, 6.3, and 6.4.

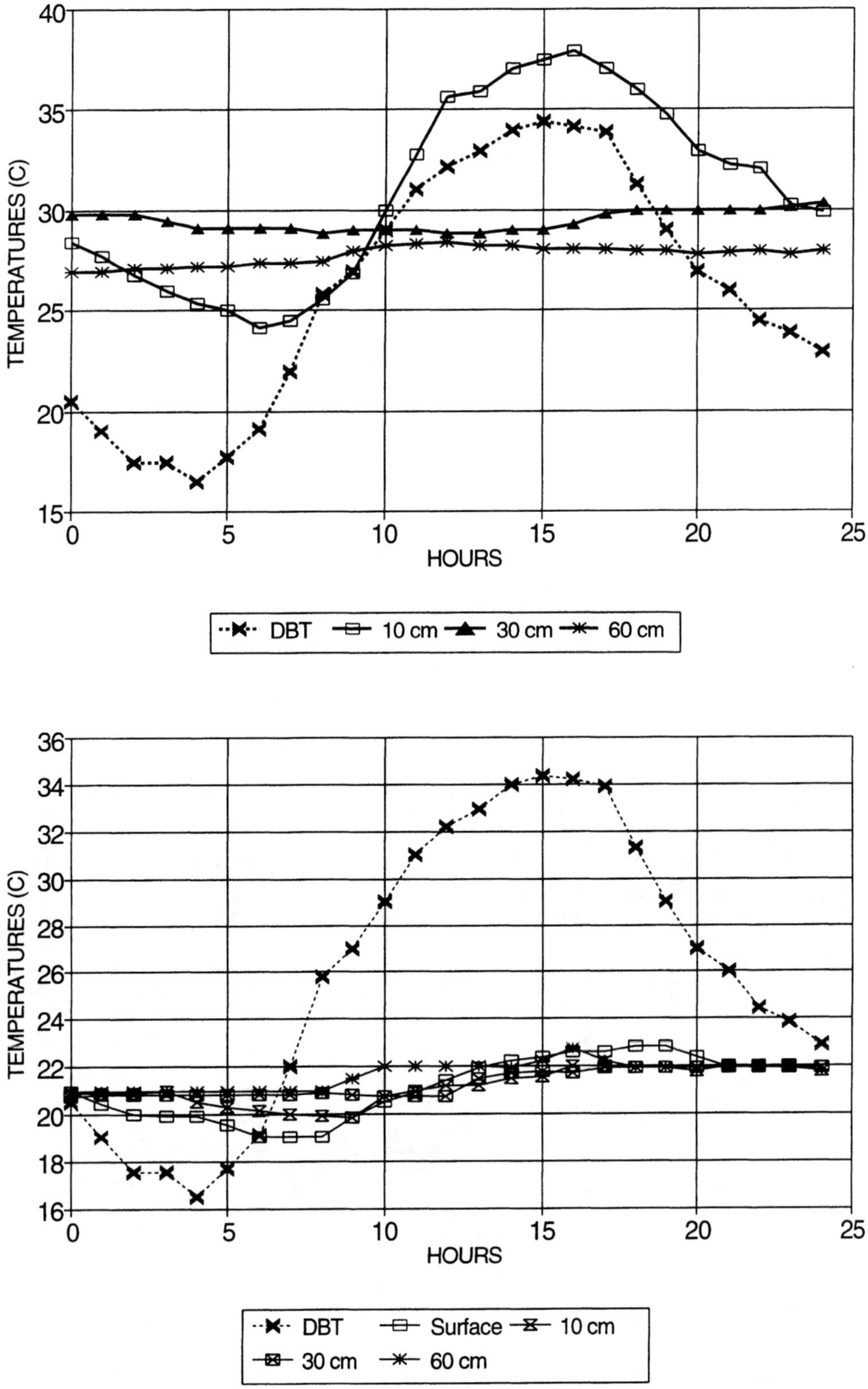

FIGURE 6.2 Diurnal temperature patterns in a desert in Israel at depth levels of 10, 30, and 60 cm, Aug. 7, 1979: (Top) untreated bare soil; (Bottom) moist soil covered by pebbles.

TABLE 6-2.

	Soil Temperatures (C)				Ambient Air		
	Surf.	*10cm*	*30cm*	*60cm*	*Max*	*Min*	*Avg*
Bare Soil	—	30.9	29.5	27.8	34.2	16.3	25.2
Cooled Soil	21.1	21.1	21.3	21.6	34.2	16.3	25.2

Figure 6.2 shows diurnal temperature patterns of the bare soil and the watered plot covered by 12 cm of gravel measured on August 7th, when the treated lot was watered during the late night hours in order to utilize the nocturnal radiant cooling of the rocks. The soil's surface, just under the pebbles, is also shown in this figure. The ambient temperature range was from 17 to 33.5°C.

The average temperatures of that period are given in Table 6-2.

The daily average temperatures of the untreated soil were all above the air average. The highest temperature was near the surface, at a depth of 10 cm. The temperatures of the treated soil were very close at all levels, and well below the ambient average temperature. The near-surface temperature of the treated soil was lower by about 10°C than the corresponding layer of the untreated soil and about 13°C below the ambient air maximum.

Figure 6.3 shows the average daily temperatures of the untreated bare soil at depths of

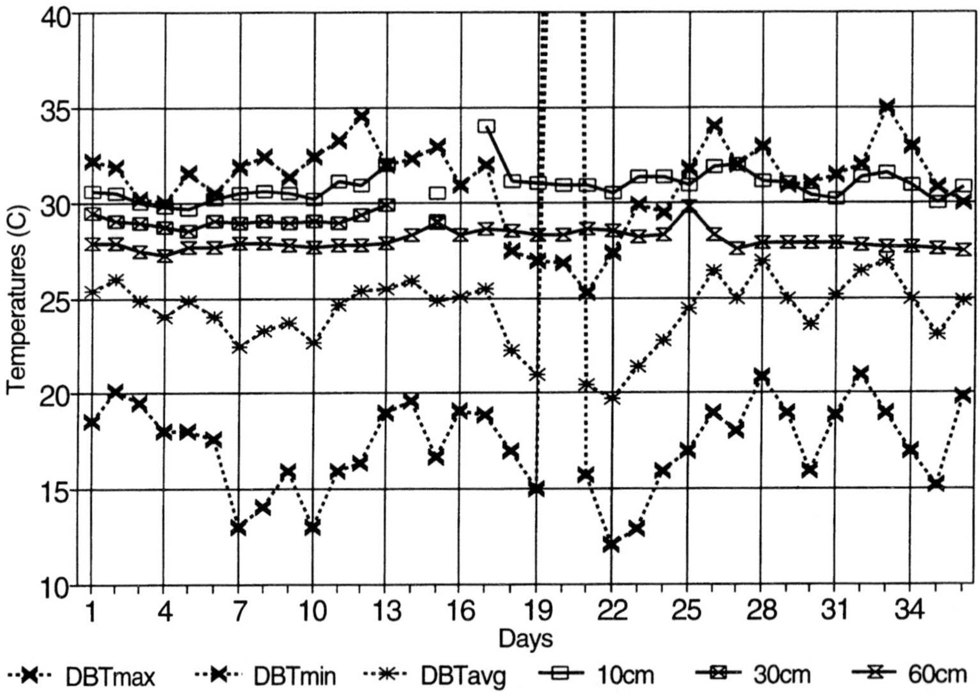

FIGURE 6.3 Maximum and minimum temperatures of bare untreated soil in a desert region in Israel, at depth levels of 10 and 60 cm, and outdoor maximum, minimum, and average temperatures, from July 27 through August 31, 1979.

TABLE 6-3.

Soil Temperatures (C)			Air Temperatures		
10cm	*30cm*	*60cm*	*Max*	*Min*	*Avg*
30.9	29.1	28.1	30.3	21.6	25.7

10 and 60 cm, together with the maximum, minimum, and average of the ambient air, from July 27 through August 31, 1979. The average temperatures of that period are given in Table 6-3.

Thus, at 10 cm depth the average soil temperature was close to, and a little above, maximum air temperature. By extrapolation it can be estimated that the maximum surface temperature was about 32°C. Even at a depth of 60 cm the average soil temperature was far too high to serve as a cooling source.

The situation was very different with the treated soil. Figure 6.4 shows a plot of soil shaded by gravel and watered, together with the maximum, minimum, and average of the ambient air, from August 8th through 31st, 1979. Although ambient temperature during this period underwent strong fluctuations—a sharp drop was followed by a sharp rise—still the soil temperatures were very stable. The average temperatures of that period are given in Table 6-4.

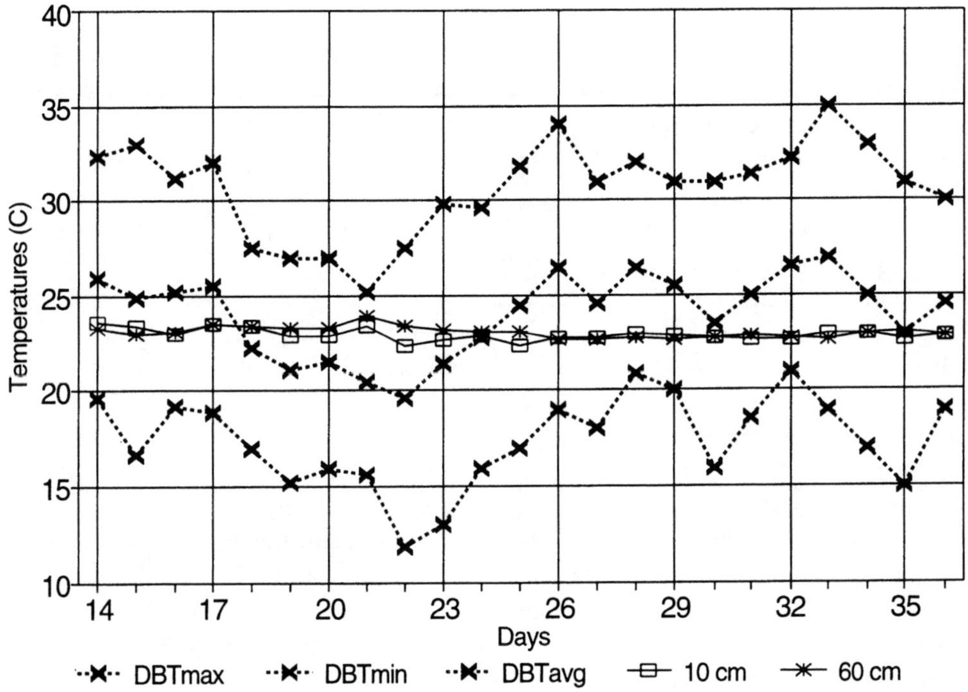

FIGURE 6.4 Maximum and minimum temperatures of moist soil covered by pebbles and outdoor maximum, minimum, and average temperatures in a desert region in Israel, August 7-31, 1979.

TABLE 6-4.

Soil Temperatures (C)		*Air Temperatures*		
10cm	*60cm*	*Max*	*Min*	*Avg*
22.9	23.1	30.7	17.4	24.1

Thus, the treated soil temperature was about one degree below the average ambient air temperature. The average difference between the treated soil and the ambient air maximum was 7.7°C. It should be noted that the soil temperature was relatively constant in spite of the sharp fluctuations in the ambient temperature. As a result, the difference between the soil and the ambient maximum temperature increased significantly during the heat wave, reaching up to 12°C during the peak outdoor temperature.

In evaluating these results it should be taken into account that this was a short term experiment and the test plots were rather small. The temperature of the deep layers under the test plot was still the same as that of the bare soil, as was the temperature of the soil surrounding the treated plot. Heat flow from the deep soil and the surrounding area obviously affected the temperature of the treated plot.

Ellis Hayeem's Study

The second study, a long-term one, was conducted by Ellis Hayeem, who later submitted the data as a M.Sc. Thesis to Trinity University in San Antonio, Texas (Hayeem 1984). Soil temperature was measured in two circular plots, 6 meters in diameter, at depths down to 2.9 meters. One plot was exposed and not irrigated, serving as a control. The second plot was covered by 10 cm of pebbles and was irrigated during the summer of 1982. The results of this study are shown in Figure 6.5, which shows the temperature patterns of the two plots, at the different depth levels, throughout 1982.

During the midsummer (June–July) the near surface temperature of the shaded and irrigated soil was lowered by 8.5°C, from about 31.5°C to about 23°C. The maximum temperatures at the various depth levels of the treated soil were about the same—there was just a small temperature drop with depth, of about 1.3°C between the surface and 3 meters, although there was a time lag of about 10 weeks in the date of the maxima at the two respective levels. In the winter there was very little difference between the exposed and covered plots.

The results from these two studies indicate that if a cooling treatment is applied to the soil surface in summer there is no need to go down to any considerable depth in order to utilize the soil as a heat sink. In fact, a depth of about 0.5 to 1 meter could provide sufficient coolth storage capacity to this end. Heat "dumped" into the soil at this depth will elevate the interface between the soil and the rocks somewhat. This in turn will elevate soil's vapor pressure and enhance the evaporative cooling of the soil, thus limiting the temperature build-up.

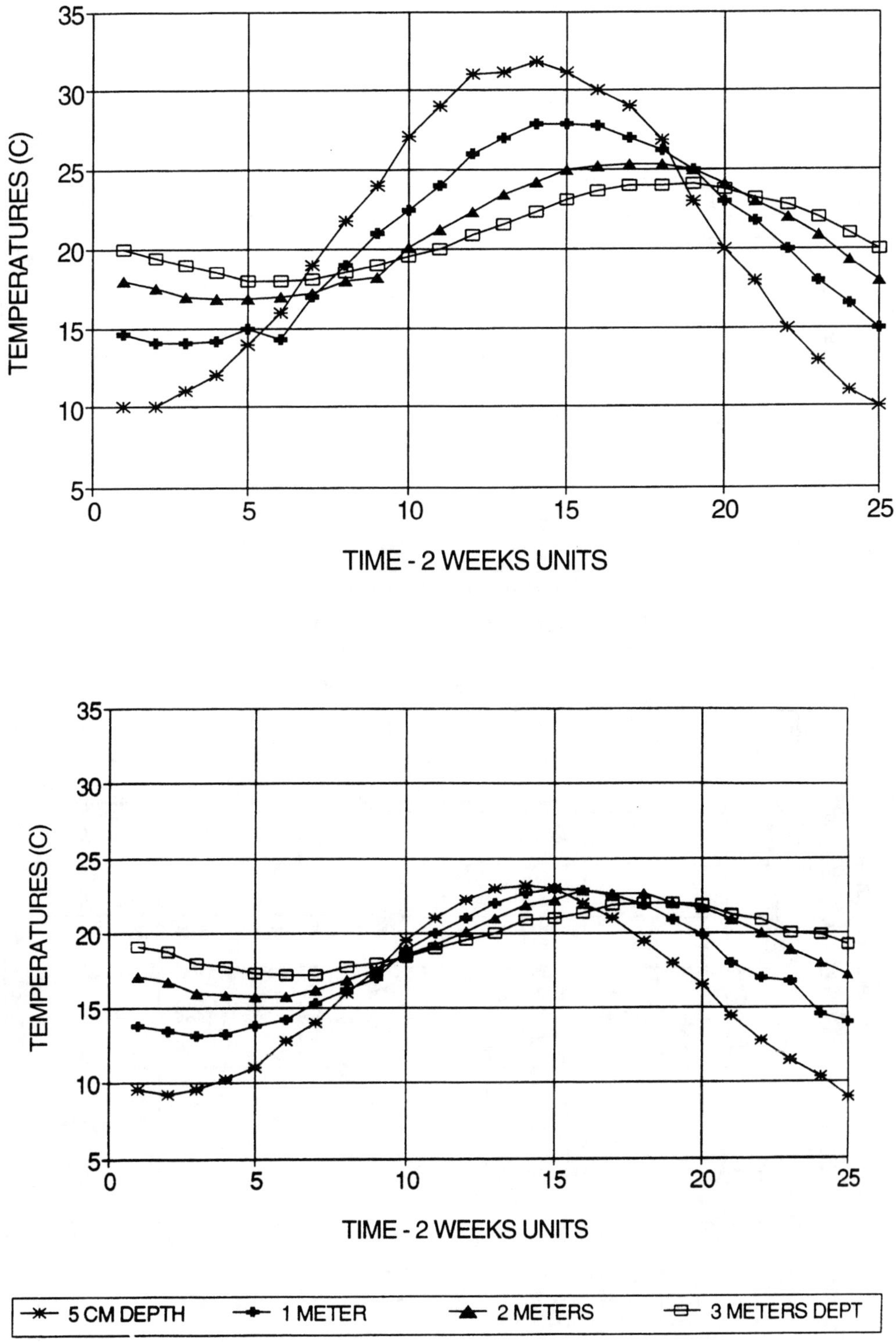

FIGURE 6.5 Annual patterns of monthly average soil temperatures in a desert region in Israel during 1981: (Top) untreated bare soil; (Bottom) moist soil covered with pebbles.

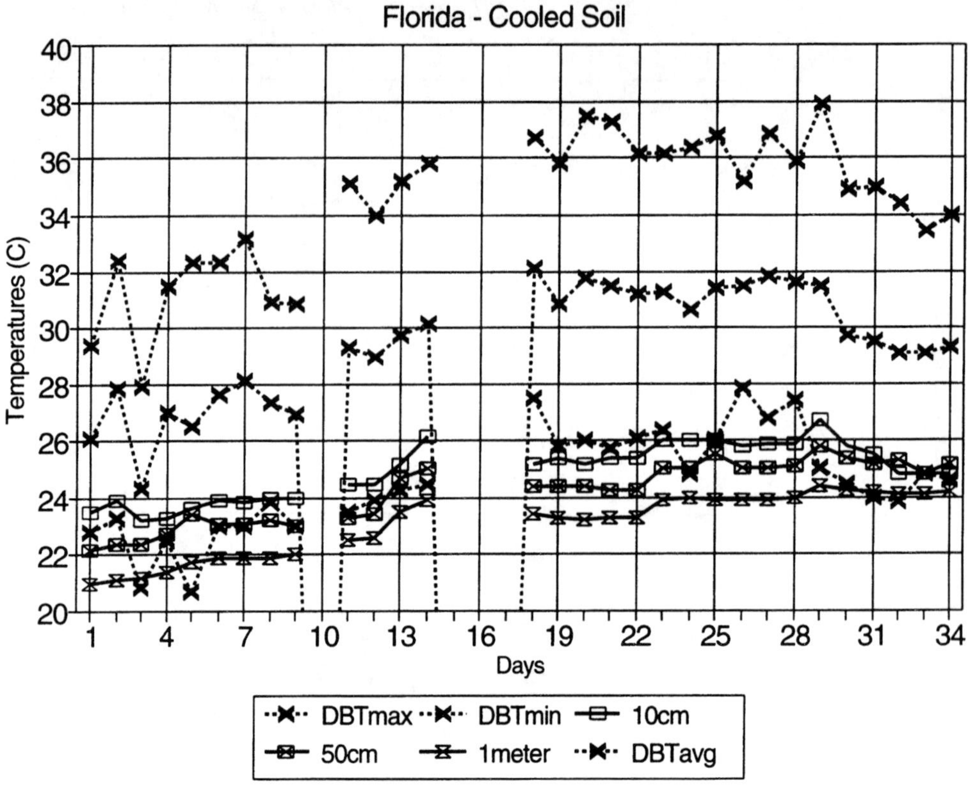

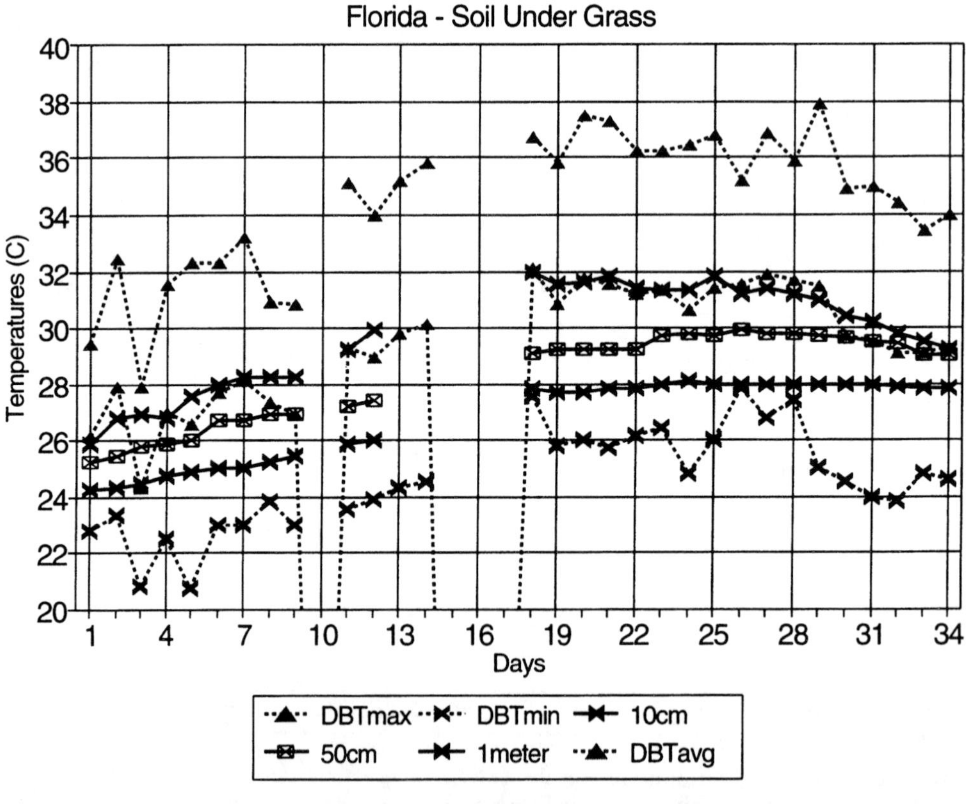

FIGURE 6.6 Daily average temperatures of cooled soil under a crawl space and of "natural" soil with grass, at depths of 5, 50, and 100 cm, in the summer of 1983 in Tallahassee, Florida.

Shading the Soil by a Building—Study in a Hot Humid Climate

Many buildings are built off the ground, either with a crawl space or at greater height above the ground. This practice is common, especially in wet regions, with buildings of wood frame construction. Such buildings provide complete shading for the soil below, and they do not obstruct the wind flow over the soil. If moisture is available in the soil in the summer, either by rain or through watering, the soil can be cooled evaporatively below the temperature of the natural exposed soil in the region.

Measurements of soil temperatures below a crawl space, and in a nearby open grass-covered field, were taken at three depth levels (5, 50 and 100 cm) by Larry Birch and Bill Winkey, with the participation of the author, at the School of Architecture of Florida A & M University (FAMU) in Tallahassee. Some results of this study are summarized below.

Figure 6.6 shows the patterns of the average temperatures of the soil under the crawl space, along with the "natural" soil covered by grass, during the hottest period, from June 21st through July 31st, together with the maximum, average, and minimum of the ambient air. At the peak of the annual cycle, the temperature near the surface soil with the grass cover was above the ambient average (about 31.5°C), while the corresponding layer under the crawl space was near the ambient minimum (about 26°C). At a depth of one meter the soil under the grass reached a peak of 28°C, while under the crawl space it reached only 24.3°C.

The difference between the outdoor air maximum and the soil temperature at a depth of one meter was about 12°C. Such a difference can provide not only sensible cooling but also some dehumidification of the ambient very humid air. The vapor content (humidity ratio) in a hot wet climate usually reaches its maximum in the early afternoon hours, as a result the ample evaporation sources are exposed to the sun.

Figure 6.7 shows diurnal patterns of the ambient and cooled soil temperatures during July 22—the peak of the summer. Between 1 p.m. and 7 p.m. the ambient air temperature was above 35°C and the WBT above 29°C, so that the dew point temperature was above 27°C. The soil temperature at a depth of one meter was about 24°C, which was below the ambient dew point. This provides the potential for partial dehumidification of ventilation air.

An interesting phenomenon was observed on rainy days, as is illustrated in Figure 6.8, which shows both temperature patterns and times of rainfall on June 24. It should be noted that the ground to the south of the building sloped down toward the cooled soil. Whenever it rained during the daytime, it caused a sharp rise in the treated soil temperature. Apparently, the water running off the hotter exposed soil was heated, and this elevated the temperature of the cooled soil.

Shading the soil by a mulching layer (for example, with gravel) allows the building and the cooled soil to be directly coupled by passive thermal conduction. The coupling is typically done by berming the walls or even covering the whole building (except for doors and windows) with the cooled soil. The gravel layer should be at least 10 cm thick, and of at least four pebble diameters, to minimize the conductive heat transfer from the surface of the layer.

The mulching layer receives substantial direct solar radiation during the day. Some of this energy is transmitted to the soil through the pebbles by conduction. To minimize this gain, the pebbles should be very light in color, and the gravel layer sufficiently thick. On

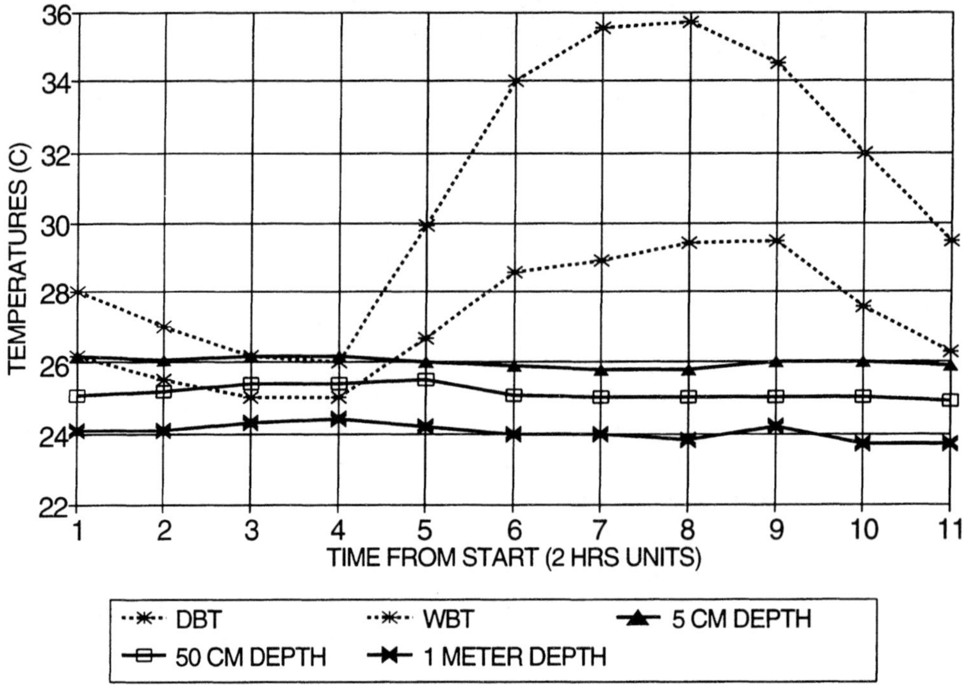

FIGURE 6.7 Diurnal temperature patterns of cooled soil under a crawl space, at depths of 5, 50, and 100 cm, on July 22, 1983 in Tallahassee, Florida.

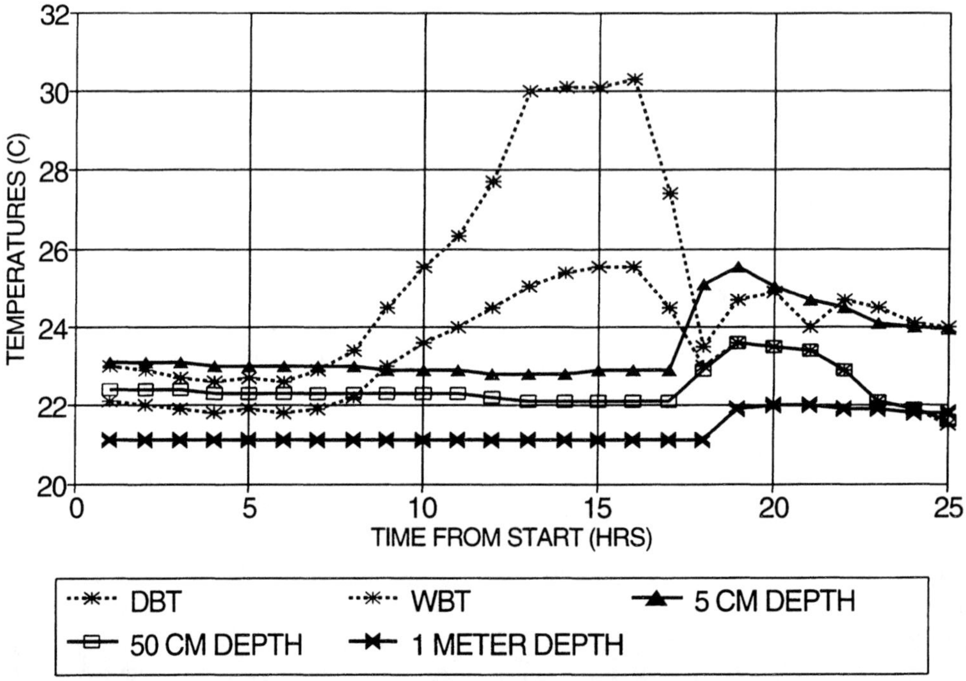

FIGURE 6.8 Diurnal temperature patterns of cooled soil under a crawl space, at depths of 5, 50, and 100 cm, on a rainy day on June 24, 1983 in Tallahassee, Florida.

the other hand, at night, the top layer of pebbles is cooled by longwave radiation. To utilize this radiant heat loss, watering should take place at the end of the night, so that the water will transport the cold from the pebble layer to the soil.

Options for Utilizing the Cooled Soil

The ways in which the cooled soil can be utilized depends in part on the relationship between the building and the soil—whether they are coupled thermally or "separated." Several examples of different relationships between a building and the cooled soil are (in order of decreasing coupling) mentioned below:

- **A** A shaded soil layer covers the roof. The walls are bermed. Coupling between building and soil can be either passive or active, depending on the climatic conditions in winter. Shading of the soil over the roof can be provided by a pergola or by a mulching cover, such as pebbles.
- **B** The walls are bermed by earth, except for the doors and windows, while the roof is insulated and is not covered by soil. Coupling between building and soil can be either passive or active, depending on the climatic conditions in winter.
- **C** The building is above ground and the walls are insulated and exposed to the environment. The roof is uninsulated and is covered by a layer of soil which, in turn, is kept moist and is covered by a mulching layer or is shaded by a pergola. The cooling of the building is passive, by conduction through the ceiling to the cooled soil. The passive coupling option in the three examples given above can be applied only in places with mild winters.
- **D** The cooled soil is under the building (for example, in a building with a crawl space). In this case no extra soil area is used for the cold storage function. Coupling between the building and the cooled soil in this case also can only be by active means.

  The most common way by which a soil area is shaded by a building is when the building is raised off the ground, leaving a crawl space under a suspended floor. The shaded area can be even larger than the enclosed floor area when outdoor timber decks, which are common in many hot humid regions, form an integral extension of the building. If irrigation is to be provided artificially, it may be advantageous for the soil to be levelled, and a small "dam" formed at the edges. Free air movement should be ensured by adequate openings on at least two opposing sides to enable the evaporation from the soil and to prevent excessive moisture buildup under the floor. Sufficient height to allow access for maintenance should be provided. Transfer of heat from the building interior to the soil acting as a heat sink may be either by air or by water flow.
- **E** The cool soil is at some distance from the building—for example, it may be under trees or a pergola in an attached garden. Coupling between the building and the cooled soil can only be by active means. The shaded cooled earth plot itself may be incorporated in the landscaping scheme, as in Japanese rock gardens. The use of high-canopy vegetation, and partial shade provided by adjacent buildings, would enhance the performance of this system.

Any soil area next to the building can be cooled and utilized as a cooling source for the building. The area should be shaded, either permanently or only during the summer. Shading can be provided by trees with a high canopy or by a vine-covered pergola, but not by a lawn or shrubs that would minimize evaporation from the soil surface. It would be advisable to cover the soil by a mulching layer even if it is shaded to minimize the convective heat gain from the ambient air. The two last coupling options are applicable also in places with cold winters.

Direct Coupling of the Building with the Soil

Direct coupling between a building and the cooled soil can be achieved only when the structural walls and/or the roof of the building are in direct contact with the soil, without any insulation installed in between them. In this situation the cooled soil forms the "environment" to which the walls and/or the roof are exposed. Direct coupling can be provided with any one of the three first design options outlined above, because the walls and/or the roof are in direct conductive contact with the cooled soil. With such conductive coupling, the indoor surface temperature of the external envelope in contact with the soil will be close to the surrounding earth temperature.

Direct passive coupling between the building and the soil is applicable only in regions with mild winters—where minimum temperatures are above about 5°C. Some distinction should be made, however, between locations that are hot or warm year round (that is, with winter average *minimum* temperatures above about 10°C) and locations that are hot in summer but have mild winters (winter average minimum temperatures are 5 to 10°C).

In places where the winter season is also warm, or even hot (for example, Maracaibo in Venezuela), it can be good to provide direct thermal coupling of the building with the cooled soil with as little thermal resistance as practical. It means that the building elements in contact with the cooled soil should be made of a conductive material, such as concrete or *dense* solid bricks, and there should be no thermal insulation. Only a waterproofing membrane should be provided where the building surfaces are in contact with the soil.

On the other hand, in regions with minimum temperatures around 5°C and a probable diurnal average temperature of about 10°C in winter, some thermal resistance of the envelope elements in contact with the soil would be advisable to minimize the heating load. Walls of hollow concrete block, for example, might be more appropriate in this case than solid concrete walls would be.

Covering the roof with earth may be much more expensive than berming the walls. (See the discussion below of wall berming versus roof coverage.) Therefore, whenever covering the roof is expensive, a reasonable design solution would be to berm the walls (or to "sink" the building below grade) for a height from 1.5 meters above the floor to the full height of the walls. It would then be possible to cool the soil surrounding the walls using one of the cooling methods discussed above while providing high thermal insulation to the roof, thus minimizing its thermal effect.

Berming the walls does not have to interfere with natural ventilation and lighting. Good ventilation and daylighting can be provided to a building with bermed walls through

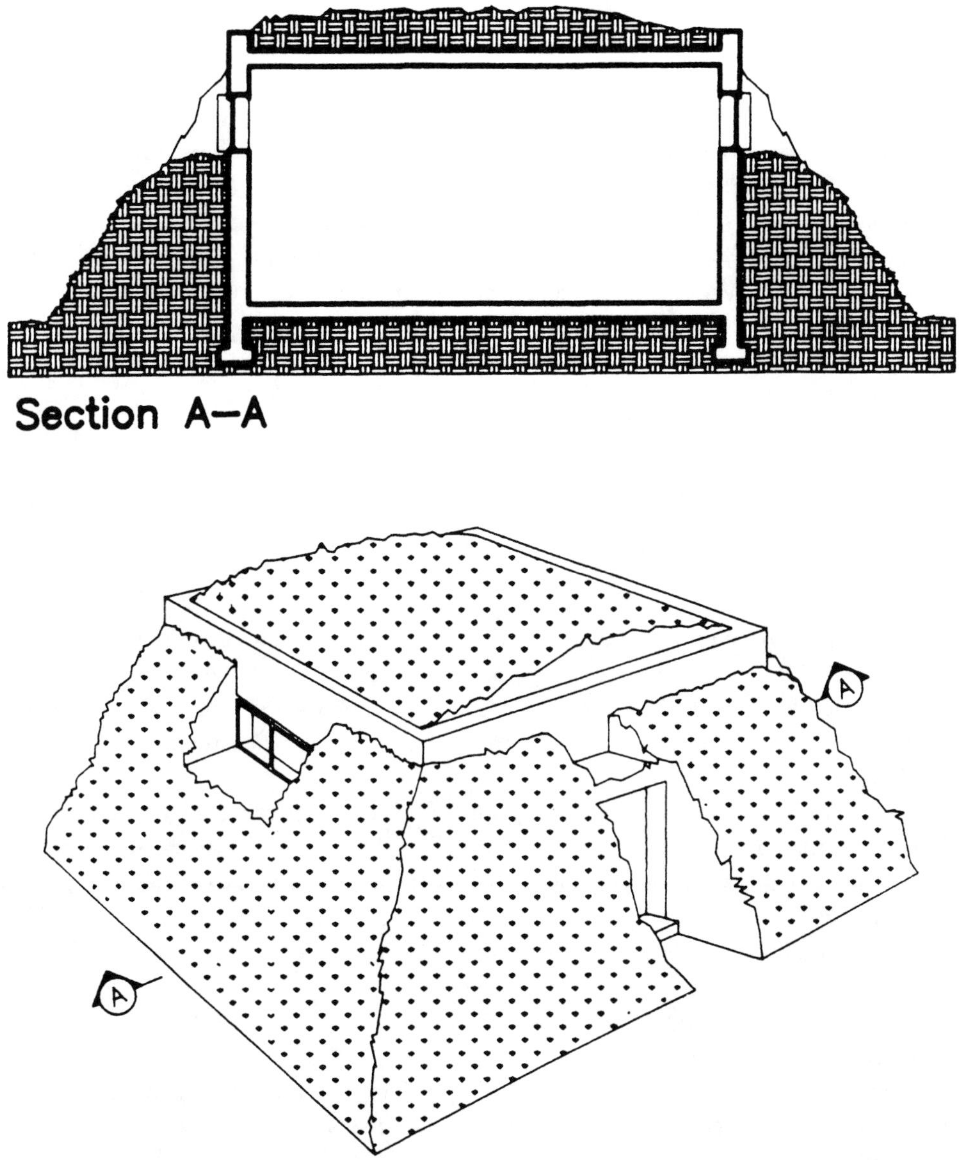

FIGURE 6.9 A scheme for daylighting a building with partially bermed walls.

high windows near the ceiling, as is illustrated in Figure 6.9. Such ventilation openings can be designed even if the walls are bermed to their full height by providing bays in the berms adjacent to these openings. With this design, almost all of the wall area is in contact with the cool soil and only the area immediately surrounding the windows is exposed to the ambient air and to the sun, if it is not shaded by a pergola or other shading device. In some cases night flushing of the indoor space by leaving the windows open for ventilation can provide additional cold storage for the following day.

Indirect (Active) Coupling Between the Building and the Cooled Soil

In regions with cold winters a building has to be insulated. Direct conductive coupling of the indoor space with the surrounding soil through highly conductive walls, floors, and roof may be undesirable because it would cause a high rate of heat loss in winter. A system of controlled coupling, which can be activated in summer and stopped in winter, would be the appropriate solution. When the heat transfer between the building and the cooled soil has to be by active means it can be provided either by forced air or by water flow. There are several technical options that can provide this coupling.

Air as a Circulating Medium

Heat exchange with the soil can be done through an array of pipes. These pipes are usually made of a plastic such as PVC. The air circulation may be indoor air moving through a closed circuit, or it may be outdoor air used for ventilation. In arid regions it may be advantageous to locate the intake within the crawl space in order to take advantage of the lower air temperature there. The air will then be supplied at a lower initial temperature, and naturally humidified. In humid regions it would be preferable to take in ambient air from a shaded area away from the crawl space because the air within the crawl space usually would be too humid. It is necessary to remove water that may condense in subsoil pipes in humid climates.

This air-cooling system is advantageous for buildings in which higher-than-normal daytime ventilation rates are required—for instance, in schools. It is particularly applicable to buildings of lightweight construction, because the thermal storage mass of the system is in the soil itself. Also, as the heat sink is separate from, but close to, the served building spaces, almost any low-rise building form may be considered, especially where the floor would be raised off the ground for any reasons.

In buildings where high ventilation rates are not required a closed circulation system would be more efficient.

Climatically, this system may offer some cooling potential for hot humid climates, where other natural cooling strategies would fail: nocturnal radiation because of the predominance of cloudy night skies, and nocturnal ventilation because of the small diurnal temperature range and the need for high daytime ventilation rates.

Limitations

Applications dealing with cooling the ventilation air are virtually unlimited, as soil temperatures will always be lower than daytime ambient temperatures during the cooling season. In the closed circulation mode, however, use is confined to regions where the soil temperature is sufficiently low to supply air at temperatures below the comfort range. In practice, if the soil exceeds about 20°C, use is limited to cooling the ventilation air.

In building design terms, the distance between the served space and the heat sink should be limited to minimize "parasitic energy" and unwanted heat-gain in the recirculation system.

Water as the Circulating Fluid

The heat transfer from the building to the cooled soil can be effected by the flow of water rather than air. The water pipes serve as heat exchangers with the soil, but there must be another heat exchanger within the building to cool the indoor air. This heat exchanger can either be a conventional fan-coil or a radiant panel. Because of the small temperature difference available, the radiator preferably should be in the form of a large horizontal overhead panel—for instance, a concrete ceiling with embedded pipes—to take advantage of the maximized downward cold flow by the combined convective and radiant heat transfer. A cooled concrete ceiling also can serve as an effective heat sink.

Comments on the Two Options

In many places with cold winters the heating will be continuous while in summer the cooling is often intermittent (mainly during the hours when the indoor space is hottest, the afternoon and early evening). On the other hand the heat dissipation from the earth surrounding the tubes to the rest of the earth's mass is a continuous process. Therefore, with relatively high soil conductivity, which is a characteristic of wetted soil, and with dissipation of the heat by increased evaporation from the soil surface, there need not be a significant heat build-up in the soil surrounding the air tubes. Winter preheating will lower the earth temperature at the deeper layers somewhat, increasing the potential for summer cooling.

Double Walls with an Air Space

The walls in contact with the cooled soil can be constructed in two layers: a highly conductive and water-proved external layer (steel or concrete), with direct contact with the surrounding soil, and an internal insulated layer. There should be an air space between the two layers. The external layer will acquire a temperature close to that of the earth surrounding it. It always will be lower in summer and higher in winter than the average ambient air, and it will provide complete solar protection in summer.

During the summer, indoor air can be circulated in a closed loop through the air space separating the two layers, thus providing convective cooling to the interior. In winter, with no air circulation, the inner layer and the air space provide an insulation from the cooler external layer. Such a solution is applicable in regions which have hot summers but cold (or even moderately cold) winters.

Condensation is likely to occur on the inner surface of the external layer, which is in contact with the soil. Treatment to prevent fungi growth is desirable. Using the air space for preheating the ventilation air can improve this situation.

STUDIES ON "ACTIVE" COOLING BY EARTH TUBES

Princeton Energy Group Study

The Princeton Energy Group (PEG) has developed a computer simulation code for predicting the cooling potential of their "Cool-Pipe" system (Glennie et al. 1979). They have concluded that, in the climate of Trenton, N.J., with a soil temperature of 10°C (50°F) at the beginning of the summer, it is possible to derive the cooling equivalent of 1136 kwh in the three-month period of June 1st through August 31st, from a pipe of 38 cm diameter, 24.4 meters long, with air-flow rate of 680 m^3/hr.

From parametric simulation the PEG has concluded that soil conductivity has little effect on the seasonal performance of the system. High conductivity enables better performance at the beginning of the cooling period, while lower conductivity ensures better performance towards the end of the summer. The air-flow rate appeared to have the highest impact on the performance, with the length of the pipe being the factor second in importance. The simulation results yielded the following regression equation for predicting the expected cooling from a pipe in the ground (depth of 1.2 m) in the climate of N.J. (initial ground temperature 10°C, summer design D.B. 31°C, W.B. 23°C, cooling degree days: 538°C days):

Expected cooling (kwh) = 0.8xF+2.54xD+16.1xL+28.5xCond+0.92xHC
where: F = air flow (m^3/hr); D = Diameter (cm); L = Length (m);
C = soil Conductivity (w/mC); HC = soil Heat Capacity (wh/m^3C)

Experimental Studies with Earth Cooling Tubes

Only a few experimental studies in which the performance of earth cooling tubes was monitored are known to the author. Four such studies will be summarized in this section:

- Goetsch, Peterson and Muehling (1983)
- Scott, Parsons and Koehler (1982)
- Francis (1981)
- Akridge and Benton (1981)

Goetsch, Peterson, and Muehling's Study

The performance of four earth tubes used to cool the ventilation air in a building used for swine farrowing in Illinois was monitored by Goetsch, Peterson, and Muehling. The details and performance of each system will be described separately.

System #1 was installed in an area with sandy soil in Mason County. It included eight runs of 25 cm diameter unperforated corrugated plastic drainage tubing, each 30.5 meters in length. The tubes were distributed in a radial fashion, leading to a central collection sump. The average depth was 2.4 m and the tubes sloped ½% to the collection sump, where a pump removed any condensed moisture from the system.

Outdoor temperature during the monitoring period (July 16–20, 1980) ranged from 19.4 to 40.6°C. The temperature of the air, after flowing through the tubes, ranged from 21.1 to 32.2°C, at an air flow velocity of 1.11 m/sec. The soil temperature, as can be inferred from the data, was about 23°C. The peak sensible cooling rate was 4800 w. The peak cooling rate per meter of length of the tubing was 19.7 w/m of tubing. The moisture content of the air was reduced almost constantly as the air was cooled to a temperature below its dew point. During the monitoring period the sump pump removed approximately 13.2 liters of water per hour from air flow rate of 1630 m^3/hr, yielding an average moisture removal of 8 grams from each m^3 of air.

It should be of interest to note the diurnal pattern of the moisture content of the ambient air. Regularly it was elevated during the daytime hours, following the temperature pattern. This is typical to many hot humid regions, where moist soil and vegetation provide plenty of surface area for evaporation, which increases with higher temperatures.

System #2 was installed in Montgomery County, Illinois. The soil was silt loam with high water content. In this system, too, the tubes were laid in a radial pattern. Three tubes of 13 cm diameter were laid in each of three trenches (9 lines altogether). Each line was 45.7 meters long and all were connected to a central collection sump constructed of 91 cm diameter concrete wall tile. The tubes were installed at a depth of 2.4 m at the outer edge and sloped to about 3.5 m at the sump.

The air flow rate was 1700 m^3/hr. The soil temperature (estimated by the author) was about 23°C. The ambient air temperature ranged from 16.7 to 35°C. The treated air temperature ranged from 21.1 to 26.7°C. The peak sensible cooling rate, as given in the report, was about 4380 watts, or 10.6 w per meter length of tubing.

System #3 consisted of 22 nonperforated corrugated plastic drainage tubes, 25 cm in diameter. The lines were 55 meters in length and were arranged in parallel, spaced 1.2 meters apart, at a depth of 2.4 to 2.7 meters. It was installed north of Macoupin, Illinois, in an area with soil of silt loam.

The flow rate was 10,195 m^3/hr and the estimated soil temperature about 22°C during the monitoring period (July 13–15, 1981). The outdoor temperature ranged from 21.1 to 34.4°C. The outlet air temperature ranged from about 23 to 24°C. The peak cooling rate was about 35,660 watts, about 29.5 w/m of tubing length. The stability of the outlet air in this system is notable, resulting from the longer length of the tubes (55 meters) relative to the air flow speed (2.54 m/s), and the high conductivity of silt loam soil.

System #4 was installed south of Central Sangamon County, Illinois. The system consists of 5 tubes, 30 cm in diameter, of nonperforated corrugated plastic drainage tubing in a parallel pattern. Each line was 79 meters long and laid at a depth of 3 meters. The lines led to a collector tube of 38 cm in diameter and then into a 76 cm culvert. At the inlet end a tee fitting and a 30cm diameter vertical pipe served as an air inlet.

The estimated soil temperature was about 17°C. The outdoor temperature ranged from 15.6 to 33.3°C, while the treated air ranged from 17.8 to 20.6°C. The peak sensible cooling rate, given by the authors, was 15,380 W. or 38.8 W/m of tubing length, with a flow rate of 3480 m^3/hr and a flow velocity of 2.67 m/sec.

TABLE 6-5. COMPARISON OF THE 4 SYSTEMS

System	*1*	*2*	*3*	*4*
Tube diameter (cm)	25	13	25	30
Tubing length (meters)	30	46	55	79
Number of lines	8	15	22	5
Tube spacing	radial	radial	1.2m	3.4m
Air velocity (m/s)	1.1	2.5	2.5	2.7
Total flow rate (m^3/hr)	1,6	1,7	10,2	3,5
Flow per tube (m^3/hr)	204	113	463	696
Peak sensible cooling (W/m)	19.7	10.6	29.5	38.8

Comparison of the Performance of the Four Systems

Table 6-5 compares the main properties and performance data of the four systems.

Scott, Parsons, and Koehler's Study

In this study a steel air tube, 45 cm (18") in diameter and 132 m (440′) long, was installed at a depth of 2.40 cm (8′) at the Cornell Poultry Research Farm in Ithaca, N.Y. Temperature measurements were taken inside the tube at 30 m (100′) intervals. In addition, soil temperatures were taken around the tube, and the undisturbed soil temperature at a considerable distance from the tube was taken as well. Measurements were taken with both continuous and intermittent operation of the fan. The authors' paper also presents data on the tube's heat exchange effectiveness (E), defined as the ratio:

$$E = \frac{\text{(inlet temp.)} - \text{(outlet temp.)}}{\text{(inlet temp.)} - \text{(undisturbed soil temp.)}}$$

This can be defined also as the temperature drop relative to the potential drop. The exchange effectiveness follows a diurnal pattern similar to that of the ambient air with maxima and minima coinciding with those of the outdoor air maxima.

During a week with continuous operation in July, 1984, the undisturbed soil temperature rose only from 52 to 53°F. The outdoor temperature diurnal swing was about 20°F and the maxima rose gradually from 90 to 97°F. The outlet temperatures rose steadily, from 60 to 70°F. The maximum exchange effectiveness underwent a steady decline during the week, from 66% to 60%.

During a week with intermittent operation in September 1984, the outdoor temperature maxima fluctuated between 78 and 98°F and the undisturbed soil temperature was about 56°F. The outlet air temperature exhibited an almost steady pattern, rising each day from about 60 to about 63°F. The maximum exchange effectiveness was about 80%, indicating that the soil next to the tube was able to regenerate its cooling potential during the night, when the fan was not in operation.

Francis's Studies

Francis (1981) has reported on the field performance of two installations of underground airtubes in the Midwest of the United States. This region is characterized by very cold winters and hot, wet summers. Because of the large annual temperature range, the underground earth temperature in that region is well below the summer ambient air temperatures, and even below the ambient dew-point. Consequently, the subsurface earth could be utilized as a heat sink, both for sensible cooling and for dehumidification of the ambient air.

The first case reported by Francis is about the annual space conditioning of a swine house with air tubes installed 3 meters below the surface. Outdoor air is blown through two tubes, 120 m long and 20cm in diameter, by a 1 hp blower. In this case the outdoor air was 35°C DBT and 31.7°C WBT (humidity ratio of 0.026, relative humidity of about 80%). The outlet air was 18.3°C DBT and 13.9°C WBT (humidity ratio of 0.0074). The flow rate per tube was 408 m^3/hr (240 cfm) and the air speed in the tubes was 3.4 m/sec (685 fpm). Thus, in addition to the sensible cooling from 35 to 183°C, about 18.6 grams of water were extracted from every Kg of air. The COP of the cooling was 18.5. It is of interest to note that almost all of the cooling was achieved at a length of 60 meters (87% of the total cooling). Thus, it was concluded by Francis, a much higher COP could be achieved by a larger number of shorter tubes.

In the second case Francis reported, ambient air was blown through an air tube of 15cm (6") diameter and 17 meters (55') long, at a depth of 3 meters under the earth surface. The flow rate was 119m^3/hr (70 cfm). No data is given in the paper about the soil temperature. The ambient air was at 26.4°C (79.5°F) DBT and 24.4°C (76°F) WBT. The outlet air was at 20.6°C (69°F) DBT and 18.9°C (66°F) WBT.

In addition to a sensible cooling of 5.2°C, 708 grams of water per hour were precipitated from the air, whose humidity ratio was lowered from about 0.019 to about 0.013 (kg/kg of dry air). COP was 12.5.

Akridge and Benton's Study

Akridge and Benton (1981) have measured the performance of underground air tubes installed in a "Thermal Envelope house" in Georgia. The two tubes were each about 50 cm diameter and 30 meters long, at an average depth of about 2 meters. Generally, the cooling of the air flowing through the tubes in this study was proportional to the elevation of the ambient air (in its diurnal pattern) above the relatively constant earth temperature. With inlet temperature of about 30°C (86°F) and soil temperature of about 21°C (70°F) the outlet air temperature was about 23°C (74°F).

The soil temperature near the tubes, within a radius of 30 cm, rose appreciably even after three days of intermittent use of the tubes (for about 8 hours per day).

General Conclusions and Discussion of Earth Cooling Tubes

All the cases reported above have demonstrated the potential of underground air tubes to provide not only sensible cooling but also, in hot humid seasons, appreciable dehumidification of the ambient humid air. The main factor that allowed such results in the American northern midwest was that region's large annual temperature range, which results in a relatively low underground summer earth temperature (about 19 to 23°C).

Analysis of the detailed data in Francis' paper shows that in all three cases about 70% of the potential cooling from the ambient air to the underground temperature was achieved at a tube length of about 12 meters. This suggests that underground cool tubes need not exceed a length of about 15 meters. The required overall flow rate can be achieved by increasing the number of the tubes or increasing their size.

A major point of interest is the extent to which underground cool tubes can be utilized in regions with not-so-cold winters. In fact, the main need for cooling exists in low-latitude regions, which have a much smaller annual temperature range than Illinois. Typical underground temperature in summer in such regions, at a depth of about 2 to 3 meters, might be above 22°C. In these regions the soil is too hot to use as a cooling source, except for precooling in very hot places. However, as has been demonstrated by the author's studies in Israel and at FAMU, it is possible to lower the earth temperature of a given lot around or under the building well below its natural level. To be of practical usefulness the earth temperature around the tubes should not exceed about 21°C, although with some forms of soil utilization, such as direct coupling to an earth covered building, even a soil temperature of 24°C would be adequate, especially in regions with maximum air temperature above 35°C. In many hot regions cooling the soil to such temperatures is achievable.

For a given length of air tubes there is an inverse relationship between the flow rate and the temperature drop of the air. Higher flow rates (and higher air speeds in the tubes and less time for the air to stay in the tubes) result in *higher temperatures* of outlet air although more *cooling energy* is obtained. Higher flow rates also cause greater flow resistance and require an increase in fan power. The considerations concerning the optimal flow rate become more complicated in humid regions if there is a potential for dehumidification of the air while it flows in the tubes.

In humid regions where the soil temperature can be kept below the daytime dew point of the ambient air (with "cold regeneration" during the nights) it might be preferable to lower the flow rate in order to cool the air in the tubes to a temperature below its dew point and thus to obtain air dehumidification. In such a case drainage of the condensed water should be provided by proper sloping of the air tubes. In places where dehumidification in the earth tubes is possible it may be advisable to increase the length of the tubes beyond the length needed for sensible cooling in order to increase the amount of water condensation.

ARCHITECTURAL ASPECTS OF EARTH-SHELTERED BUILDINGS

In many hot regions direct coupling the building with the earth, either only by berming the walls or also by covering the roof with earth, can greatly reduce the heat load of the building. When the earth itself is cooled below the natural summer soil temperature, further improvement—and in some cases complete natural cooling—is possible. Therefore this type of building deserves special discussion in the context of passive cooling. In this section the architectural aspects of earth-sheltered buildings are discussed, and particular attention is paid to such buildings in hot regions.

General Architectural Considerations

Different design approaches can be applied to earth-sheltered buildings located at sites with different topographical characteristics (flat land or slopes with different orientations) in order to optimize energy conservation and solar utilization potential. Designers must also consider such related "human factors" as the need for visual contact with the outdoors and for natural lighting and ventilation. When and where it is desirable, underground or earth-covered buildings should have natural lighting and ventilation through windows opened to central courtyards, peripheral light walls, or skylights. Views of the outdoor landscape can be best provided in an above-ground earth-covered building that has an exposed wall oriented towards the desired direction.

Earth-sheltered buildings may also have an impact on the structure of cities and the quality of the urban environment. As cities grow, pressure on the downtown area increases, and the demand for high-density development and high-rise construction comes into conflict with the need for open spaces for recreation and the desire to preserve environmental quality. In densely built up cities, underground development can provide open spaces right at the center of the city. There are several examples of large underground parking lots that sit below downtown central parks—for example, the Boston Common. In many cases, open-pit mining and quarries are destroying the natural landscape, and the owners are required to fill back the excavated space. As has been suggested by the University of Minnesota team (USGS 1982) it is possible to construct a variety of facilities in the excavated space, and then cover the space up again. The area can either be reconstructed as it was, or the builders can shape a new land form.

The design of an earth-sheltered building should deal with several aspects specific to this type of building. Even general architectural design problems may have special implications in the case of earth-integrated buildings. A systematic procedure followed during the design stage can help in arriving at optimal solutions. This procedure could follow the steps proposed below:

- **A** Decisions concerning human psychological needs must be made. These decisions should take into consideration the desires of the occupants of the building, as well as the building's planned use (as a residence, an office, a school or so on). People's desire for natural lighting, natural ventilation, and visual contact with the outdoors should also be considered.
- **B** An analysis of the climate during both the summer and the winter should be made, and the designer should determine the building's requirements for winter heating, potential for solar heating, and need for summer cooling and natural ventilation. The choice between the different basic types of earth-sheltered buildings will be based on this analysis as well as the site analysis and human aspects analysis.
- **C** An analysis of the site should be performed, with the special requirements of an earth-sheltered building kept in mind. Such an analysis can help in determining the relationship between the building and the earth.

D Decisions concerning the means of access into the building for people and vehicles must be made.

E Structural and technical problems, such as created by the stresses imposed by the earth, waterproofing, and drainage must be considered. These problems are outside the scope of the present discussion, and they will not be dealt with here. Relevant information can be found in such books as Labs (1989), Golani, and UGSC (1982).

Human Factors in the Design of Earth-Sheltered Buildings

The term "human factors in design," in the context of the present chapter, is related to the psychological and physiological needs whose satisfaction depends on a building's design details.

Specifically, such factors as the human requirements for natural lighting and ventilation, which affect both thermal comfort and psychological satisfaction, should be considered. Also the need for visual contact with the outdoor environment, which is of a more purely psychological origin, may have special importance in earth-sheltered buildings. The design solutions of these problems may determine its occupant's satisfaction with these kinds of buildings, and the larger public's acceptance of them. In the long run, adequate design solutions to these factors may determine the prospects of large-scale habitation in earth-sheltered buildings.

Natural Lighting and Ventilation

The provision of natural lighting and natural ventilation should get serious consideration, because of its psychological importance and because of the help it provides in conserving energy, especially in hot regions. The role of natural light and ventilation in earth-sheltered buildings may even be more important than in ordinary above-ground buildings, mainly because of the psychological and public acceptance implications. From the psychological point of view, the provision of daylight, even to sections of the buildings that are relatively far away from the exposed parts, may alleviate the danger of the gloominess and feelings of being in a cave that are often associated in the public's opinion with earth-covered buildings.

Natural ventilation is important in earth-sheltered buildings, not only for thermal comfort and to maintain an adequate indoor air quality, but also (and, in some cases, mainly) to remove any moisture that may penetrate into the indoor space through the earth-covered building, in spite of all precautions.

From the energy point of view, the situation is more complicated and dependent on climatic conditions. In cold regions, the openings that are necessary to provide natural ventilation usually increase the heat loss of earth-sheltered buildings. Even sealed apertures, which serve only for lighting and not for ventilation, can become a source of heat loss in winter (if the glazing is not of sufficient thermal resistance) and of inadvertent solar heat gain in summer. On the other hand, in a warm climate the cross ventilation that can be provided through such openings may ensure indoor comfort and reduce—or even eliminate—the need for cooling. Fortunately, there are usually good designs that can provide natural lighting and ventilation in most types of earth-sheltered buildings without undue penalties from the energy conservation viewpoint.

Climatic Analysis for Earth Sheltered Buildings

The "environment" to which an earth-sheltered building is exposed is much more stable than the corresponding daily outdoor meteorological conditions, and seasonal differences in the climatic conditions are much smaller. However, basic differences may exist between thermal control options and the design details of earth-sheltered buildings in different types of climates.

In this respect, a distinction should be made between hot and cold climates, when the terms are related to the overall annual and seasonal characteristics more than to short-term variable conditions. As a result, the climatic analysis can be done mainly on the basis of the annual average temperatures. Nevertheless, seasonal conditions (average rather than diurnal patterns) should also be considered in determining the human comfort needs and design details of the building, although they are less important than the annual averages.

Through long-term thermal control it is possible to modify the underground earth temperature in the vicinity of the building and to bring it closer to the human comfort range. These unique opportunities presented by earth-sheltered buildings do not exist in above-ground ones.

In a cold climate the main problem is heating, and one of the design objectives for earth-sheltered buildings would be to raise the underground temperature around and below the building to above its natural level. In summer the natural temperature of the subsurface soil may minimize the cooling needs. In hot climates the main problem is cooling, and the objective would be to lower the underground temperature below its natural level. In winter passive solar heating may provide most of the heating needs. In places with a mild climate some heating in winter and cooling in summer may be needed in earth-integrated buildings.

As an earth layer of about 30 cm is sufficient to suppress the diurnal temperature wave to an insignificant range, only the average monthly temperatures are of interest in estimating the expected conditions inside earth-integrated buildings. Even heat waves in summer and cold waves in winter that last for as long as several days have only a small effect on the underground conditions.

Comparing Wall Berming and Roof Cover in Earth-Sheltered Buildings

From the structural point of view it is much easier to support the earth load on walls than over the roof. Therefore, in practice the depth of soil coverage over the roof, even when it is provided, may be limited to about 40 to 60 cm. But even such a shallow layer of earth involves extra expense and the question arises about the effects of such earth cover on the comfort and energy conservation of the building. These effects and costs should be compared to that of additional insulation over the roof.

Because of the limited thickness of an earth cover over a roof the annual temperature range at the bottom of the earth layer is about 80% of the surface's temperature range. (See the appendix to this chapter.) It means that the average minimum temperature is above the surface minimum by about 10% of the annual surface range. In regions with cold or even cool winters it means that the roof has to be insulated to minimize heat loss in winter.

Once the roof is insulated, the additional resistance of the soil above it is relatively small. For example, a layer of 5 cm expanded polystyrene has a thermal resistance of about 1.5 (m^2.C/W). A layer of 0.5 meters of soil, with conductivity of about 1.2, has a resistance of only (0.5/1.2) about 0.4. Nevertheless, an earth cover over the roof stabilizes the temperature fluctuations—both the diurnal and the day-to-day changes. It means that average conditions, rather than extreme "design" climatic condition, are affecting the indoor thermal conditions. Whether this contribution, which is a real one, justifies the expense involved in putting such a layer of soil over the roof, is questionable.

The opinion of the author is that roof coverage by soil is justified, from the cost/benefit viewpoint, only in regions where the winter is comfortable or warm, so that insulation is not needed then. An earth layer cooled in summer by shading and watering can be kept in the summer in most regions at temperatures close to the comfort range, so that insulation is not needed, and in fact not even desirable, during the summer.

Orientation Considerations

Orientation, in earth-sheltered buildings, concerns mainly the orientation of openings—windows and doors. This orientation may be chosen by considering energy conservation and solar utilization. However, in many cases it will be dictated by topography, or be influenced by the view and landscape seen in the different directions.

Earth-sheltered buildings, by definition, have a limited exposure to the outdoor air, as a significant area of the walls are bermed, or even completely covered by the earth. This is often the case with the roof as well. Therefore, the aspects of solar energy availability, natural ventilation, and lighting should be considered when deciding which walls, or parts of walls, will be exposed.

The orientation and topography of the site, and the building that is designed on it, determine the potential and design options for solar heating of the earth-sheltered building in winter, and its exposure to prevailing winds in both winter and summer. In addition, the landscape and views that can be seen from the exposed part of the building may also affect the choice of orientation.

For better energy conservation, in the northern hemisphere the preferred orientation of the exposed side of a building is towards the south. Such an orientation allows inexpensive passive solar heating in winter and simple shading of the exposed wall and openings in summer. A building having a southern orientation can therefore be planned with more freedom in detailing than with other orientations. When the topography of the site slopes southward, all types of earth-integrated buildings can be considered. The importance of a southern orientation increases in colder regions, especially in areas that are cold but sunny in winter.

When the site properties are favorable—for example, if they offer both a southern orientation and a southern view—any type of design and any detail may be considered. The topography and climatic conditions, however, may render some types and design details more suitable than others. For example, in a flat area with a cold climate the building could be earth-covered on all sides and the roof, with only the southern side exposed. A one-story building may provide the optimal design.

However, in a hot climate with similar site conditions the potential for cross ventilation in summer is important. A building with an earth-covered roof, an exposed southern wall, full-height berming on the eastern and western walls, and partial berming on the northern and southern walls to a height of about 1.5 meters may be a better solution, because it allows cross ventilation through southern and northern openings above the berms.

Another design option in hot regions for ventilation and natural lighting is to use vertical skylights (often called roof monitors) that project from the roof through the earth layer, some facing the prevailing wind and some the opposite direction. This arrangement would ensure effective cross ventilation to the earth-covered building.

Groundwater Considerations

The groundwater level at the site has a significant impact on the design details of earth-integrated buildings. The highest expected level of the water-table should be taken into account. In places where the groundwater level is deep and well below the floor of the earth-integrated building, it is possible to contemplate an underground building, fully or partially below the natural earth surface, without high expenses for water- and moisture-proofing.

On the other hand, in places where the water table may be high and close to the floor of the building, a safer approach would be to design a bermed or earth-covered building above the natural ground level. In that case, an artificial hill would be formed around and above the building.

Ventilation and Condensation Control in Hot Humid Regions

Earth-covered buildings in hot humid regions and seasons may face special humidity and condensation problems. This is because the surface temperatures of the walls, the floor, and even the ceiling, may be below the dew point of the outdoor air. As a result, the ventilation air, which in hot-humid regions or seasons has a high water-vapor content, cools down when it comes into contact with the cooler surfaces and may precipitate excess moisture. In addition, the high relative humidity resulting from the cooling of the outdoor air may be too high for comfort, and may cause mildew and fungi growth. The problem, of course, is aggravated by the water-vapor produced inside the building as the people inside it engage in such activities as breathing, cooking, and washing.

However, the root of the problem—the relatively low temperature of the earth surrounding the building—may also provide the solution. By blowing the ventilation air through air pipes embedded within the cooler soil, it is possible to obtain cooling and partial dehumidification of the air, thus enhancing thermal comfort and reducing the risk of internal condensation.

To prevent corrosion and fungi growth inside the air pipes, they should be made of materials that do not provide favorable conditions for fungi development, and provisions should be made to drain the condensate.

The same air pipes and ventilation system can be utilized during the winter, of course, to preheat the cold outdoor ventilation air, using the warmer soil as the heat source.

STORING WINTER COLD UNDERGROUND FOR SUMMER UTILIZATION

Different design options for storing winter cold in insulated soil and utilizing the cooled soil during the summer as a cooling source for buildings was suggested by Givoni in 1979 in a paper presented at the conference on earth-covered settlements in Fort Worth. One such scheme was also presented in a Research Proposal submitted in 1984 to the Department of Energy by the school of Architecture of Florida A&M University (FAMU—by Givoni) in Tallahassee. Although the project was not funded, the concept underwent further analysis and various design options were considered. The analysis and ideas developed in the proposal are presented in this section.

Different design options can be considered for storing winter coolth in the subsurface layers of the soil. All of them have four basic elements:

- A Increasing the cooling rate of the "deep" soil layers (for example, at a depth of 2 to 10 meters) in winter.
- B Minimizing the heat flow from the soil surface to the deeper layers in summer.
- C Minimizing the heat flow from the surrounding warmer soil into the cooled soil.
- D Utilizing the cooled soil for cooling the building in summer.

Increasing the Cooling Rate of the Deep Soil in Winter

The major physical process of heat transfer between the near-surface and the deeper soil layers is by conduction, which is mainly a function of the soil's thermal resistance. Even in moist soil the rate of cooling of the deep layers in winter, especially in regions where the average minimum winter temperature is above zero degrees C, is rather slow, because of the thickness of the soil layer.

The cooling rate of the deep layers can be significantly increased if they will be cooled by convection—that is, by the percolation of a large quantity of cold water from the surface. It should be noted that the volumetric heat capacity of water (1.0 cal/cm^{3}C) is about 2.5 times as great as that of soil (approximately 0.4 cal/cm^{3}C). Considering the limited area of the soil that can be considered for winter coolth storage (about the same area as the floor area of a low rise building), cooling it by cold water percolation would not involve excessive waste of water, even in arid regions.

When a thin layer of water is exposed to the sky at night, it is cooled by a combination of evaporation and long-wave radiation. It reaches temperatures lower than the minimum of the outdoor air. In regions with clear nights the water temperature may even be below the minimum dew point, because of long wave radiant heat loss. Therefore, when using water as a cold transport medium, the climatic parameter of interest is not the average winter temperature but the *average minimum* for the winter period.

When cooled water is drained to the soil under an insulation layer, the soil attains a temperature close to the water's. Daily "flushing" of the insulated soil can cool the deeper layers, by convection, much faster and to lower temperatures than is possible by conduction. Of course, the convective cooling process is accompanied by augmented conduction as a result of the high water content of the flushed soil.

As the soil is cooled gradually during the winter, the use of water for soil cooling can continue as long as the minimum water temperature at dawn is lower than the temperature of the deep layers of the soil. As mentioned above, this concept of storing winter cold underground for summer use was developed in 1984 for FAMU in Tallahassee, Florida. Therefore the climatic conditions of this region will serve to illustrate the concept's potential. The following discussion is taken from the above research grant request.

Typical winter minimum temperatures in northern Florida are around 3 to 8° C. The winter is the dry season, skies then are generally clear. Consequently, a thin layer of water exposed to the sky at night could be cooled by longwave radiation to a temperature about 3 to 4° K below the air minimum.

If water will be cooled every night and then drained into an insulated soil, either under a building covering a large ground area or into "open" soil covered by an insulated layer, the soil below the insulation, to a depth of several meters, could be cooled throughout the winter to a temperature well below its natural level. The drained water will return to the underground aquifer, and thus it will not be wasted.

The soil under a large building, or under an insulated "open" layer, at a depth of a few meters, undergoes a relatively small annual swing around the annual average temperature. In many warm regions this annual average is about 18 to 20° C. Assume a layer of water 5 cm deep with an average temperature of 6° C drained each morning into insulated soil with an initial average temperature of 18° C. Over 100 winter nights it is equivalent to a water layer of 5 meters in height, cooling by percolation the insulated soil to a depth of several meters. The cooling effect is equivalent to about 72 Kwh/m^2.year.

Part of the stored cold is of course dissipated into the surrounding soil. Assuming an annual storage efficiency of 50%, for a large insulated area of cooled soil, summer utilization of the stored coolth yields about 36 Kwh/m^2.year.

Minimizing the Heating Rate in Summer

An exposed soil reaches higher average surface temperatures than the outdoor air during the summer. Even if it is shaded, the soil surface temperature approximates the air average. Lower temperatures at the deeper layer would increase the conductive heat flow downward, greatly reducing the efficiency of the cold store achieved during the winter. Therefore, an integral component of any system for underground winter coolth storage should include insulation of the area above the cooled soil.

Depending on the situation several design options can be considered.

A **A large building with a concrete slab floor.**

Many commercial buildings are one story tall and have flat roofs that cover a large floor area (about 30×30m or more), with a concrete slab floor on grade. When such a building is a candidate for utilizing stored winter cold a relatively simple design solution can be suggested. (This was the scheme presented in the Research Proposal submitted to the DOE.)

The winter coolth can be stored in the soil under the building while a shallow pond can be placed over its roof. Thus no specialized area would be needed for storing the winter coolth.

An insulation layer (perhaps of expanded polystyrene) can be applied directly under the concrete floor slab. This insulation will minimize cooling of the building in winter. In fact, as the building is maintained during the summer at a comfortable temperature—about 20° C—it forms the upper external boundary for the cooled soil. This reduces the summer temperature difference, and the resulting heat flow into the soil, as compared with an open area.

The shallow pond over the roof should be filled every evening during the winter months (mid-November through February) and drained each morning, through embedded pipes, into the soil under the insulated floor. Thus the cold transfer in winter bypasses the thermal resistance of the insulated floor.

In fact it is desirable to provide a layer of rocks about 50 cm thick above the soil and under the insulation. This layer will enable the soil to expand into the air spaces between the rocks when it is wet.

The water can be distributed to the whole soil area under the building through perforated pipes.

B Cold Storage Under an Open "Green" Area or Parking Lot

When the winter cold storage is planned under an open land area, such as a park, or under an open parking area, the insulation layer design has to take into account the use of the area as well the protection of the insulation. To minimize the summer surface temperature and the associated downward heat flow, the best approach would be to keep the area green—to use it as a lawn.

With this configuration the insulation layer should be covered by some soil, enough thickness to transfer the load of people who are walking—or even running—without damaging the insulation material. This soil cover could then be planted with grass and shallow-rooted shrubs. The cold storage area would then double as a utilized space. In this case the cold collection functions—the shallow ponds—should be located elsewhere, at some distance from the cold storage area. Perhaps they could be put on the roofs of nearby buildings, presumably the ones where the stored winter cold will be used for cooling during the summer.

If the area is to be used for parking the insulation would have to be thicker to compensate for the much higher surface temperature.

Minimizing the Heat Flow from Surrounding Warmer Soil

While the downward heat flow from the surface to the undersurface cooled soil can be minimized by a specialized insulation layer, it is more difficult to insulate the cooled soil mass against lateral heat flow from the warmer soil around the cooled area. Even if a vertical "belt" of insulation were to be placed in ditches surrounding the storage area, it would be practically limited to a depth of about 2 meters while the cooled soil may be 4 to 6 meters deep, or even deeper.

In the absence of any lateral insulation an "edge effect" should be taken into account. This reduces the effective volume of the cooled mass. As an approximation, a plane sloping inward at 45 degrees from the edges of the cooled soil can define the effective volume that can be utilized in summer as a cooling source.

Horizontal insulation several meters wide, surrounding the cooled soil and covered by such materials as soil or concrete, also can reduce the heat flow from the surrounding soil, effectively reducing the edge effect.

This edge effect is a major factor in establishing the minimum dimensions of the area of the cooled soil. From this viewpoint the shortest dimension should not be less than about 15 meters.

Utilizing the Precooled Soil as a Cold Source for Buildings

The fact that the soil is being precooled in winter practically excludes the option of direct passive conductive heat transfer between the building and the soil, even in cases where, when the soil is at its natural temperature, such an approach would be possible (for example, in earth-covered buildings in warm regions).

The thermal coupling between the cooled soil and the building interior had to be accomplished by controlled means, either by air tubes or by water pipes embedded in the soil. Cold can then be transferred from the soil to the building only when it is needed while the building would remain insulated from the cooled soil.

Another option for utilizing the cooled soil is by coupling it with a heat pump. An Earth Coupled Heat Pump is an established technology. Lowering the temperature of the soil that serves as a heat sink can improve the performance of the system significantly.

The choice between water or air systems depends not only on considerations of cost effectiveness but also on the inherent advantages and limitations presented by each option, depending on the building type, its use, and the climate.

APPENDIX 1: *Mathematical Models of the Natural Soil Temperature*

Soil temperature, as a function of depth and time, is usually expressed as a sum of one or more exponentially damped temperature waves. The formulae of the temperature waves are solutions to the heat equation in which the ground is considered to be a semi-infinite solid with a plane surface at a uniform temperature. The chosen boundary condition at the surface can take two forms, depending on the available data. If ground surface temperature is known or can be estimated, its temporal variation is the boundary condition. If this data is not available, meteorological data can be used in a boundary condition that states the energy balance at the surface.

Moreland (1978), Kusuda (1975), and Labs (1989) have used models that incorporate surface temperature data. Since their equations are similar, only Labs' model will be discussed here. His equation is:

$$T_{z,t} = T_m - A_s \exp[^{-z((\ /365a))}] \mathrm{Cos}\{(2\ /365)[t - t_o - (z/2)((365/\ a))^{1/2}\ ]\}\} \tag{1}$$

Where: $T_{z,t}$ = **soil temperature at depth z and day t**

T_m = **Mean annual ground temperature**

A_s = **Annual surface temperature amplitude**

t = **Time of the year in days**

t_o = **Date of the minimum surface temperature**

a = **Soil thermal diffusivity**

This is a solution to the heat equation for a semi-infinite solid whose surface temperature varies sinusoidally. Diurnal temperature variations have been ignored, since, for usual values of soil diffusivity, these variations are negligible at the depth levels below 30 centimeters. The relative decrease in the annual temperature amplitude with depth is shown in Figure 6.10.

A drawback to this formula is the reliance on the knowledge of the soil temperature annual amplitude at the surface, and of its diffusivity. The soil surface amplitude data usually is not available, and the issue of the soil's effective diffusivity is a complex one. (See the comments below.) Without these data this formula cannot be used.

Comments on Soil Diffusivity

Thermal diffusivity is defined as conductivity (k) divided by the product of density (ρ) and specific heat (c) of the material in question.

In any mathematical analytical model the soil diffusivity (k/ρc) has to be specific as a constant property of the soil. If the boundary conditions (annual pattern of the surface temperature) are known then the solution to the general formula is straightforward on the basis of the specified thermal diffusivity.

However, it should be pointed out that soil diffusivity is an illusive and highly variable property that changes greatly with both time and depth. The two main factors that determine the conductivity of the soil, and to some extent its heat capacity as well, are its water content and its sand/loam proportion. Soil conductivity increases greatly with water content. Also, a soil composed mainly of loam or clay has higher conductivity than a sandy soil.

The sand/loam proportion, or the presence of any other soil ingredient that affects its water-holding capacity and thermophysical properties, may vary greatly with depth, even at layers close to the surface.

The water content of any given soil exposed to the natural environment varies greatly, according to the annual patterns of precipitation. In wet regions with a water table close to the surface (a few meters below the surface) the soil is saturated below while having a variable water content at the layers above the water table. The level of the water table itself changes with the precipitation pattern.

In regions where the water table is relatively deep (more than 20 meters below the surface) the upper layers have a higher water content than the deeper layers during the rainy season and a lower water content during the dry season. As a result of the temporal and spatial variations of the water content of the soil it is impossible in most regions to define the diffusivity of the soil or to measure it using laboratory procedures. The only practical way to estimate the effective diffusivity is by inference from actual measurements of the annual ranges of the soil temperature at different depth levels.

From the annual patterns and the ratio of the annual ranges at any two depth levels an effective diffusivity of the soil between these two layers can be extracted, as is described below. Once such measurements from different regions are available, it is possible to develop a simple formula for estimating directly the annual patterns of the soil, at various depths, for regions with various soil types and wetness characteristics.

For any given type of soil, when experimental data on the soil's temperature patterns are available two numbers can be derived. One characterizes the dumping of the annual range with depth, and the other specifies the progressive delay in the date, when the maximum temperature occurs, with increasing depth. When such information is available it is possible to use it as a basis for the development of a generalized simple formula for predicting the soil temperature at any given depth and date, as is described below.

Estimating the Soil's Effective Diffusivity

When the annual patterns of soil temperature are known for two depth levels, the soil's effective diffusivity can be estimated from two parameters:

- **A** The ratio of the annual amplitude at the two levels, A_1 and A_2.
- **B** The time lag of the peak temperature with increasing depth.

Diffusivity Estimate Based on Amplitude Ratio

From formula (1) the damping of the annual amplitude at level 2, relative to level 1, is given by:

$$\mathbf{A_2/A_1 = exp\ (-R(z_1 - z_2))} \qquad (2)$$

$$\mathbf{z = depth\ level,\ meters}$$

$$\mathbf{R = (w/2a)^{1/2}}$$

$$\mathbf{w = 2\pi/365}$$

$$\mathbf{a = diffusivity = k\rho c}$$

From this formula the soil's diffusivity, **a** can be extracted:

$$\mathbf{a = (w/2\pi)[(z_2-z_1)/(lnAz_1-lnAz_2)]^2}$$

$$\mathbf{= (\pi/365)[(z_2-z_1)/(lnAz_1-lnAz_2)]^2} \qquad (3)$$

Diffusivity Estimate Based on Time Lag

The time lag with depth, L, between two layers with depth of z_1 and z_2, is given by:

$$\mathbf{L = T_2-T_1 = R(z_1-z_2) = (w/2\pi)^{1/2}\ \ (z_1-z_2)} \qquad (4)$$

Where:

$$\mathbf{R = (w/2a)^{1/2}}$$

Hence:

$$\mathbf{(w/2a)^{1/2} = (T_1-T_2)/(z_1-z_2)}$$

and

$$\mathbf{a = (w/2)[(z_1-z_2)/(T_1/T_2)]^2} \qquad (5)$$

Accuracy of the Two Procedures for Estimating Soil's Diffusivity

There is a great difference in accuracy between these two procedures for estimating effective soil diffusivity on the basis of experimental annual temperature patterns.

From the practical point of view, it is best to take a depth of at least 0.5 meter—preferably 1.0 meter—as the upper level (z_1), because at shallower levels the temperatures fluctuates too much for clear determination of the annual maximum and minimum, as well as their date.

During the periods of the maximum and the minimum temperatures the rates of change at a depth of 1 meter and below are very small. Therefore it is easy to determine the actual temperatures (maximum and minimum) with reasonable accuracy. On the other hand the determination of the actual dates of the maximum and the minimum is quite difficult and may require a several harmonics Fourier analysis. Therefore, the errors inherent in estimating soil diffusivity on the basis of the time lag are much greater than in estimating it on the basis of the amplitudes.

APPENDIX 2: *Simplified Model Based on Experimental Studies*

In several studies the annual patterns of the earth's temperature were measured at different depths. The climatic conditions and soil properties were different in the different studies. Nevertheless, all the experimental data can be compared when brought to a common basis: the relative range, at a given depth, relative to the range of the surface temperature.

In this way, data from studies reported by several authors were reduced to this common basis. The experimental conditions are summarized below.

Main Conditions Reported in Experimental Studies

Kusuda's Study (1975)

Kusuda's study was discussed in more detail above. The soil in this study was silty sand to a depth of 5 meters with density of 1850 kg/m^3 and water content of 11% to 16%. Below this depth the ground contained brown and gray disintegrated rock.

Bligh's Study (1975)

T. Bligh took soil temperatures in St. Paul, Minnesota, to a depth of 8 meters. No data on the soil properties were reported.

Penrod et al's Study

In this study of soil temperatures, down to 3 meters were measured in Lexington, Kentucky, over a period of five years. Normal yearly precipitation in the area is 1091 mm.

Fluker's Study

Fluker took soil temperature measurements down to 3 meters in College Station, Texas, over a period of five years. The soil consisted of five different layers.

Carson's Study

Carson took soil temperatures down to about 9 meters in Argonne, Illinois. The soil was a thin layer of black soil over sandy clay.

Kristenson's Study

Kristenson took soil temperatures down to 7 meters near Copenhagen, Denmark. The soil was glacial deposits.

Ashbel's Studies

Ashbel has reported soil temperatures in various places, including the arid region of Tel Amal in Israel and in Helwan in Egypt, down to about 2 meters.

Relative Range and Time Lag in Experimental Studies

Relative Range

The relative ranges computed from the above studies are shown graphically in Figure 6-10. In this figure the relative range is plotted as a function of the depth below the surface, with each study marked differently.

It should be pointed out that some studies were conducted in regions that can be characterized as humid (the eastern United States and Denmark), while other studies were conducted in drier regions (Texas, Israel, and Egypt).

At any given depth the relative range of the annual wave of earth temperature was greater in humid regions than in dry regions.

This phenomenon is expected because of the higher thermal diffusivity of the soil, which brings the surface and depth temperatures closer.

The experimental data from the different studies could actually be divided into two groups: one representing humid places (St. Paul, Minn.; Washington, D.C.: Lexington, Kentucky) and the other representing drier places (College Station, Texas; Tel Amal, Israel; Helwan, Egypt).

Most of the experimental observations are bounded by two curves, both exponential. The upper boundary curve of the relative ranges of the subsurface temperature for humid places is expressed by the formula.

$$T_{(z)}/T_{(o)} = e^{-0.3z} \tag{6}$$

and the lower boundary curve of the relative range in arid regions can be expressed by:

$$T_{(z)}/T_{(o)} = e^{-0.5z} \tag{6a}$$

where:

$T_{(o)}$ = the earth surface temperature range
$T_{(z)}$ = temperature range at depth z
z = depth below the surface, meters

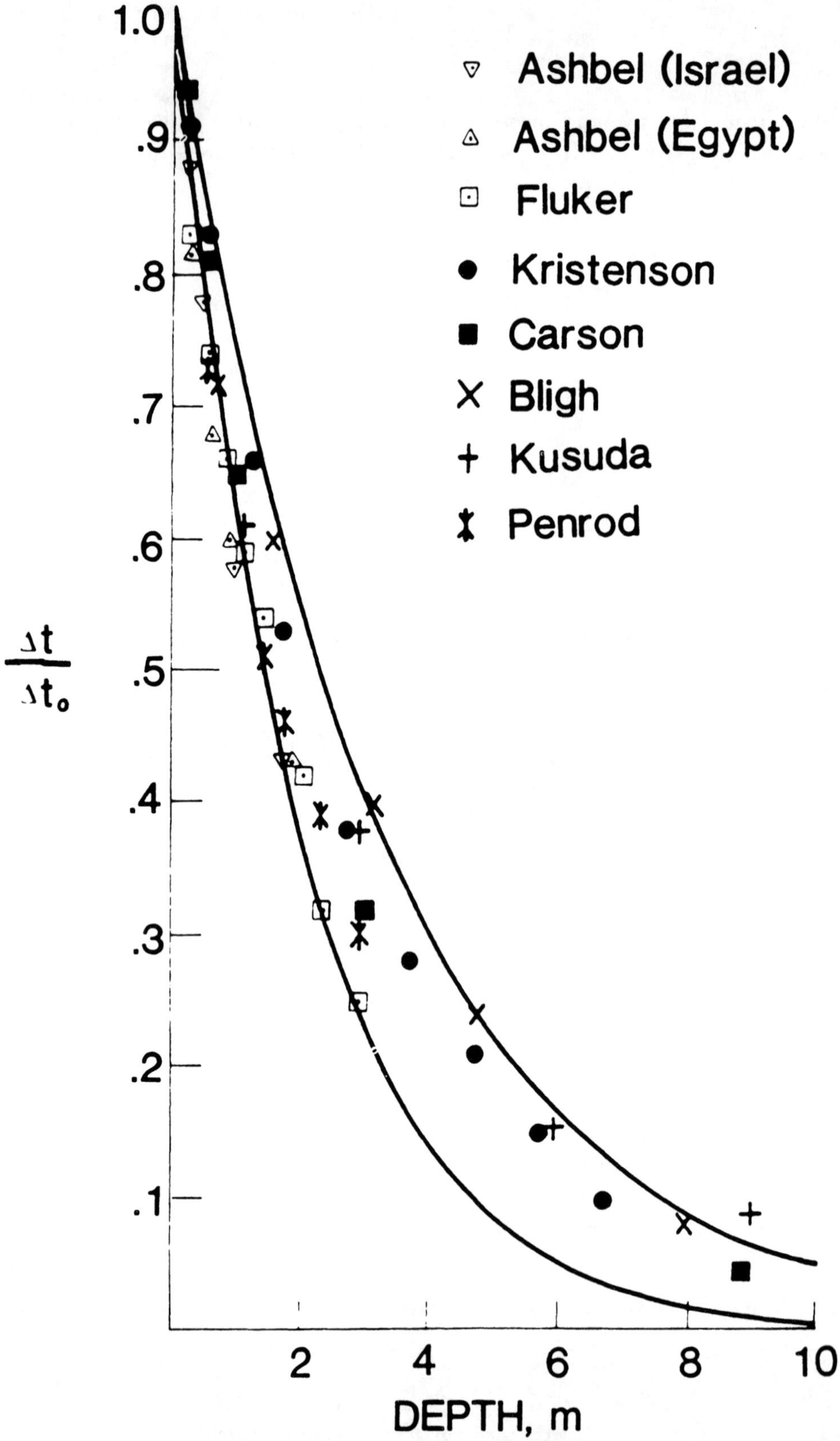

FIGURE 6.10 Relative temperature range (dT/dT_0) as a function of depth. Various experimental studies.

Different types of soil have different conductivities. Soil with a high loam and clay content can have higher water content than sandy soil. Therefore, in a given climatic region the soil diffusivity of loam and clay soils would be higher than that of sandy soil.

On the basis of these experimental studies, which were conducted in regions with different climatic characteristics, it seems possible to suggest the values shown in Table 6-A1 for the range damping factor (F) for soils in different climatic regions.

TABLE 6-A1. DAMPING FACTOR (F) FOR DIFFERENT CLIMATES AND SOIL TYPES

	Soil Type		
Climate	*Loam/clay*	*Intermediate*	*Sandy*
Desert	0.45	0.50	0.55
Arid	0.40	0.45	0.50
Intermediate	0.35	0.40	0.45
Humid	0.25	0.35	0.40
Wet	0.20	0.30	0.35

The expected annual range of any depth can then be estimated, from the known or estimated surface range, by the formula:

$$\textbf{Range (D)} = \textbf{Range}_{(o)} {*} \mathbf{e}^{-FD}$$

For quick reference Table 6-A2 shows the expected relative ranges (T_z/T_o) for the different climatic zones:

TABLE 6-A2. ANNUAL RELATIVE TEMPERATURE RANGES (°C FOR INTERMEDIATE SOIL IN DIFFERENT CLIMATES)

	Depth (m)						
Climate	*0*	*.5*	*1*	*2*	*3*	*4*	*5*
Desert	1.00	.78	.61	.37	.22	.14	.08
Arid	1.00	.80	.64	.41	.26	.17	.11
Intermediate	1.00	.82	.67	.45	.30	.20	.14
Humid	1.00	.84	.70	.50	.35	.25	.17
Wet	1.00	.86	.74	.55	.41	.30	.22

The suppression of the annual temperature with depth, for intermediate soil in arid and in wet climates, is shown in Table 6-A3.

TABLE 6-A3. TIME LAG (L) FOR DIFFERENT CLIMATES AND SOIL TYPES

	Soil Type		
Climate	*Loam/clay*	*Mixed*	*Sandy*
Desert	24	25	26
Arid	22.5	23.5	24.5
Intermediate	21	22	23
Humid	19.5	20.5	21.5
Wet	18	19	20

Time Lag

From the annual temperature patterns of these experimental studies the time lag of the occurrence of the maximum temperature could also be expressed as a function of depth, as is shown in Figure 6-11. In this case there seems to be a nearly linear relationship between the time lag and depth, ranging from about 19 to 25 days per meter depth.

From these experimental data it was not possible to clearly observe the effect of climate on the variability in the relationship between climate type and time lag. On the basis of theoretical consideration the time lag should be inversely related to the soil diffusivity, particularly to its water content, and hence also to the climate. Therefore the following formula is suggested for the time lag:

$$\text{Time Lag} = L.z$$

where the value of L depends on the climatic and soil characteristics as given in Table 6-A3.

Simplified Formula for the Soil's Temperature Patterns

Combining together the formulae for the range damping factor, F, and the time lag, L, a generalized formula estimating the soil temperature, T_s, for any depth and time is suggested:

$$T_s = T + A_o * \exp^{(-F*z)} * \mathrm{Sin}(0.986*N - 125 - L*z)$$

where:

T_s = Soil temperature on day N at depth D; (°C)
T = Annual average of the soil's temperature; (°C)
A_o = Annual amplitude of the soil's surface temperature (annual range/2); (°C)
F = Range damping factor, depends on climate and soil type.
z = Depth under the surface; (meters)
L = Time lag per meter depth, depends on climate and soil type; (days)
0.986 = Days of the year expressed in degrees (360/365)
N = Day's ordering number (January 1st = 1)
125 = April 25. According to the experimental observation, April 25 was chosen as the day number which would put the peak surface temperature at July 25th. In different regions this date may vary a little.

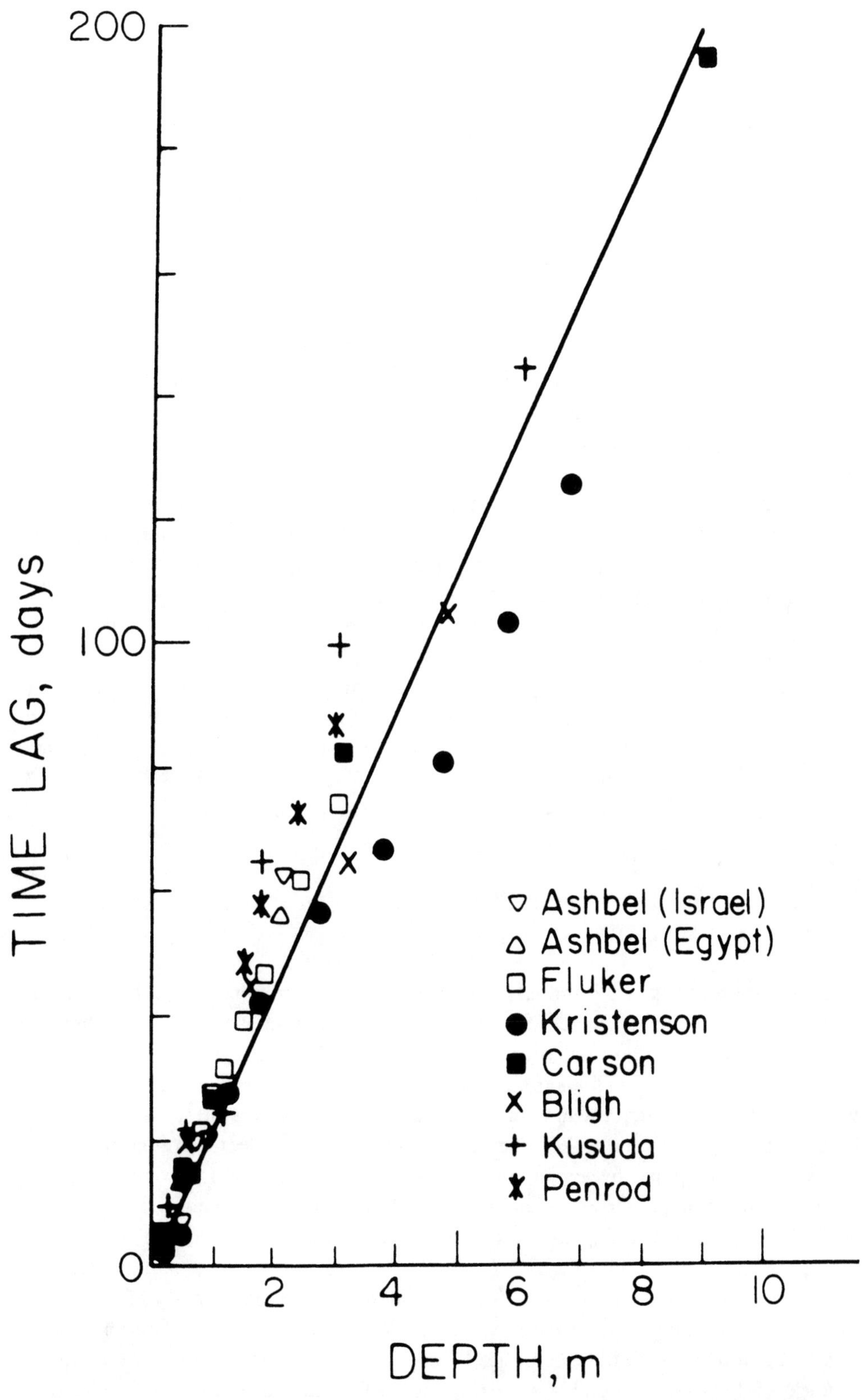

FIGURE 6.11 Temperature time-lag as a function of depth. Various experimental studies.

Comments

A When the surface temperature range is not known it can be estimated using the air temperature range and adding a number of degrees, based on the surface cover—for example, 5°C for bare soil, 0°C to completely shaded soil.

B The values for F and L given in Tables 6-A1 and 6-A3 should be considered to be temporary. In its present form the values of the parameters characterizing a given soil in this formula are very crude. More experimental information about the damping and time lag characteristics of soils in various regions will provide more specific values than those suggested above.

C The value of the annual average of the soil temperature can best be determined from annual patterns at depth levels below 2 meters, which are less affected by deviations from the normal patterns that may occur at any given year.

D The values for the range damping factor, F, and the time lag, L, can best be determined for any given region from the temperature patterns at depth levels of about 1 meter and 2 to 3 meters. The surface pattern is too erratic for reasonable accuracy.

E Any fundamental sine wave assumes symmetry around the maximum temperature. In fact, the rainy season may produce deviations from such symmetry that might increase the cooling rate of the soil. Thus, for example, in regions with maximum surface temperature at the end of July and rains from November through March, the soil temperature in November would be lower than in April, although the months are equidistant from July 31. Typically a sharp drop occurs in the soil temperatures at the start of the rainy season and there is a sharp rise after its end.

F In spite of deviations of the real soil temperature from the symmetrical patterns assumed in sine functions containing only the fundamental curves, such approximate formulae are useful in estimating the soil temperatures during the periods of high and low temperatures. For many applications this is the main interest.

Effect of Year-to-Year Climate Changes on Underground Temperatures

Year-to-year variations in the annual climatic patterns are common. The surface temperature is the most sensitive level and responds instantly to changes in the ambient air temperature, solar radiation, and precipitation. With increasing depth the effect of year-to-year changes are damped. Thus if a cooler summer with a lower maximum temperature, follows a warmer summer, the maximum soil temperatures during the second year will be lower, but the difference between the years will be smaller at deeper soil layers.

As an initial approximation, assume that the drop or elevation between successive years is in the same proportion as the damping of the relative temperature range given in Table 6-A2. For example if, in an arid region with mixed soil, the maximum annual of the surface temperature is elevated by 1°C above the long term average than at 1, 2, and 3 meters depth the elevations of the maximum would be 0.6°C, 0.4°C and 0.25°C.

By this procedure, the year-to-year variations in the temperature patterns at the different depth levels could be computed according to the actual variations in the surface conditions. This procedure allows a comparison of predicted temperatures with those measured over the course of several years.

REFERENCES

Akridge, T. M., and C. C. Benton. 1981. *Performance Study of A Thermal Envelope House. Phase II—Cooling Performance.* Atlanta, Georgia: Southern Solar Energy Center.

Akridge, J. M. 1982. *"Investigation of Passive Cooling Techniques for Hot-Humid Climates.* Research Report, College of Architecture, Georgia Institute of Technology, Atlanta, Georgia.

Ashbel, D. 1942. *Intensity of Solar Radiation and Earth Temperature,* Hebrew University Press. In Hebrew.

Bligh, T. 1975. "A Comparison of Energy Consumption in Earth-Covered vs. Non-Earth-Covered Buildings." In Moreland 1975, pp. 85–105.

Brown, M. B., and B. Givoni. 1980. "Cooling the Sub-Surface Earth." Research paper, UCLA School of Architecture and Urban Planning.

Cardon, J. E. 1963. "Analysis of Soil and Air Temperatures by Fourier Techniques." *Journal of Geophysical Research,* V-68, No. 8, pp. 2217–2232.

Cook, J., Ed. 1989. *Passive Cooling.* Cambridge, MA: MIT Press.

Fluker, B. J. 1958. "Soil Temperatures." *Soil Science,* July/Dec., pp. 35–46.

Francis, C. E. 1981. "Earth Cooling Tubes—Case Studies of Three Midwest Installations." Presented at *1st International Passive and Hybrid Cooling Conference,* Miami Beach, Florida.

Givoni, B. 1981. Earth Integrated Buildings—An Overview. *Architectural Science Review,* V–24, No. 2, pp. 42–53.

——— 1991. "Performance and Applicability of Passive and Low Energy Cooling Systems." *Energy and Buildings,* V–17, 177–199.

Givoni B., and L. Katz. 1985. Earth Temperatures and Underground Buildings. *Energy and Buildings,* V–8, pp. 15–25.

Glennie, W. L., P. S. Brock, and W. B. Hallagan. 1979. *A 'Cool-Pipe' Experiment and Design Method.* Princeton University: Princeton Energy Group.

Goetsch, Peterson, and Mueling 1983. "Evaluation of Earth-Tube Heat Exchangers as Air Tempering Systems." *ASAE Transactions,* paper #4073.

Golani, G. *Earth Sheltered Habitat.* New York: Van Nostrand Reinhold.

Hayeem, E. 1984. "An Experimental Study of Earth Temperature Modification in a Semi-Arid Desert Region." M.Sc. Thesis, Trinity University, San Antonio, Texas.

Kristenson, K. J. 1959. "Temperature and Heat Balance of Soil." *Oikos* 10:I, pp. 103–120.

Kusuda, T. 1975. "The Effect of Ground Cover on Earth Temperature." In Moreland 1975, pp. 279–303.

Labs, K. "Earth Coupling." In J. Cook, ed., 1989.

Moreland, F. L., ed. 1975. "The Use of Earth-Covered Buildings." Proceedings of a conference in Fort Worth, Texas. U.S. Government Printing Office.

Moreland, F. L., F. Higgs, and J. Shih, eds. 1980. Earth-Covered Buildings: Technical Notes. Proceedings of a conference in Fort Worth, Texas, May 1978. Washington, D.C.: D.O.E.

Penrod, E. B., and J. M. Elliott, and W. K. Brown. 1960. "Soil Temperature Variation (1952–1956) at Lexington, Kentucky." *Soil Science,* vol. 90, pp. 275–283.

Scott, N. R., R. A. Parson, and T. A. Koehler. 1965. "Analysis and Performance of an Earth-Air Heat Exchanger." Presented at ASAE Winter meeting, Chicago, Dec. 1965.

USGS. 1982. *Underground Space Center: Earth Sheltered Residential Design Manual.* New York: Van Nostrand Reinhold.

* *

7

COOLING OF ATTACHED OUTDOOR SPACES

This chapter deals with the possibilities of lowering the air and radiant temperatures in outdoor spaces adjacent to buildings—spaces such as front and back yards and patios. There is an interaction between the thermal conditions in a building and in its adjacent outdoor space. Different sources that can be used for cooling outdoor spaces are discussed.

ENVIRONMENTAL FACTORS AFFECTING COMFORT IN OUTDOOR SPACES

Many human activities take place in bounded outdoor spaces attached to homes, such as yards, patios, and courts. The duration and intensity of the use of an outdoor space may be greatly affected by the comfort (or discomfort) conditions existing in the area. Minimizing the duration and level of uncomfortable conditions can greatly increase the benefits derived from the outdoor space.

In hot regions during the summer, exposure to the sun and high temperatures outdoors often causes heat discomfort and stress. Thus people are less likely to use outdoor areas, and even less likely to enjoy using them. In such places the possibility of improving the comfort conditions by shading the area and, if possible and practical, also lowering its air and radiant temperatures, should be considered.

There are some basic differences between the environmental factors affecting human comfort in outdoor and indoor spaces. Inside buildings people usually are protected from direct exposure to solar radiation. Also, the longwave radiant temperature of an indoor space usually is not too different from the indoor air temperature. Indoors, the radiant heat load is not a significant factor affecting comfort.

On the other hand people staying outdoors are exposed to several direct and indirect forms of solar radiation. When an outdoor space is not shaded, the people in it are exposed both to the direct solar beam and to diffused solar radiation from the sky. Reflected solar radiation from the surrounding ground, which depends on the reflectivity (color) of the ground, may add significant heat load, especially in arid regions that often have scant vegetation and light-colored soil.

During the daytime the ground surface temperature, especially in arid regions, is significantly higher than the ambient air level. This temperature elevation imposes an extra longwave radiant load on the people who are exposed to it in addition to the direct, diffused, and reflected solar radiation. During periods of still air, when there is no wind, the air temperature above a given ground area is also affected by the temperature of the ground. A higher ground temperature can elevate the local air temperature at the "occupied" level, about 1 meter above the ground, and thus further increase the heat discomfort.

The physiological severity of the solar and longwave radiant loads depends on the ambient wind speed, temperature, and humidity, as well as on the clothing people wear. The higher the temperature and the humidity and the lower the wind speed, the greater the physiological heat stress imposed by the radiant load. The direct solar radiation falling on people in an outdoor space can be reduced by shading. The effectiveness of any shading device depends on geometric factors, in conjunction with the annual and diurnal patterns of the sun paths in the sky, and on the partial transparency of the shade material.

Common manufactured shading elements over outdoor spaces are thin and lightweight. Their surface temperature during the day would usually be higher than the ambient air, depending on their color. In many cases the underside of the shade element would be at a higher temperature than the ambient air as well, imposing a radiant heat load on people staying in the shade. The design details of the shading should aim to minimize this extra heat load.

Shade provided by vegetation often does not provide complete shading—some fraction of the solar radiation is transmitted directly through the foliage of the plants. The effective shading provided by plants changes with the age of the plants and, in the case of deciduous plants, can be very different in summer and in winter.

Any cooling scheme in an outdoor space should start, therefore, with minimizing the solar and the longwave radiant loads. Although such treatments do not comprise a "positive" cooling of the space, they are equivalent from the comfort aspect to actually lowering the space air temperature.

Differences Between Cooling Methods Applicable to Buildings and to Outdoor Spaces

The difference between cooling methods possible in buildings and those that may be used in outdoor spaces is caused by the very different conditions of mixing between the air confined in the space and the general ambient air, and in the temperature difference achievable between the confined and the ambient air masses. A building can be closed, minimizing the flow of hot outdoor air through it. The conductive heat gain can also be controlled by the thermal resistance of the building envelope. Consequently the indoor maximum temperature, even without any cooling, can be lowered by appropriate building design, especially in an arid region, below the outdoor level by up to about 8°C. (See Chapter 2.)

On the other hand an outdoor space, by definition, is open to flow of the ambient air, and the degree by which its temperature can be lowered below the surrounding level is quite limited. Therefore, a cooled outdoor space, even when shaded by an impermeable material, has a much higher rate of advective heat gain by wind flow across the space than a building heat gain by infiltration. Any cooling scheme of an outdoor space should take this factor into account.

Another difference between a building interior and an outdoor space concerns the radiant load. Within a building solar radiation can be eliminated and the mean radiant temperature is usually close to the indoor air temperature. Outdoors the reflected solar radiation and emitted longwave radiation from the surrounding hot ground can cause significant radiant heat load and therefore should be minimized by the treatment of the outdoor space.

Minimizing the temperature elevation of surfaces above the ambient air level in an outdoor space often does not involve an expenditure of energy. Planting trees and shrubs within and around the periphery of an outdoor space can effectively reduce the reflected solar radiation and the longwave radiation emitted from the surrounding area. On the other hand, lowering the air temperatures of an outdoor area *below* the ambient air level during daytime involves, directly or indirectly, some form of water evaporation and/or some expenditure of energy.

GENERAL CONSIDERATIONS IN COOLING OUTDOOR SPACES

When the design of a comfortable outdoor space in a hot region is considered, the first decision is whether to limit the treatment of the area only to shading it or whether, in addition to shading, also to attempt to lower its temperature below the ambient air level. In order to lower the temperature a water source would be needed.

A prerequisite for lowering the air temperature in an outdoor space is to minimize the wind flow through that space. A major source of heat transfer into a *cooled* outdoor area is by lateral advection of the warmer outdoor air. This heat advection can be minimized by the elements surrounding the boundary of the area, such as fences, walls, and shrubs, as well as by the details of the shading element over the cooled area. Of course, the wind blocking minimizes the direct physiological cooling effect of the wind. This conflict between "ventilation cooling" and temperature lowering, and the conditions under which each approach is advisable, are discussed in more detail below.

Once an area is partially isolated from the surrounding environment several cooling sources can be applied to lower the air and surface temperatures in the area, as discussed below.

Shade as a Prerequisite for Cooling Outdoor Spaces

Before considering any means for lowering the air temperature in an outdoor space shade from the sun should be provided. There are two reasons for the primacy of shading:

- **A** Protection from solar radiation has a larger physiological effect in reducing heat stress than the effects that can be expected from lowering the air temperature without providing shade in outdoor spaces.
- **B** Shading does not involve any expenditure of energy or water for irrigation, as do almost all the systems that can lower the temperatures in an outside area.

Materials that commonly are used to shade outdoor spaces are not solid and, in effect, transmit some fraction of the solar radiation. The transmitted radiation raises the ground surface temperatures in the shaded space. In practice, even with "tight" shade materials about 20% of the radiation falling on the shade can be assumed to be transmitted. The geometry and dimensions of the area affect the effectiveness of the shading.

Even in cases where the whole treated outdoor area is covered by shading elements, solar radiation can enter from the sides. The narrower the shaded area the greater is the fraction that is affected by this lateral sun penetration, especially if the area extends in the north–south direction.

The Temperature of the Shade's Underside

The temperature of the shade's underside during the daytime, except when shade is provided by plants, is usually above the ambient air level, as a result of the solar radiation that is absorbed by the shade. The temperature of the underside of the shade affects the longwave radiant field to which people staying in the area are exposed, and consequently it affects their thermal comfort. The shade's underside temperature depends on the external color of the material and its thermal conductance. The lighter the color and the lower the conductance the closer will the shade's underside temperature be to the ambient air level.

An exception is when shade is provided by a thick layer of leaves, such as that given by trees or vines with dense foliage. The lower leaves are protected from the sun by the upper part of the canopy and cooled by evapotranspiration, so their temperature is very close to the ambient air level—in fact, it might even be lower.

Tradeoffs between Physiological Cooling by Wind and Outdoor Air Cooling

In dealing with the issue of comfort in outdoor spaces a conflict may exist between two approaches:

- **A** Providing comfort through the cooling effect of the wind.
- **B** Lowering the air temperature in the outdoor space.

Any lowering of the air temperature in an outdoor space requires that the (warmer) wind penetration be minimized. The conflict between the two approaches exists because minimizing the wind penetration in an open space usually is accomplished by fixed devices. The wind blockage can not be turned on and off, as is possible in a building.

There are no windows to open and close outside. Therefore, once the space is semi-enclosed by such devices as surrounding walls, shrubs, or earth berms, the wind cooling effect is minimized even during periods when wind would be desirable.

In some cases the state of comfort at the ambient air temperature with moderate wind speed may be better than at somewhat lower temperatures but with lower wind speed. Quantitative comparison of the comfort conditions in the two cases can be done by the use of "comfort indices." Generally, it can be expected that in humid regions the wind will have greater effect on comfort than will any achievable reduction in temperature. In this case, reflected solar radiation and emitted longwave radiation from the surrounding area can be reduced by the treatment of the area; for instance, by covering it with low but dense vegetation so that the surface temperature would be lowered while the wind would not be blocked.

In arid regions the choice between the two options depends on the local typical maximum air temperature. In places where the maximum temperature is below about 30 degrees Centigrade, wind can be very effective as a comfort element. In such regions shading the space, while leaving it open to the wind, might be the best approach to minimize heat stress. In places where the daytime temperatures are higher, the humidity is low (that is, where the relative humidity is below 50%) and water is available, lowering the temperature and blocking the wind may be the preferable approach. This can be accomplished by the various means discussed below.

Lowering the Surface Temperature in Shaded Outdoor Spaces

Two "treatments" can be applied in outdoor spaces to lower the surface temperature in a shaded space to below the ambient level. They are based, directly or indirectly, on evaporative cooling and include:

- **A** Lowering the temperature of elements surrounding the cooled areas, and
- **B** Cooling the pavement of the area itself.

Lowering the Temperature of Elements Surrounding the Space

In order to minimize the radiant heat load from reflected solar radiation and emitted longwave radiation from the ground around the area, the cooled area should be surrounded by elements that can be kept at temperatures below that of the exposed surrounding ground. These can be elements such as walls, high shrubs, or earth berms. The area also can be sunken below the surrounding ground level.

When the elements surrounding the cooled area are plants, walls covered by vines, or earth berms covered by vegetation, their temperature closely follows the ambient air level, and the excess radiant load thus is minimized. This point is discussed further in the section dealing with courtyards adjacent to buildings. Moreover, if the surfaces of the surrounding elements are wetted, *while shaded*, their temperature will drop toward the ambient WBT and, in arid regions, these temperatures can be significantly lower than the ambient air.

In practice, if the surrounding elements consist of built walls or earth berms covered by concrete or stones, it is possible to wet them or even to have a water film running over their surfaces—for instance, in the form of cascades—thus providing constantly wet surfaces around the cooled area. The water flowing over the surrounding walls need not be of high quality. Where brackish water is available it can be used in this application, as a high rate of water flow can minimize salts and mineral deposits.

Cooling the Pavement

Even if an outdoor dry surface is shaded effectively, its temperature would be somewhat elevated above the ambient air level. Cooling the pavement of a relatively wide shaded area thus can lower the radiant temperature to which the people in the area are exposed, although the effect of the air temperature, at the level of the "occupied" zone, would be negligible unless the cooled area is very large.

There is a point in cooling the pavement in a given space only if the dimensions of that area are large enough so that the change in the mean radiant temperature will be significant. This means that the smaller dimension of the area (its width) should be above a given minimum—above about 3 to 4 meters. Consequently, there is no point in lowering the temperature of pavements in narrow strips, where the effect on the mean radiant temperature would be minimal. An exception is a paved private "confined" yard surrounded by walls, where cooling the pavement can be accomplished by very simple means and can have a measurable physiological effect even in such a small area.

In a private paved yard the pavement can be cooled simply by wetting it from time to time during the daytime, using a sprinkler or even a garden hose. These methods of wetting are not practical in a public outdoor space, however, as they disrupt the use of the space.

A method for cooling a pavement, developed by Professors Asiain and Dominguez and applied at the '92 EXPO site in Seville (Asiain 1988, Dominguez 1989), is to circulate cool water under the surface in thermal contact with the material. The source of the cool water can be a pool that is cooled by a continuous spray of water, day and night, for example, or it can be a shaded pond, as is discussed in more detail later on. Water flowing over walls surrounding the rest area (as discussed above) can continue to flow under the pavement into a central pond.

LOWERING THE AIR TEMPERATURE IN OUTDOOR SPACES

Once an area is effectively shaded and partially separated from the surrounding air by being sunken and/or surrounded by walls, trees, or shrubs, the air and radiant temperatures in it can be lowered below the level of the outdoor air. The air within a semiconfined space can be cooled either by the introduction of precooled air or by direct evaporative cooling of the air inside it. Air can be precooled by a passive evaporative cooling tower or by a convective "shower" cooling tower, systems that are discussed below.

To maximize the physiological benefits from the cooled air it is advisable to introduce it near ground level. In this way temperature stratification over the area would be enhanced, with the cooler air at the level where people are standing or sitting.

Untreated river water, brackish water, or even sea water can be utilized for some of the cooling systems possible in outdoor spaces, as will be noted in the discussion of the specific systems. Such water sources are available in many arid and desert regions.

The depression of the wet bulb temperature (WBT) below the air temperature (DBT) in summer during the hot hours of the day is the most direct quantitative criterion for assessing the potential of evaporative cooling in a given location. The expected temperature drop, below the ambient air level, of air exiting from an evaporative cooling system would be about 60 to 80% of the WBT depression (DBT–WBT).

Mechanical Evaporative Cooling

Mechanical evaporative coolers can cool large quantities of air to a temperature very close to the ambient WBT with relatively low expenditure of energy as compared with compression air conditioners. It is a conventional technology for cooling buildings, and it is used often in arid regions. Because of its relatively low cost and low energy consumption it can also be applied for cooling open spaces.

It should be noted, however, that mechanical evaporative coolers use high-quality water. Brackish water, which is available in many arid and desert regions, cannot be used in mechanical evaporative coolers because it will clog the wetted pads through which the air is drawn. This is in contrast with some other techniques of evaporative cooling, in which brackish water *can* be used.

Mechanical evaporative coolers are noisy. To minimize noise in the outdoor area it is preferable to install the evaporative cooling machine at some distance from the area. The cooled air can be directed by ducts to the area. To derive the greatest physiological benefits from the cooled air it is advisable to duct the air under the pavement, introducing it near ground level, at the space's "occupied" zone.

Artificial Fog

Artificial fog can be generated by injecting water at high pressure through very minute orifices. If the water droplets are small enough, and the WBT depression below the air temperature is large enough, the droplets evaporate almost instantly. The evaporation of the droplets cools the air without forming liquid water over people staying nearby, although the fog is visible.

Fog generators consume a large quantity of a pump's energy relative to their cooling effect, and they need water with low mineral content or water that has been treated chemically (Mee Inc., personal communication).

Even with a very light wind the fog is carried away downwind. When the winds in a given region are light and the direction is more or less constant, fog generators set out in a line, located upwind of the area to be cooled, can provide a measurable cooling effect. Artificial fog is an especially attractive cooling technique for local "spot" cooling of small areas where people congregate, such as in front of kiosks along walkways.

The Tucson Passive Evaporative Cooling Towers

A passive evaporative air cooling tower attached to a building has been developed by Cunningham and Thompson at the Environmental Research Laboratory of the University of Arizona in Tucson (Ref. 2). It consists of a downdraft tower topped by vertical wetted cellulose pads. Water is distributed at the top of the pads, collected at the bottom by a sump, and recirculated by a pump. A detailed description of this system, its performance, and a mathematical model developed by Givoni that predicts its performance, is presented in Chapter 5. Such a passive evaporative cooling tower can be used to provide cool air for semiconfined outdoor areas.

Inertial/Convective "Shower" Air Cooling Tower

This cooling tower was proposed by the author for the '92 EXPO and constructed in the Rotunda of the '92 EXPO site in Seville, Spain.

When fine drops of water are sprayed vertically downward from the top of an open shaft the falling water, which has a very large surface area, entraps a large volume of air. The momentum of the falling water is thus transmitted to an air stream, creating an inertial air flow. The evaporation of the fine drops cools the air to a level close to the ambient wet bulb temperature. As any evaporation from the droplets takes place in the free air stream, brackish or sea water, if it is available, could be used in this system. The performance of the system has been tested by one of my students at UCLA (Al Hemiddi 1991, 1992). The test facility configurations included both an enclosed room and a patio.

A detailed description of this system, applied to cooling a building, and the performance of a model of a cooled room, is presented in Chapter 5. However, when cooling a patio some modifications of the model are needed to take into account the solar penetration and mixing with the outdoor air through the open "roof." The basic formulae of the model and these modifications are presented below.

Modelling the Temperature of an Enclosed Patio Cooled by the "Shower" or the Arizona Cooling Towers

The flow rate and temperature of the air exiting from an evaporative cooling tower—either the shower type or the Arizona type—are the same in the case of a cooled patio as in the case of a cooled building as presented in Chapter 5. Here the operational formulae are summarized:

For the Shower Tower:
The Flow Rate of the Air Exiting From the Tower (Flow)

Flow = 45*WF*H^0.5; (m^3/hr), (A tower without a wind catcher)
Flow = (45*WF*TH^0.5)+440*(WS–1); (Tower with a wind catcher)

Temperature of the Exiting Air

$$T_{exit} = DB-(DB-WB)*(1-\exp(-.8*H))*(1-\exp(-.15*WF))$$

When:

WF = Water Flow Rate (Liters/minute)
TH = Height of the shower

For the Arizona Tower:

$$\text{Flow} = 0.033 * \text{Aevap} * (\text{HT}*(\text{Ta}-\text{WBT})^{0.5};\ (\text{m\^{}3/hr})$$
$$T_{exit} = \text{Ta} - 0.87 * (\text{Ta} - \text{WBT});\ (^\circ\text{C})$$

When:

Aevap = Area of the wet pads

The Air Temperature of the Patio

The basic difference between the indoor temperature of a building and the temperature of a patio is due to their exposure to the sky. While a cooled building is usually closed and protected from the sun, a patio is fully or partially roofless and thus exposed to the sun and to mixing with the ambient air flowing above the building. These factors make it necessary to modify the formulae predicting the space temperatures in these two cases.

The heat gain and loss of an enclosed patio partially or fully open to the sky can be calculated as follows:
Let:

Tin = Average air temperature in the patio, °C
Tout = Temperature of air flowing out of the patio, °C
Ta = Outdoor air temperature, °C
WS = Wind speed next to the patio, m/s.
Ar = Area of the roof opening, m^2.
Qsol = Solar heat gain through the roof, W/m^2
Conv = Convective gain from mixing with the air flowing above the patio,; Conv = Ar*10*WS, (W/C.m^2).
0.333 = Volumetric heat capacity of the air, (Whr/C.m^3)
UA = Hourly heat gain coefficient of the wall surrounding the patio, (W/C).
Tout = Texit+2*(Tin–Texit), (°C) (Assuming the patio air temperature is half way between the cooled exit air and the outflowing air.

Then:

$$\text{Gain}=\text{UA}*(\text{Ta}-\text{Tin}) + \text{Ar}*10*\text{WS}(\text{Tout}-\text{Tin}) + \text{Ar}*\text{Qsol}$$
$$\text{Loss}=0.333*\text{Flow}*(2*(\text{Tin}-\text{Texit})) = 0.67*\text{Flow}*(\text{Tin}-\text{Texit})$$

Hence:

$$\text{Tin} = \frac{(\text{Ar}*\text{Qsol}+\text{Ar}*10*\text{WS}*\text{Ta}+\text{UA}*\text{Ta}+0.67*\text{Flow}*\text{Texit})}{(0.67*\text{Flow}+\text{Ar}*10*\text{WS}+\text{UA})}$$

(Instantaneous patio temperature, disregarding any thermal mass.) Comment: The mixing of the patio air with air flowing above depends greatly on the details of the patio. For a cooled patio this mixing should be minimized. The convective coefficient is thus assumed as 10 W/C.m^2. In any specific case this coefficient should be estimated according to the details of the patio and the building by which it is enclosed.

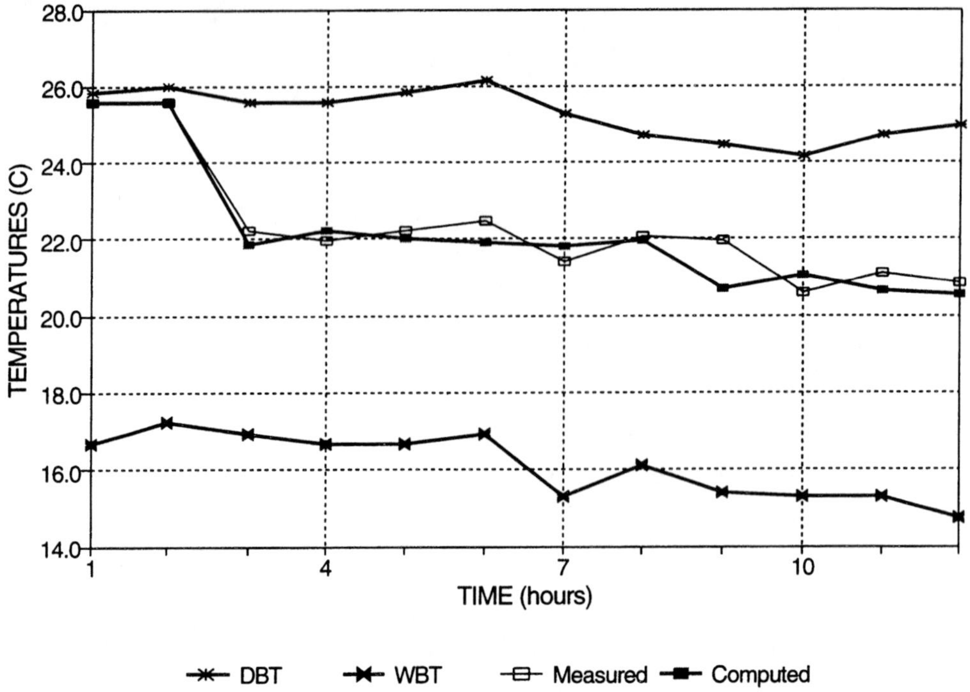

FIGURE 7.1 Measured and Computed Patio Temperatures, January 14, 1991.

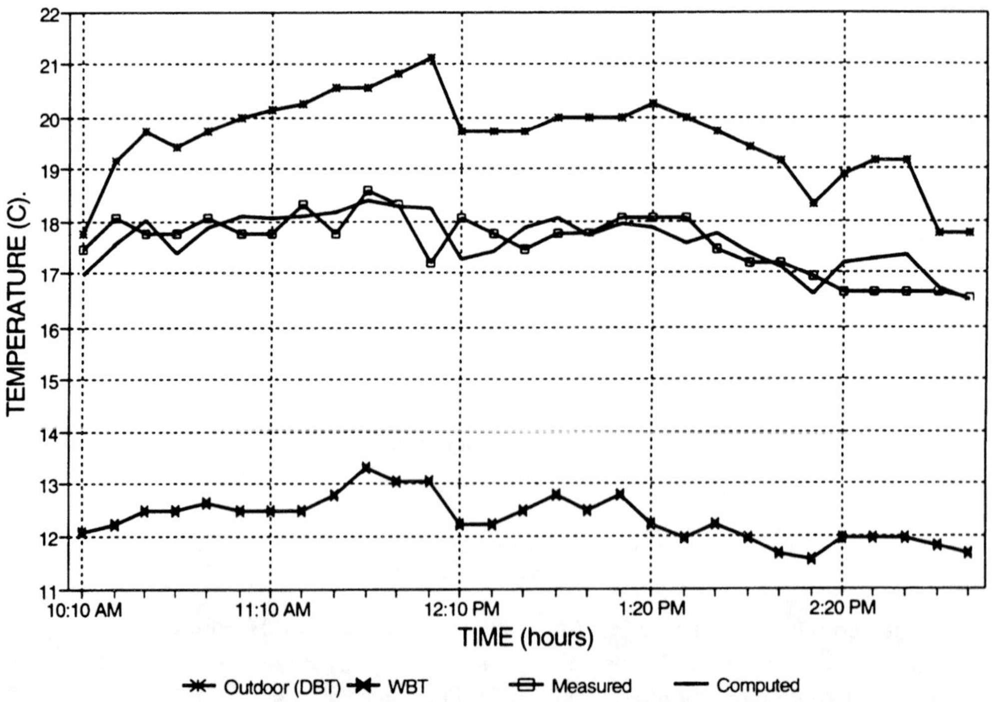

FIGURE 7.2 Measured and Computed Patio Temperatures, March 7, 1991.

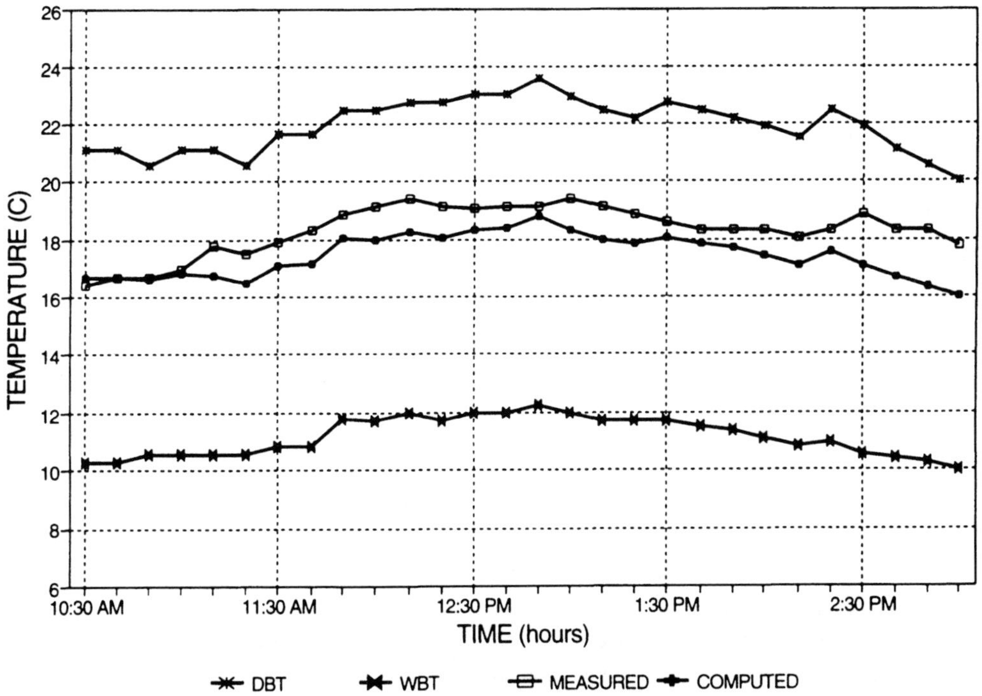

FIGURE 7.3 Measured and Computed Patio Temperatures, March 8, 1991.

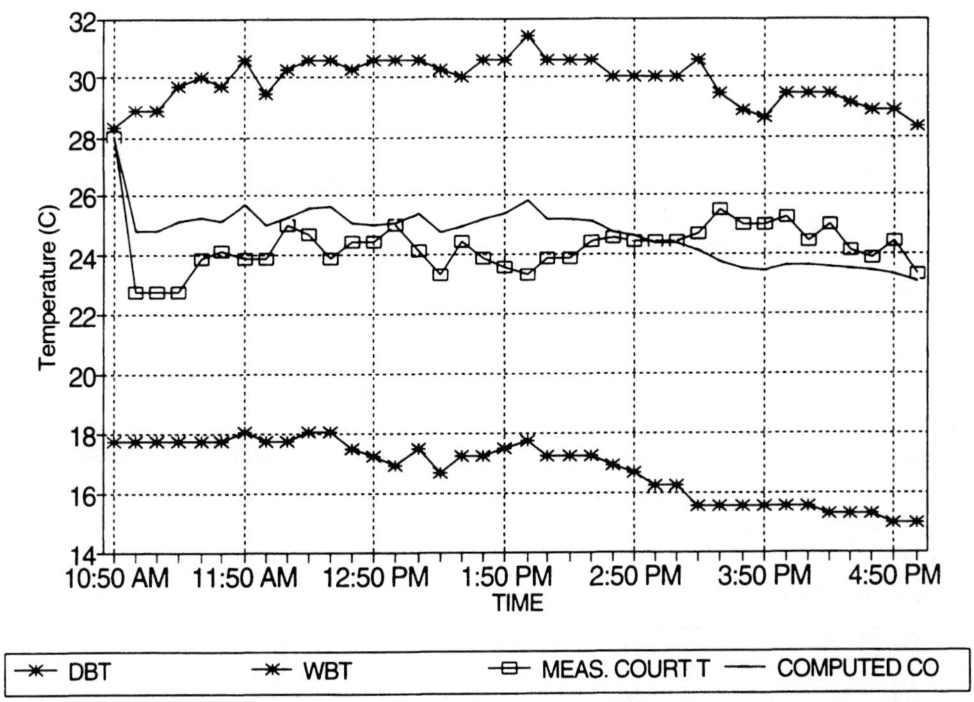

FIGURE 7.4 Measured and Computed Patio Temperatures, May 6, 1991.

TABLE 7-1. MEASURED AND COMPUTED TEMPERATURES IN A COOLED PATIO

Date	*DBT*	*WBT*	*Tin-Meas.*	*Tin-Comp.*	*dTmeas/dWBT*
Jan. 14	25.1	15.9	21.7	21.5	0.37
March 7	19.6	12.3	17.6	17.6	0.27
March 8	21.9	11.1	18.3	17.4	0.33
May 6	28.1	17.3	24.0	24.0	0.38
Averages	23.7	14.1	20.4	20.1	0.34

Figures 7.1–7.4 show measured and computed temperatures during four days with the same configuration (Jan. 14, March 7, March 8 and May 6, 1991) when 60% of the patio roof was open to the sky.

Table 7.1 summarizes the measured and computed average temperatures during the experiments done during the four days, along with the measured temperature drop of the indoor temperature relative to the Wet Bulb Depression (dTmeas/dWBT).

The measured average relative indoor temperature drop, for the courtyard model, was about 35% of the WBT Depression. It is, of course, less than it would be in a room cooled by the same system because of the direct solar gain and the mixing with the outdoor air through the roof opening. These two heat gain factors can be reduced greatly by suitable design details of the patio. For instance, trees with a wide and high canopy in the patio could provide shade and minimize the cooled patio air from mixing with the warmer air flowing above it.

Effects of Shower Height and Solar Load on the Patio Temperature

With the availability of the model for the cooled patio air temperature it is possible to evaluate the effect of various design parameters on the patio temperature under given climatic conditions. Three such parameters, in the case of a shower system, are the water flow rate, the height of the tower, and the solar load (which can be controlled by shading of the patio roof opening).

Figure 7.5 shows the relative exit temperature drop (relative to the WBT depression), computed by the model for different shower heights (1 to 5 meters) and for various water flow rates (10 to 30 liters per minute). In fact, with a height of 2 meters and a water flow of about 14 l/min, the relative temperature drop actually measured was above 70%, and with three meters height and water flow of 12 l/min the measured temperature drop was near 80%. Therefore, it is not possible to lower the exit air much further, even with greater heights and water flow rates. The main benefits from such increased heights and water flow may be in the higher air flow rates.

Figure 7.6 shows the calculated temperatures in a small patio, assumed to have a floor area of 15 m^2 and a roof opening of 10 m^2, when the ambient air temperature is 35°C and the WBT 17°C, with shower heights of 2 to 5 meters and solar loads (controlled by shading) of 100 to 400 w/m^2. Figure 7.7 shows the temperatures in the same patio with water flow rates of 10 to 30 liters/min and shower heights of 2 to 5 meters, when the solar gain is constant at 200 W/m^2.

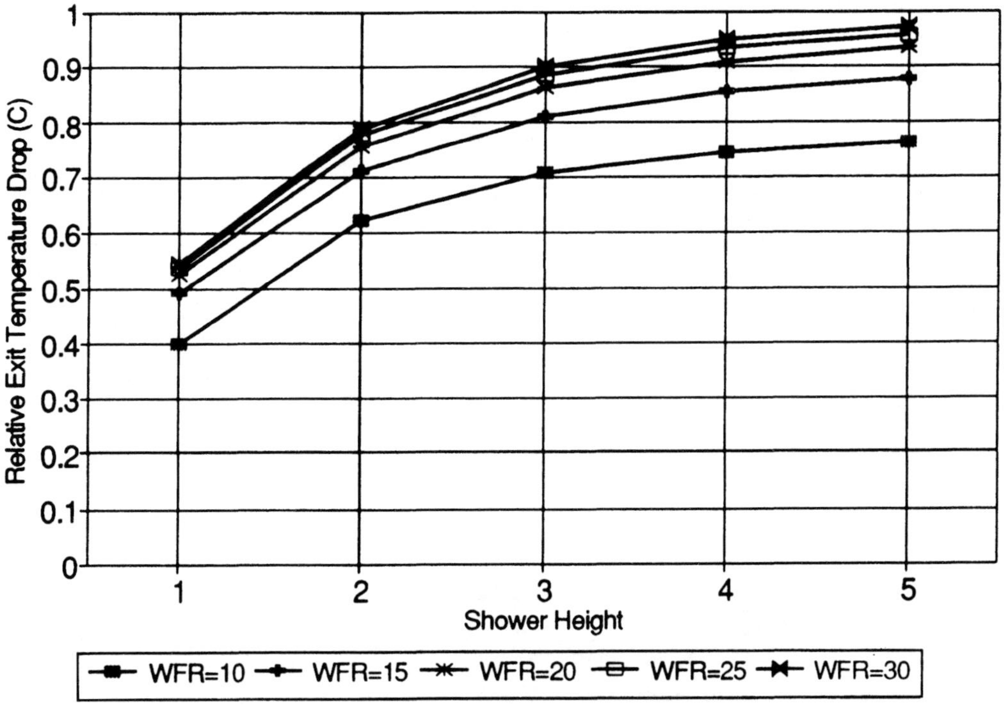

FIGURE 7.5 Effects of Shower Height (m) and Solar Load (Watts) on the Exit Temperature Drop (relative to WBT depression).

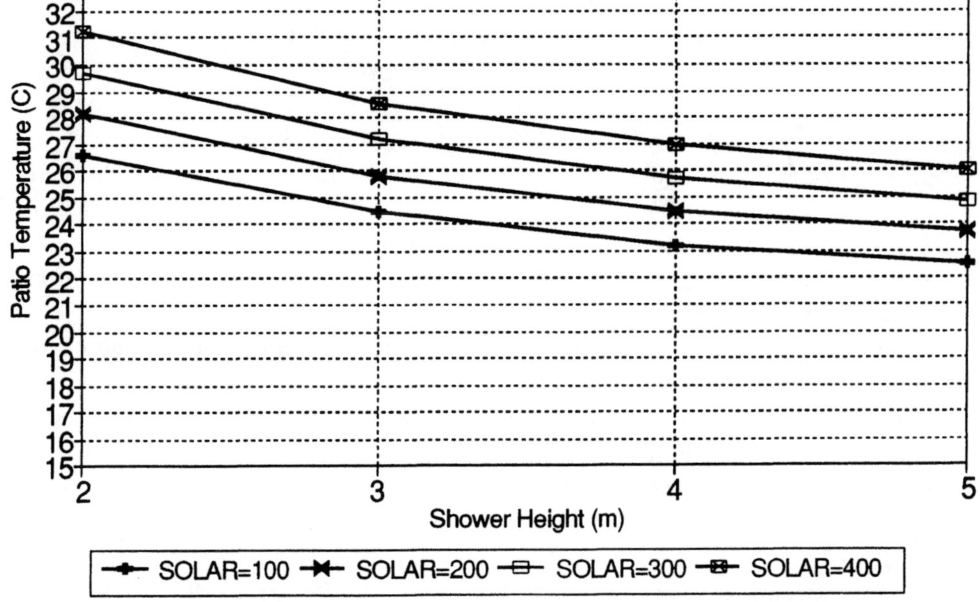

FIGURE 7.6 Calculated Air Temperature of a Patio as a Function of the Solar Heat Load and the Shower Height.

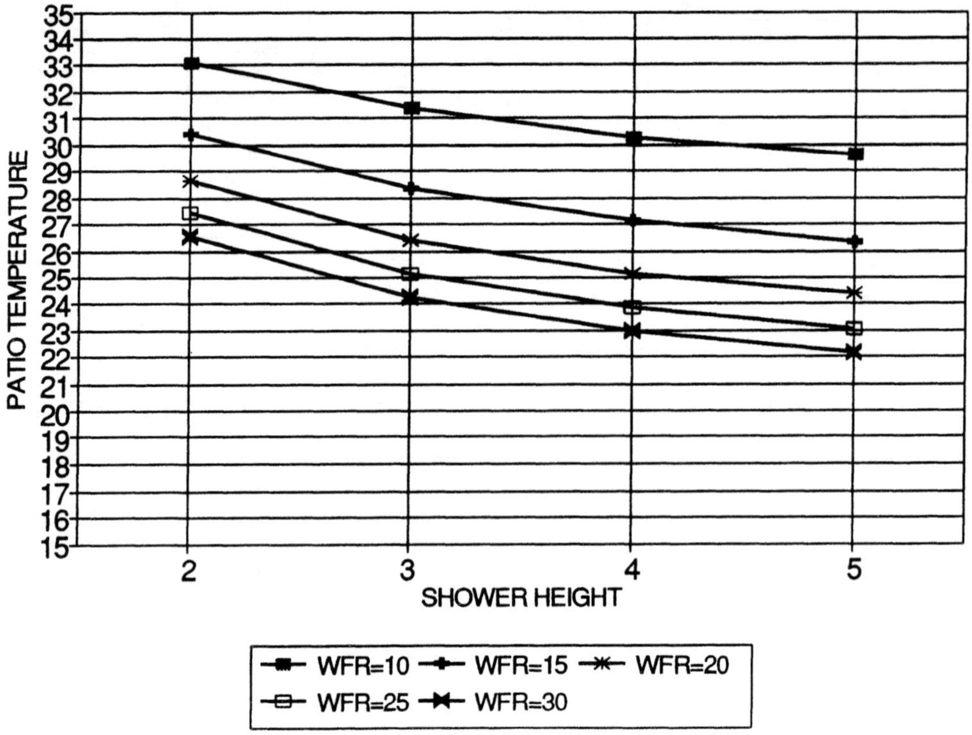

FIGURE 7.7 Calculated Air Temperature of a Patio as a Function of the Water Flow Rate (in l/min) and the Shower Height.

DISCUSSION OF ISSUES SPECIFIC TO ADJACENT OUTDOOR SPACES

Open spaces adjacent to buildings and enclosed by walls—that is, courtyards and internal patios—are very common in many hot regions. Houses with courtyards are considered the most "appropriate" in many hot regions, especially in deserts. This notion is based on the fact that such courtyards are very common in such regions, mainly in developing countries, and have been so for centuries.

When analyzed from the viewpoint of their impact on thermal comfort and energy consumption (Givoni 1994), it is found that the performance of courtyards depends greatly on the details of their design. With some details courtyards can provide a pleasant outdoor environment and also improve the indoor thermal conditions. With other details (for example, paved and/or bare ground) internal patios actually may elevate the indoor temperature of the building around them and cause poor ventilation in the rooms located on the leeward side.

This has been demonstrated in a study done by Etzion (1991) who measured air temperatures at one meter height in two courtyards. Both had concrete pavement over their whole areas. The temperatures of the two courtyards was very similar and in both cases the temperatures were higher, day and night, than the ambient air in the open space nearby at the same height. The average minimum temperature was higher by about 0.5°C and the maximum by about 2.3°C.

However, by applying some specific design details and cooling systems to courtyards adjacent to buildings, during the summer it is possible to lower the air and radiant temperatures within the courtyard below the level of the ambient air. Of course, a cooler courtyard provides a more comfortable environment for children's play and other outdoor activities in an area adjacent to the buildings and also lowers the heat gain by the building through the common walls.

The use of plant shading—trees and pergolas—makes it possible to minimize the radiant temperature within a yard, as well as to reduce the radiant load on the adjacent windows and walls. But just by shading alone it is not possible to lower the air temperature in an open area significantly, unless the planted and shaded area is very wide.

In winter, of course, a sunny courtyard is much more pleasant and healthy than a shaded one. In regions with rainy winters solar radiation is also desirable because it speeds up the drying of wet walls. The trees should not completely shade the building's walls, particularly the windows that face the courtyard. Therefore, deciduous trees and vines, which have thick foliage in summer but become bare in the winter, should be used for shading.

When adequately treated, such adjacent open spaces can modify the climatic conditions of the ambient environment next to the skin and openings of the building and thus also improve the indoor climate. Such improvements may include:

- **A** Shading the building walls adjacent to the open space with the shade elements in the courtyard without blocking the sun in winter.
- **B** Increasing the insulation value of adjacent walls by dense and high shrubs and vines, which create a semidead airspace next to the wall surface. This last effect can be especially significant and beneficial in both summer and winter when walls have very low thermal resistance, particularly in the metal or asbestos-cement buildings that are common in many developing countries.
- **C** Lowering the air temperature with one of the various forms of evaporative cooling that can be installed in the courtyard—for example, wetted pavement, wet walls, or a small pond with a fine spray.

It should be stressed again that modifications of the climatic conditions next to the skin of the building can be achieved only if the courtyard is separated from the environment outside it. Separation is required not only laterally, by high walls around the yard, but also from above, by built elements or by tree canopies, which separate the confined cool air in the yard from the warmer air above.

Thermal Interaction Between a Building and a Courtyard

The possibilities of thermal interaction between a building and an adjacent courtyard, and of the utilization of the cooling system of the building to cool the adjacent courtyard as well, depend to a large extent on whether an evaporative cooling system or a compression air conditioner is used to cool the building.

Evaporatively Cooled Buildings

Evaporative cooling offers the best possibilities for utilizing the cooling system of the building for cooling the courtyard, because it cools very large quantity of air and exhausts the cooled and humid air from the building in a once-through flow. The exhausted cool air can be directed into a confined courtyard, where it can lower the temperature before it is discharged to the general environment.

Integrating the cooling systems of a building and its courtyard works best with an interior courtyard, which is surrounded on all sides by the building. If exhaust air is to be used to cool a courtyard the building's duct system, which distributes the cooled air within the building, should be designed accordingly.

In this way the cooling system is utilized twice: it cools first the building interior and then the outdoor space, which can serve as a more pleasant area for rest and outdoor activities. The cooled courtyard, in turn, reduces the heat gain of the building across the adjacent walls and windows.

Buildings Cooled by Compression Air Conditioners

Buildings cooled by compression air-conditioning systems have very low rates of air change, and such cooling systems are not capable of cooling both a building and that building's adjacent open space. Therefore, when a building is cooled by a compression air conditioner the courtyard has to be cooled by a different system.

Un-conditioned Buildings

Even when a building that is located in a hot arid region is not cooled by a mechanical system, an appropriate design can still keep it at a significantly lower temperature than the outdoors during the day. Such a design should include solar control. The building's walls and roof should be a reflective (near white) color, and the building should be made of materials that can provide suitable thermal resistance and heat capacity.

An adjacent cooled courtyard should then use separate cooling systems that are applicable to outdoor spaces, such as mist generators, wet walls, cooling towers, and fine-spray fountains, as discussed above. The temperature of the air in the courtyard would then be lower than that of the outdoor air.

The cool air from the courtyard can be used as the source for building ventilation through openings in the separating walls, either by natural (thermosyphonic) ventilation or by such forced ventilation devices as an exhaust fan placed at the other side of the building.

Cooling Options for Courtyards

In a private courtyard that is internal or adjacent to a building, practically all of the cooling systems discussed above can be applied to lower the temperature of the air and the surfaces that enclose it.

When comparing direct evaporative cooling of a building itself with evaporative cooling of an adjacent courtyard, it should be noted that a courtyard can be cooled by systems that may not be used indoors, including wet walls, fine droplet fountains, and mist generators. Furthermore, a mechanical evaporative cooler has to use high-quality water to minimize clogging of the wet pads where the evaporation takes place. On the other hand, some of the systems suitable for courtyard cooling can use water that is not suitable for mechanical evaporative coolers. For example, they can use brackish water, which is often available in arid regions.

Shading

The function of shade elements in a cooled courtyard is not only to block the sun but also to provide some separation between the cooled air volume below the shading and the hotter ambient air above it. Shading of courtyards, which fulfills these two functions, can effectively be provided both by trees with high trunks and wide canopies and by pergolas made of vines with relatively thick foliage.

Such shading, together with the building itself and the walls surrounding the courtyard, can "isolate" the air mass within the yard and keep the wind from mixing the cooled air with the warmer ambient air as it otherwise would. With such a partial separation a desirable thermal stratification can be created in the courtyard during the daytime, with the cool air below and the warmer air above preventing thermal air currents. At the same time water vapor, generated by various evaporative cooling systems that may be applied in the courtyard, can diffuse through the tree canopies to the ambient environment even in the absence of air flow. In this way evaporative cooling can proceed continuously without making the courtyard excessively humid.

During the night the tree canopy blocks direct radiant heat loss from the courtyard to the sky, because radiation is emitted only from the upper layer of leaves. However, whenever the air below the canopy is warmer than the air above, it rises, and the air that has been cooled by contact with the upper leaves can sink down to replace it, creating natural convective cooling.

Deciduous trees and vines can shade the courtyard in summer while admitting the sun in winter, rendering the courtyard comfortable all year long.

Cooling a Courtyard Adjacent to or Enclosed Within a Building

Because of the effect of landscaped but otherwise uncooled courtyards on the exposure conditions and thermal performance of buildings, the impact of cooled courtyards can be quite significant. When a courtyard is shaded by trees and cooled by some of the cooling systems discussed above, the air temperature would be lower than the ambient temperature. This, in turn, can reduce the energy use of buildings and the peak demand for electricity for air conditioning.

Several features of an enclosed courtyard adjacent to a building can be designed as cooling elements. Among them are:

- **A** The walls surrounding the yard
- **B** A shaded pond with spray fountains
- **C** The pavement of the courtyard

The Walls Surrounding the Courtyard

When walls surrounding a courtyard are exposed to the sun, the temperature of their surfaces that face the courtyard may be elevated above the ambient air temperature. This imposes a radiant heat load on the people who stay in the courtyard and reduces the possibility of cooling the courtyard air temperature below the ambient level.

The simplest solution to this problem is to cover the wall surfaces that face the courtyard with climbing vines or high shrubs. The leaves of the vegetation will be kept at temperatures close to the ambient air, thus lowering the contribution of the walls to the courtyard's radiant field, although not providing a "positive" cooling effect.

The surfaces of the walls facing the courtyard can be cooled below the ambient level by a closed cycle flow of water that flows down over the surfaces, eventually falls into a pond in the courtyard, and then is recirculated. The air next to the cooled surfaces of the walls facing the courtyard is also cooled, and it flows down by free convection. However, the effect of this convective cooling on the courtyard air temperature would be rather small because of the very small "thickness" of the falling cool air stream. Still, in this way the wetted walls can lower the radiant temperature in the courtyard, and at least to some extent it can lower the air temperature as well.

Water Ponds

In a private courtyard adjacent to a house it is possible to design the entire landscape, including a pond that is surrounded by trees and effectively shaded. A shaded pond would be cooler than an unshaded one and can maintain the water average temperature slightly above but close to the diurnal average WBT. Deciduous trees will allow sunshine into the courtyard during the winter.

A cool pond by itself will have a very small effect on the air temperature in the courtyard, because there is just a limited heat exchange between the pond's cool water and the warmer air above it. This heat exchange can be greatly increased if the water in the pond can be circulated through a fine-drops fountain. The semiconfined air in the courtyard can then be cooled, mainly by convection, by the large surface area of the fine drops.

If a pond with still water is not shaded its temperature will be higher than that of a shaded pond. It is possible to lower the temperature of such a pond by circulating the water through a fountain that generates a fine-drop spray. This cooling effect will be accomplished at the cost of an increased rate of evaporation and water use as compared with the case of a shaded pond.

The Courtyard's Pavement

Many courtyards have areas that are paved with such material as stone, tile, or bricks. The temperature of the pavement affects the radiant field in the courtyard and, if the yard's air volume is partially isolated from the general environment, it can also have some effect on the temperature of the air above.

The temperature of the pavement in a private courtyard can easily by lowered if the pavement is regularly wetted during the hot hours of the day, because the evaporation from the surface of a moist pavement draws almost all the energy for the vaporization from the pavement.

If the walls of the courtyard are cooled by a water cascade, the water flow can continue under the pavement of the courtyard into the pond, thus further lowering the environmental radiant temperature in the space.

REFERENCES

Al Hemiddi, N. 1992. "A Cooling Tower for a Passive Evaporative System in Hot Dry Climate." ISES Congress. Cocoa Beach, Florida.

——— 1991. "Preliminary Investigation of the Effect of a Passive Direct Evaporative Cooling System on a Courtyard at UCLA." Proc. PLEA '91. Seville Spain.

Asiain, J. L., J. M. C. Lainez, A. L. B. Rguez, and J. P. L. Halcon. 1988. "Climatic Control for the Open Spaces of the 1992 World Fair, Seville, Spain." Proc. PLEA '88 Conference. Oporto, Portugal, pp. 527–532.

Cunningham, W. A., and T. L. Thompson. 1986. "Passive Cooling with Natural Draft Cooling Towers in Combination with Solar Chimneys." Proceedings, Passive and Low Energy Architecture (PLEA). Pecs, Hungary.

Dominguez, A. S. 1989. "Simulation, Monitoring and Evaluation of a Recent Experience: the EXPO '92 Project in Seville." Proc. Workshop on the Interaction Between Physics and Architecture, International Center of Theoretical Physics. Trieste, Italy.

Etzion, Y. 1991. "The Thermal Behavior on Non-Shaded Closed Courtyards in Hot-Arid Zones." Accepted for publication in *Architectural Science Review.*

Givoni, B. 1994. Building and Urban Design in Different Climates. In print, Van Nostrand Reinhold.

——— 1989. "Cooling of Outdoor Spaces." Proc. Workshop on the Interaction Between Physics and Architecture, International Center of Theoretical Physics. Trieste, Italy.

——— 1969. *Man Climate and Architecture,* 2nd ed. London: Applied Science Pub.

Hoyano, A. "Climatological Uses of Plants for Solar Control and the Effect on the Thermal Environment of a Building." 1988. *Energy and Buildings,* V-11, pp. 181–199.

McPherson, E. G. "Effects of Orientation and Shading from Trees on the Inside and Outside Temperatures of Model Homes." 1981. Proceedings, International Passive and Hybrid Cooling Conference. Miami, Florida, pp. 369–373.

McPherson, E. G., L. P. Herrington, and G. M. Heisler. 1988. "Impacts of Vegetation on Residential Heating and Cooling." *Energy and Buildings,* V-12, pp. 41–51.

Parker, J. H. 1981. "A Comparative Analysis of the Role of Various Landscape Elements in Passive Cooling in Warm Humid Environments." *Int. Passive and Hybrid Cooling Conference,* Miami Beach, pp. 365–368.

——— 1983. "The Effectiveness of Vegetation on Residential Cooling." *Passive Solar Journal,* V-2, No-2, pp. 123–132.

——— 1987. "The Use of Shrubs in Energy Conservation Planting." *Landscape Journal,* V-6, pp. 132–139.

Shlachtman, P. J., and J. H. Parker. 1985. "The Effect of Vegetative Landscaping on the Time Dynamics of Residential Heat Gain in Warm-Humid Climates." Proc. 10th Passive Solar Conference.

INDEX